STP 1261

Volatile Organic Compounds in the Environment

Wuncheng Wang, Jerald L. Schnoor, and Jon Doi, Eds.

ASTM STP Publication Code Number (PCN)
04-012610-16

ASTM
100 Barr Harbor Drive
West Conshohocken, PA 19428-2959
Printed in the U.S.A.

Library of Congress Cataloging-in-Publication Data

Volatile organic compounds (VOCs) in the environment / Wuncheng Wang,
Jerald L. Schnoor, and Jon Doi, Eds.
(ASTM special technical publication; 1261)
The Symposium on Volatile Organic Compounds (VOCs) in the
Environment was held 11-13 April 1994 in Montreal, Canada"-
-Foreword.
Includes bibliographical references and index.
ISBN 0-8031-2048-6
1. Volatile organic compounds—Environmental aspects—Congresses.
I. Wang, Wun-cheng. II. Schnoor, Jerald A. III. Doi, Jon.
IV. Symposium on Volatile Organic Compounds (VOCs) in the .
Environment (1994 : Montréal, Québec) V. Series.
TD196.073V55 1996
628.5′2—dc20 95-13347
CIP

Peer Review Policy

Each paper published in this volume was evaluated by three peer reviewers. The authors addressed all of the reviewers' comments to the satisfaction of both the technical editor(s) and the ASTM Committee on Publications.

The quality of the papers in this publication reflects not only the obvious efforts of the authors and the technical editor(s), but also the work of these peer reviewers. The ASTM Committee on Publications acknowledges with appreciation their dedication and contribution to time and effort on behalf of ASTM.

Printed in Fredericksburg, VA
May 1996

Foreword

The Symposium on Volatile Organic Compounds (VOCs) in the Environment was held 11–13 April 1994 in Montreal, Canada. ASTM Committee E47 on Biological Effects and Environmental Fate sponsored the symposium. Wuncheng Wang, U.S. Geological Survey, presided as symposium chairman and Jerald L. Schnoor, University of Iowa, and Jon Doi, Roy F. Weston, Inc., presided as symposium cochairmen. They are all editors of this publication.

Contents

Overview

Volatile organic compounds (VOCs) are an empirically defined group of compounds that: (1) typically have a low molecular weight, small specific gravity, low water solubility, and low boiling point; (2) include chemical classes such as aliphatic and aromatic hydrocarbons, halocarbons, aldehydes, ketones, and alcohols; and (3) are commonly analyzed using purge-and-trap gas chromatography coupled with mass spectrometry. Acetaldehyde, benzene, carbon tetrachloride, chloroform, propane, and vinyl chloride are but a few examples of VOCs. They are used extensively by many industries and are ingredients in many consumer products, such as fuels, paints, aerosols, cosmetics, disinfectants, refrigerants, and pesticides.

Many VOCs can adversely affect human health and the environment. Twenty-five VOCs, such as benzene, carbon tetrachloride, and chloroform, are known or suspected carcinogens; concentrations of these compounds in public water supplies are regulated by the U.S. Environmental Protection Agency (USEPA) and various state regulatory agencies. Public health concerns related to releases of VOCs into the atmosphere include depletion of the atmospheric ozone layer and, as a consequence, an increase in the incidence of human skin cancer, an increase in the mutation or die-off of plants and animals, and a decrease in food crop production. Accidental spills of VOCs, such as methyl isocyanate, have been known to be extremely disastrous to humans and animals. Because of their widespread use, their toxicity, and their environmental effects, VOCs are scrutinized closely by regulatory agencies, the scientific community, and concerned citizens.

VOCs have been detected in municipal and industrial wastewater discharges, the atmosphere, surface and ground waters, soil and landfills, municipal sewage sludge, and some foods. Their presence in the environment has been widely reported in scientific journals, the mass media, and in special technical reports. The concentrations of VOCs in urban air are of particular concern and, in March 1994, the USEPA issued stringent regulations aimed to reduce current VOC emissions by 80%. VOCs such as multihalogenated benzenes have been detected in shrimp in the United States. Xylene has been detected in tissue from fish in areas as remote as the Arctic Region.

VOCs are known to migrate into shallow and deep aquifers. VOCs, especially xylene from petroleum hydrocarbons and gasoline, have been detected in many shallow ground-water systems. After the prolonged flooding during the summer of 1993 in Iowa, many VOCs were detected in water from more monitoring wells (number of wells): benzene (1), carbon tetrachloride (1), dichloroethane (1), ethylbenzene (3), tetrachloroethylene (1), and xylene (6). The depths of these wells ranged from 9 to 48 m below land surface. Methyl tert-butyl ether is an oxygenated fuel additive for air-quality improvement mandated by the Clean Air Act Amendments of 1990. According to a 1993–1994 survey of national shallow ground water from 211 wells in 8 urban areas and 524 wells in 20 agricultural areas, this compound was detected in 27% of urban wells and 1.3% of the agricultural wells. Concentrations ranged from less than the detection level of 0.2 μg/L to as high as 23 000 μg/L.

Many VOCs currently are regulated in drinking water, industrial and municipal effluents, solid and hazardous wastes, and air quality programs by Federal and state agencies. For example, 33 of 126 chemicals regulated under the Safe Drinking Water Act, 29 of 297 toxic substances regulated under the Clean Water Act, and 77 of 418 toxic substances regulated under the Resource Conservation and Recovery Act can be classified as VOCs. Other statutes that are the basis for regulation of VOCs include the Toxic Substances Control Act and the Occupational Safety and Health Act.

VOCs in air are regulated by provisions of the Clean Air Act. More than 50 of 189 toxic substances in air are classified as VOCs. According to the regulation, each industry is required to release less than maximum air toxic concentrations, the so-called technology-based criteria on the basis of maximum achievable control technology. Many states have adopted air toxic regulations more stringent than the Federal regulations. For example, South Carolina has mandated that 257 compounds should meet air quality standards—a large number of them are VOCs.

The unique characteristic of all VOCs in the environment is their volatility, which makes the study of VOCs more challenging than all other types of contaminants. It is difficult to close mass balance calculations and recovery of VOCs are low. There is a distinct negative bias in monitoring VOC concentrations. For example, the conventional grab sampling method, for obvious reasons, is inadequate. VOC samples need an environment in which there is no opportunity for volatilization into air. Alternatively, VOCs can be captured in a solvent for later analysis. In addition, VOC studies involve complex questions such as their distribution, source(s), transport, and fate in the environment. Other equally important issues include ecological and health risk assessments, control, and environmental remediation strategies. The number of studies related to VOCs in the environment has been increasing steadily in recent years as public awareness of the occurrence and potential hazards of these compounds has increased.

As part of the effort to increase the understanding of VOCs in the environment, an International Symposium on Volatile Organic Compounds in the Environment was sponsored by the Committee E47 on Biological Effects and Environmental Fate, American Society for Testing and Materials (ASTM). The purpose of the Symposium was to bring together scientists, engineers, regulators, and concerned citizens in an open forum to exchange information on VOCs. The objectives were to evaluate the current information and to help guide future research directions. The Symposium was intended to be an international forum because the transport of VOCs is not restricted by national boundaries and the understanding of VOCs has international applications.

The Symposium was held 11–13 April 1994, in Montreal, Canada, and included contributions from attorneys, scientists, engineers, and environmental managers from Austria, Brazil, Canada, Czech Republic, Great Britain, Korea, and the United States. The keynote address, given by Professor Yoram Cohen of the University of California, Los Angeles, set the tone of the Symposium by describing the multimedia perspective of VOCs. Professor Cohen's presentation was followed by a lecture by Carol Boman, an attorney formerly with Howard, Rice, Nemerovski, Canady, Robertson and Falk of San Francisco, on the international regulation of VOCs, with special reference to the North American Free Trade Agreement.

The Symposium was divided into sessions that focused on regulation and assessment, air quality, environmental fate, environmental measurement, environmental monitoring, and control and remediation of VOCs. The ultimate objective of VOC research is to improve environmental quality through control or remediation, both of which were heavily emphasized in this Symposium.

The Symposium Committee wishes to thank all speakers who participated in the Symposium and contributed to the collective information set forth in this *ASTM Special Technical Publication 1261*. Session chairs, who responded to the call from the Committee by serving in various sessions and recruiting speakers, are: William Sonzogni (University of Wisconsin, Madison, Wisconsin), Susan Wilson (Lancaster University, Lancaster, Great Britain), Michael Harrass (Amoco Corporation, Chicago, Illinois), Mark Hinman (Exxon Biomedical Sciences, East Millstone, New Jersey), Donald Goolsby (U.S. Geological Survey, Lakewood, Colorado), William Kelly (University of Nebraska, Lincoln, Nebraska), Virginia Rohay (Westinghouse Hanford Company, Richland, Washington), and Jae Park (University of Wisconsin, Madison, Wisconsin). Many technical reviewers assisted greatly by providing numerous constructive comments. Thomas Doane, Chairman of the ASTM Committee E47, provided leadership and support that were essential to the success of this Symposium.

Wuncheng (Woodrow) Wang
U.S. Geological Survey,
Columbia, South Carolina;
symposium chairman and editor.

Jerald L. Schnoor
University of Iowa,
Iowa City, Iowa;
symposium cochairman and editor.

Jon Doi
Roy F. Weston, Inc.,
West Chester, Pennsylvania;
Symposium cochairman and editor.

Keynote Address

Yoram Cohen[1]

Volatile Organic Compounds in the Environment: A Multimedia Perspective

REFERENCE: Cohen, Y., "**Volatile Organic Compounds in the Environment: A Multimedia Perspective,**" *Volatile Organic Compounds in the Environment, ASTM STP 1261,* W. Wang, J. Schnoor, and J. Doi, Eds., American Society for Testing and Materials, 1996, pp. 7–32.

ABSTRACT: The environment is a complex system of interacting environmental media. Pollutants do not stay in the medium where they originate but move across environmental phase boundaries. The distribution of pollutants throughout the various environmental compartments (for example, air, water, soil, and biota) is the result of complex physical, chemical, and biological processes. The resulting environmental and human health risks depend upon the degree of exposure of human and ecological receptors, via multiple pathways, to these chemicals. Thus, environmental pollution is a multimedia problem. Volatile organic compounds (VOCs), in particular, are very mobile in the environment. For example, VOCs which are initially present in the soil or water media can readily volatilize to the atmosphere where they can be transported over significant distances from the source location. In this paper an overview is presented of VOC sources, VOC ambient levels, the multimedia distribution of VOCs in the environment, and multipathway exposure to VOCs.

KEYWORDS: VOCs, sources, emissions, volatilization, multimedia distribution, multimedia exposure, wet deposition, air/water exchange, soil transport

The development of multimedia-based risk reduction strategies, designed to minimize the adverse effects of environmental pollution on human and ecological receptors, requires a clear understanding of the sources of pollutants and their pathways to the receptors of concern. Knowledge of pollutant fate and transport is essential in order to assess the potential impact on the environment and to determine the need for risk reduction strategies. For example, the assessment of the state of various ecosystems (locally, regionally, and globally) requires an evaluation of exposure of multiple receptors, which are in contact with various media directly or indirectly (for example, plants which are in contact with soil and air), to environmental contaminants. Volatile organic compounds (VOCs) are of particular concern since they readily volatilize to the atmosphere and thus can be distributed over large regions leading to a population-wide exposure to these chemicals. VOCs are present in liquid and solid process and waste streams, in consumer products, and in fossil fuels. VOCs can be emitted to the environment from a variety of controlled and uncontrolled sources and as a result there is a large variation in the ambient VOC concentrations and composition on both local and regional scales. Because of their relatively high vapor pressures, VOCs are found predominantly in the atmosphere. However, VOCs have also contaminated surface water, ground water, municipal drinking water, and the subsurface soil in various local environments. In general, VOCs in the environment present problems that are associated with: (a)

[1] Department of Chemical Engineering, University of California, Los Angeles, CA 90095.

human health problems, (b) the formation of ozone and urban aerosol precursors, and (c) odor.

In most urban regions the concentration of VOCs contribute significantly to the total cancer risk associated with toxic air pollutants. In California, the Air Resources Board has estimated the population-weighted statewide risk from exposure to average ambient concentrations of air toxics as shown in Table 1 for selected VOCs. In Los Angeles, with a typical average benzene concentrations of about 4 ppb (Venturini et al. 1991), the associated individual lifetime cancer risk is about 220 per million. An even higher cancer risk estimate of about 500 per million results for 1,3-butadiene at the ambient concentration level of about 0.44 ppb reported by Loscutoff and Poore (1993). Exposure to VOCs is also associated with the emissions of VOCs from treatment, storage and disposal facilities (TSDFs) (Shen et al. 1993; Govind et al. 1992). It is estimated that emissions from TSDFs result in about 140 excess cancer incidences per year (*Federal Register* 1991).

The potential health risks associated with VOCs and their role in the formation of photochemical smog have led to increasing public concern regarding the presence of VOCs in the environment. With the passage of the 1990 Clean Air Act the United States Environmental Protection Agency has been charged with the task of regulating the emissions of hazardous air pollutants with the goal of reducing their emissions from a variety of stationary sources. Also, Section 313 of the Superfund Amendments and Reauthorization Act (SARA) has set a requirement for the reporting of toxic chemical releases resulting in an extensive database of toxic chemical release inventory for VOCs and other chemicals (Friedman 1989; Bowman 1989). There are other federal laws that regulate VOC emissions directly or indirectly through the National Emission Standards for Hazardous Air Pollutants (NESHAPS), the New Source Performance Standards (NSPS), the Occupational Health and Safety Administration (OSHA) Standards, and other state and local regulations. States such as California, for example, have passed other specific legislation such as the 1987 "Hot Spots" Act and the 1988 Motor Vehicle Air Toxics Bill in order to accelerate the control of toxic air contaminants associated with stationary and mobile sources.

The presence of VOCs in the environment is of a significant environmental concern. Mitigation of environmental problems associated with VOCs will require careful consider-

TABLE 1—*Estimated population-weighted statewide risk in California from exposure to average ambient concentrations of selected VOCs.*[a]

Substance	Potential Risk (Lifetime Cancer Cases per Million)
Benzene	220
Ethylene oxide	80[b]
Carbon tetrachloride	10
Methylene chloride	10
Vinyl chloride	10
Ethylene dibromide	4
Trichloroethylene	2
Chloroform	1

[a] Source: Venturini, P. D., Denton, J. E., Howard, K., Huscroft, S. V., Shiroma, G. S., and Suer, C. B., "California Air Toxics Program: Review of Progress and Future Direction," presented at the Air and Waste Management Association, 84th Annual Meeting and Exhibition, Vancouver, BC, 16–21 June 1991.

[b] Based on data for South Coast Region.

ation of sources, monitoring data of their ambient levels in various environmental media, and mass transfer rates across environmental phase boundaries. VOCs in the environment must be treated from a multimedia perspective and, accordingly, this paper presents an integrated view of the major issues associated with VOCs.

Sources of VOCs

Volatile organic compounds (VOCs) are released to the environment from a variety of outdoor and indoor sources (Fig. 1). Indoor releases of VOCs result from indoor activities such as cooking, use of office machines, releases from building materials, consumer products, as well as tobacco smoke (EPA 1989, 1993). In some cases, vehicular emissions can also lead to indoor contamination directly in houses that have attached garages. Outdoor releases are due to combustion of fuels, fugitive emissions from petrochemical and chemical facilities, VOC emissions from public-owned treatment works (Dunovant et al. 1986; Loehr and Ward 1987; Govind et al. 1991), mobile sources, and product use and disposal. In urban areas, VOC emissions typically correlate with the population density as is illustrated, for example, by the similar VOC emissions per person in San Francisco and Los Angeles (Table 2). EPA estimates that about 32% of the nationwide VOC emissions are from mobile sources, while 17% is attributed to direct industrial processing and production activities (Fig. 2*a*). In urban regions mobile sources contribute significantly to the total emission rate of VOCs as is shown

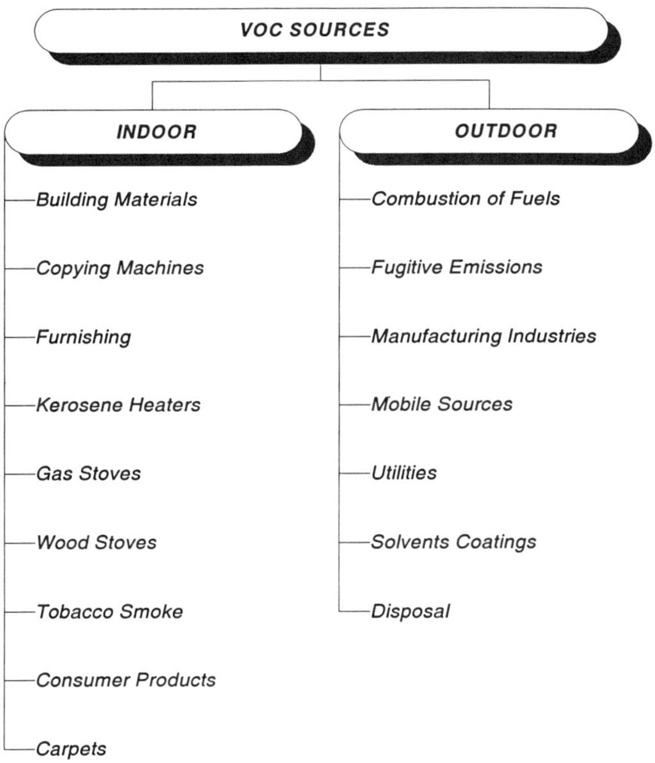

FIG. 1—*Outdoor and indoor VOC sources modeling.*

TABLE 2—*Comparison of VOC emissions for urban and rural locations.*[a]

			VOC Emissions		
Location	Population	Area, sq mi.	tons/day	lb/day/person	lb/day/person
San Francisco[b]	5 800 000	5 500	1 700	0.6	600
Los Angeles[b]	12 875 000	6 500	2 100	0.3	650
North Coast, CA[c]	300 000	12 300	120	0.8	20
Valdez (only)[d]	~3 600	100	1.0 to 1.4[b]	0.5 to 0.8	20 to 28
Valdez (plus Terminal)	...	...	120	66	2 400

[a] Source: adapted from Cohen et al. 1992.
[b] Source: California Air Resources Board, 1991, Total Organic Gas Emissions.
[c] Estimated from North Coast, CA, lb/person and lb/sq mi. North Coast Air Basin includes Del Norte, Humboldt, Mendocino, Trinity and portions of Sonoma County.
[d] Goldstein et al. 1992.

in the example of the VOC emission inventory for Chicago (Fig. 2*b*). In this particular example, vehicular emissions account for about 35% of the total VOC emissions in the region (Scheff and Wadden 1993) which is consistent with the nationwide profile. In contrast, VOC emissions in rural areas are significantly lower as is shown in Table 1 for the North Coast Air Basin in California (Cohen et al. 1992).

There are cases where stationary sources are a significant contributor to total VOC emissions. For example, VOC emissions at Port Valdez of about 120 tons/day (Goldstein et al. 1992), associated with crude oil loading operations at the Alyeska Marine Terminal, are the dominant VOC emissions in the Port Valdez region (Table 2). As shown in Fig. 3, the chemical-specific emission rate due to tanker loading operations correlates with the simple emission index which is defined as $E_i = W_i Y_i$, where W_i and Y_i are the VOC mass fraction in the crude oil and mole fraction in the vapor phase at saturation conditions (or the pure VOC), respectively (Cohen 1992). Clearly, the emission rates increase with increasing concentration or vapor pressure, or both, of the VOC present in the crude oil. Emissions from offshore platforms is another example of VOC emissions from crude oil production. Recently, Countess and Browne (1993) reported that VOC emissions from a typical offshore platform range from about 7 tons/year to 25 tons/year. Emissions of VOCs are also associated with the handling and processing of crude oil in refineries as revealed by the detailed study of emissions from the AMOCO Yorktown Oil Refinery (Klee et al. 1994). In the Yorktown

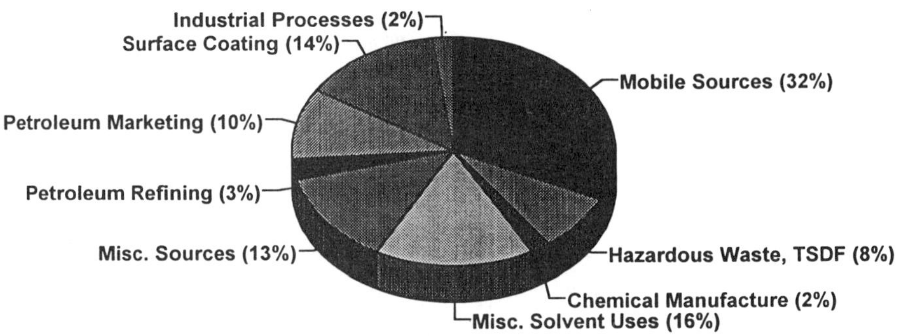

FIG. 2(*a*)—*Sources of nationwide VOC emissions (EPA 1992).*

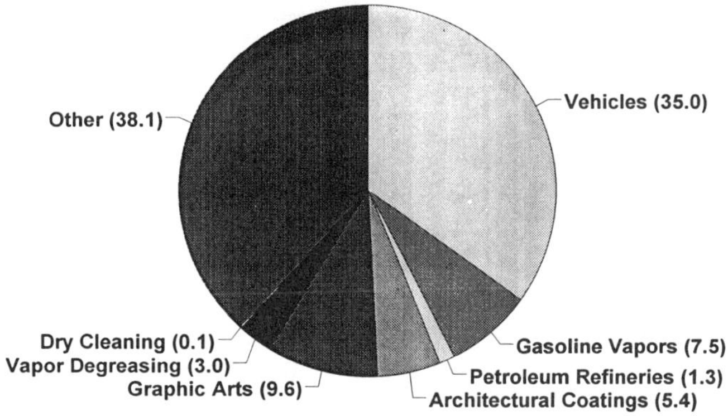

FIG. 2(*b*)—*VOC emission profile (%) by source category for Chicago, IL (source: Scheff and Wadden 1993).*

region, however, emissions of the eleven VOCs analyzed by Cohen et al. (1991) revealed that the Yorktown refinery contributed to 42% of these VOC emissions while vehicular emissions comprised 57% of the total emissions of these eleven VOCs (Fig. 4). The above examples suggest that the control of VOC emissions is complex given that in different locations the dominant VOC emission sources may vary in their character. In the case of Port Valdez, reducing emissions from the Marine Terminal is a logical approach, while in the vicinity of the AMOCO Yorktown refinery, reducing emissions from both the refinery and stationary sources is needed to have a significant positive impact. It is important to note that in most urban regions motor vehicles are a significant contributor to emissions of hydrocarbons, suggesting that an integrated approach that addresses emissions from both mobile and stationary sources is required.

In viewing the national flows of hazardous waste streams it appears that a significant fraction of hazardous waste generated is either emitted to the atmosphere or discharged as

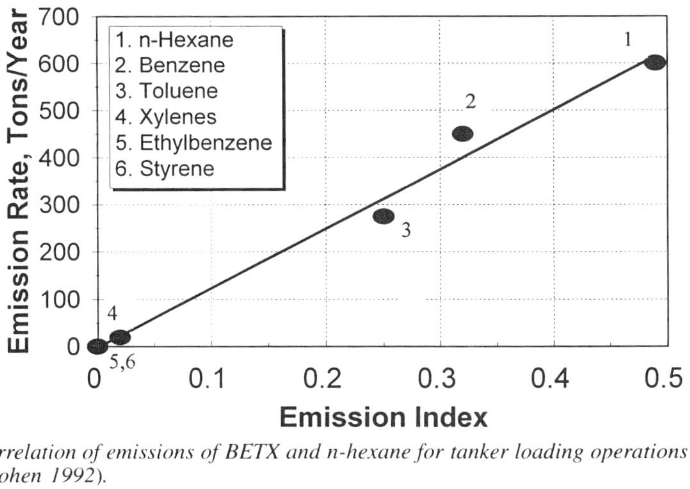

FIG. 3—*Correlation of emissions of BETX and n-hexane for tanker loading operations at Port Valdez, AK (source: Cohen 1992).*

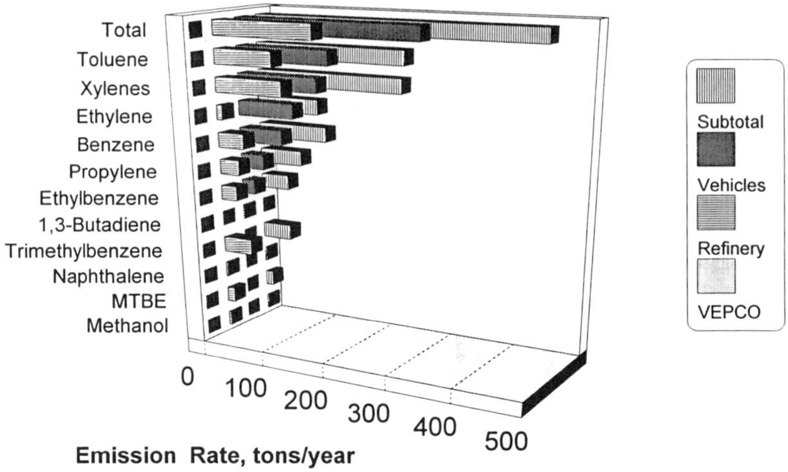

FIG. 4—*Emissions of selected VOCs in the Yorktown Region (source: Cohen et al. 1992).*

aqueous waste, albeit at low concentrations (Cohen and Allen 1992). With the development of the toxic release inventory (TRI), information regarding VOC emissions has been mounting. Although this database does not provide a complete reporting of all VOC sources (for example, it does not include mobile sources and small generators), it can be queried to provide an assessment of the relative distribution of the various emitted VOCs. In a recent review, Doa (1992) summarized TRI data on VOCs with the largest total releases and transfers in the U.S. and their corresponding air and water releases (Fig. 5). Accordingly, VOCs with the largest TRI total releases and transfers comprise about 32% of the total TRI releases and transfers. At the same time, VOCs with the highest reported TRI release to the air comprise about 60% of the total air releases, while VOCs amount to only about 7% of releases to surface water. Perhaps, more illustrative is the fact that VOCs constitute about 73% of the total TRI releases and transfers of carcinogens. It is also noted that of the 25 chemicals with the largest TRI total releases and transfers, nine were classified as VOCs, while of the 25 carcinogens with the largest TRI releases and transfers, 15 were classified as VOCs; thus, it is no surprise that there is a serious public and regulatory concern with the presence of VOCs in the environment.

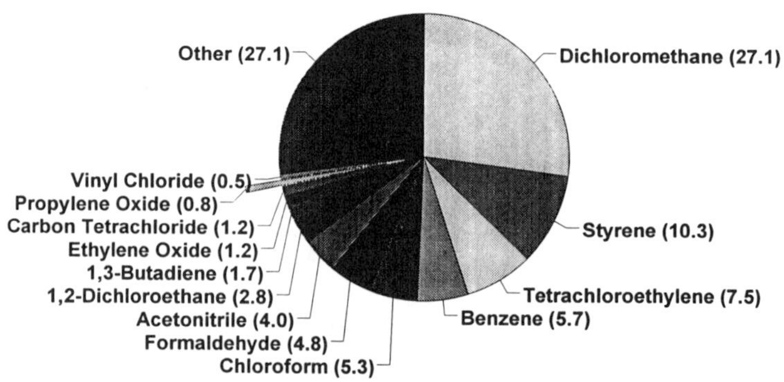

FIG. 5—*Profile of total (%) TRI releases and transfers for carcinogenic VOCs (source: Doa 1992).*

From a public health concern toxic VOCs are of particular importance. However, the variety of VOC sources makes it difficult to implement a uniform strategy to control the release of VOCs to the environment. For example, in Southern California, the emission profile for benzene, which is a known carcinogen, by source category indicates that mobile sources contribute about 58% of the total benzene emissions (SAQMD 1988). Therefore, even if all other sources were eliminated, the exposure to benzene would only be reduced by 40%. On the other hand, perchloroethylene, used mainly as a solvent, is emitted primarily from three major activities with dry cleaning being the major source with over 80% of the emission rate (SAQMD 1988). Another illustration of the complexity of VOC's emission control is the case of fugitive emissions from large chemical manufacturing facilities. The breakdown of the contribution of various plant components to fugitive emissions from a typical organic chemical manufacturing facility is shown in Fig. 6 (USEPA 1982). It is important to recognize that in these facilities fugitive emissions occur as slow leaks from literally thousands of plant components.

The above examples illustrate that the releases of VOCs and associated exposures are due to multiple different industrial, residential, and municipal sources. Thus, monitoring and control of VOC emissions is indeed a technical and regulatory challenge.

Ambient Levels of VOCs

It is expected that chemicals which are emitted from various industrial facilities will contribute to the ambient levels of these VOCs in the atmosphere. In recent years increasing efforts by industry have resulted in a rich information base on VOC emissions and ambient concentration levels in industrial zones (Klee et al. 1994; LaGrone 1991; Baker and Warren 1992). For example, the Houston Regional Monitoring Corporation (HRM), a consortium of 45 companies located in the Houston Texas slip channel industrial complex, embarked on a study designed to assess the level of VOC emissions from the various facilities in relation to the ambient VOC levels (LaGrone 1991). Based on the HRM study, the measured ambient concentrations of selected VOCs versus the estimated annual emission are shown in Fig. 7. The general trend is that the measured ambient concentrations correlate positively with the industrial emission rates. This does not necessarily mean that emissions from the industrial complex solely account for the observed ambient VOC levels. In fact, the observed ambient levels of benzene, ethylbenzene, toluene, and xylene (BETX) in this example, are of the same order of magnitude or even lower than the measured ambient levels of these same VOCs in other urban centers (USEPA 1988c; Goldstein et al. 1992; Cohen et al. 1992; Lostcoutoff and Poor 1993). Ambient levels of BETX are similar in most urban centers with higher concentrations in the more congested cities since the emissions of such aromatics by

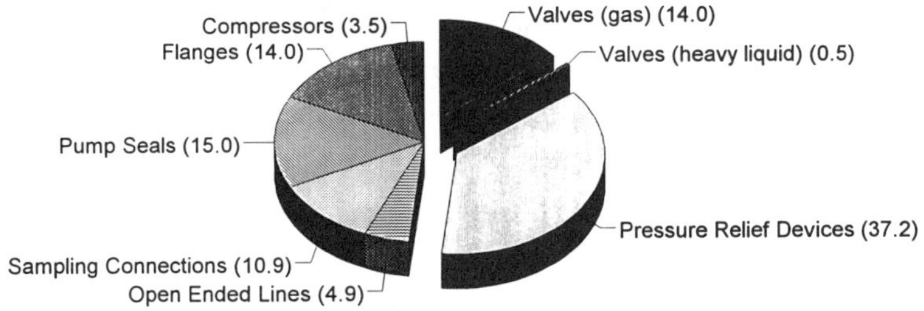

FIG. 6—*Fugitive emission profile (%) for a typical organic chemical manufacturing industry (SOCMI) plant (source: EPA 1982).*

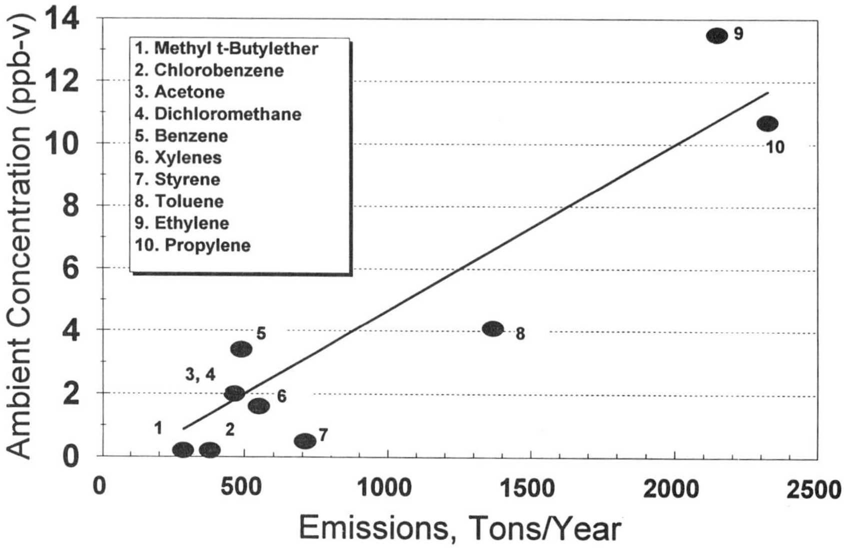

FIG. 7—*Ten highest SARA emissions totals for an industrial complex in Houston, TX (source: LaGrone 1991).*

automobiles is significant. Note that in a recent extensive toxic air contaminant monitoring program (TACM) in California (Loscutoff and Poore 1993), a toluene:benzene concentration ratio of about 2 was found in both the atmosphere and in automobile exhaust, suggesting that the source of these VOCs in California urban centers is mainly due to vehicular emissions. Various VOCs are also formed in the atmosphere. For example, formaldehyde and acetaldehyde are formed in the atmosphere due to photooxidation of hydrocarbons and during combustion of fossil fuel (Grosjean 1991; and references therein). Note also that in a recent study, Zhang et al. (1994) gave a detailed account of the indoor sources of various aldehydes. Seasonal variations in the atmospheric concentrations of some VOCs can be significant in urban regions. This is the case for contaminants for which there is a seasonal source variation (for example, additional emissions due to burning of fossil fuels in the winter). The ambient concentrations of VOCs with constant emission rate are also expected to be affected by the seasonal change in the atmospheric mixing height and wind dispersion. In addition, greater atmospheric transformations in the summer months will also affect the resulting VOC concentrations especially in urban centers.

Over the last decade it has become apparent that VOCs are also generated indoors due to a variety of household activities (Ott 1985; USEPA 1983, 1989; Wallace et al. 1991; Wallace 1989; Austin et al. 1992; see Fig. 1). The celebrated total exposure and monitoring (TEAM) study by Wallace et al. (1986) nearly a decade ago has conclusively shown that the concentration of certain VOCs can be higher in the indoor environment and personal air. As an illustration, Fig. 8 depicts the concentrations of selected VOCs measured outdoors and in personal air in Los Angeles, during the summer of 1987 (Wallace et al. 1991), and from the 1989 California South Coast Basin Toxic Air Contaminants Monitoring Study (Lascutoff and Poore 1993). Clearly, VOC concentrations in personal air can be higher than outdoor VOC concentrations. There are a variety of reasons for the high VOC content in indoor air (Wallace et al. 1991; Wallace 1989; Lioy 1990; Naugle and Pierson 1991) including sources from tobacco smoke, fugitive emissions from various household solvents, venting of dry-

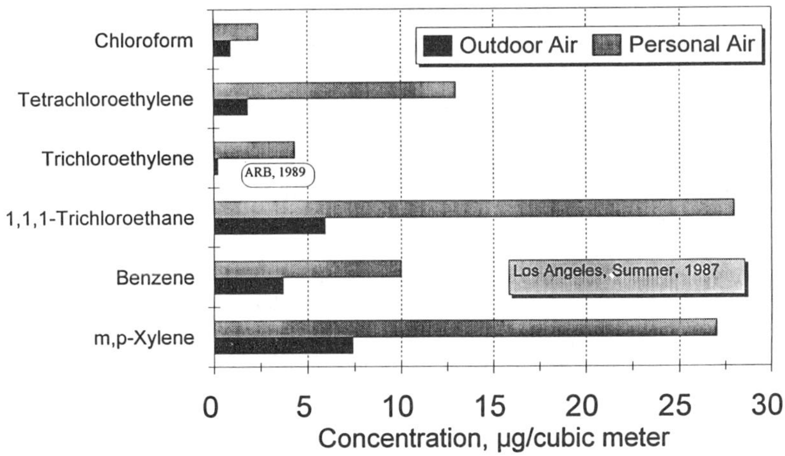

FIG. 8—*Concentrations of selected VOCs in the outdoor and personal air in Los Angeles (source: Wallace 1991; Loscutoff and Poore 1993).*

cleaned clothes, and various household products (for example, cleaning fluids, insecticides). Indoor exposure to VOCs can also occur due to consumption of contaminated drinking water and exposure to contaminated water while showering and bathing (McKone 1987; Lioy 1990; Andelman et al. 1990). Figure 9 illustrates the concentration of selected VOCs measured in drinking water in both New Jersey and Los Angeles regions during the 1982 and 1984 phases of the pioneering TEAM study of Wallace et al. (1986). It is noted that the chlorinated hydrocarbons which were detected in drinking water are due, in part, to the formation of halocarbons during water chlorination (Bull 1985), as well as to the contamination of groundwater supplies.

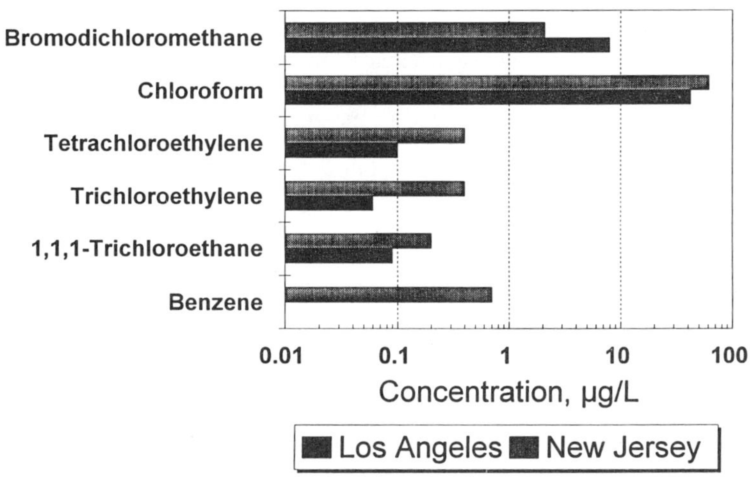

FIG. 9—*Concentrations of selected VOCs in drinking water in New Jersey and Los Angeles (source: Wallace 1986).*

In recent years it has become apparent that the thousands of hazardous and municipal landfills are potential sources of VOCs. VOCs in the landfills leach to ground water and volatilize to the atmosphere. Chemical waste treatment, storage and disposal facilities (TSDF) are believed to contribute significantly to the total emission inventory of toxic VOCs (Shen et al. 1993). In a study of 400 landfills in California, ten different VOCs were tested (Lagarias et al. 1993) revealing that at 70% of the sites at least one of the listed VOCs was present with average concentrations of landfill gas samples ranging from about 4 to 480 000 ppb. While it is accepted that VOC air emissions and leaching to ground water from TSDFs can lead to local exposure problems, there are cases where contamination of aquifers has been reported to occur over a large area. In the San Gabriel Valley there is great public concern regarding VOC contamination of the San Gabriel Aquifer which provides drinking water to nearly one million people (*LA Times,* 14 March 1994). Contamination of the aquifer by VOCs is believed to be the result of disposal activities linked to the operation of nearly 48 000 firms that have set up their operations in the valley after World War II, to household septic tanks, and to past agricultural activities.

Distribution of VOCs in the Environment

When considering the distribution of VOCs in the environment, it is often convenient to distinguish between the nonbiological and biological environmental media as shown in Table 3 and the human receptor as a separate environmental compartment. The major environmental compartments in Table 3 can be further subdivided into subclasses as deemed necessary. The distribution of VOCs in the multimedia environment will vary with topographical and meteorological parameters (wind conditions, mixing heights, etc.), the physicochemical properties of the environmental media, and properties of the VOCs. Although the properties of the environmental media (Table 3) are not pollutant specific, their physicochemical properties can affect the pollutant interaction with these compartments. For example, soil properties such as fraction of organic carbon, moisture content, and pH affect the sorption of organic pollutants by the soil matrix. Another example is the increase in bioconcentration of hydrophobic organic chemicals in biota with increasing lipid content.

The partitioning of VOCs among the different environmental compartments are also governed by the physicochemical properties of the pollutants of interest including partition coefficients, biotransfer factors, and intermedia transport parameters (for example, mass transfer coefficients). Some of the major physicochemical properties and partition coefficients that are normally required in multimedia analysis are listed in Table 4*a* and 4*b*. Detailed reviews of environmental properties that are needed in environmental assessment can be found in various sources in the literature (Lyman et al. 1990; Howard 1989, 1990, 1991; Conway 1982; Novotny and Chesters 1981; USEPA 1988a; Cohen 1986; Cohen et al. 1990; Swann and Eschenroeder 1983; and references therein).

TABLE 3—*Major environmental media.*

Nonbiological Compartment	Biological Compartment
Atmosphere	Aquatic animals
Surface water	Terrestrial animals
Sediment	Vegetation
Surface water	Food (solid and liquid)
Soil	
Ground water	

TABLE 4(*a*)—*Physicochemical properties for intermedia transport analysis.*

- Molecular weight
- Liquid density as a function of temperature
- Molar volume at the boiling point
- Vapor pressure as a function of temperature
- Critical properties (critical temperature, critical pressure, critical volume, critical compressibility factor)
- Water solubility as a function of temperature
- Boiling temperature
- Heat of vaporization; heat of sublimation; heat of fusion

The rates of VOC transfers among the different environmental compartments are governed by complex intermedia transport processes. A summary of the intermedia transfer pathways among the various environmental media for VOCs are given in Tables 5*a* and 5*b*. Among the various modes of transport, the transport of particle-bound chemicals is not believed to be of major significance for VOCs since the major fraction of VOCs is found in the vapor phase. However, exposure to soil-bound VOCs can affect the accumulation of contaminants in vegetation and terrestrial organisms, as well as contribute to the exposure of humans to VOCs via soil ingestion and dermal absorption.

It is emphasized that the propensity of VOCs to accumulate in the environment is directly related to their biotic and abiotic transformation rates. In general, the most significant degradation process for VOCs in the atmosphere is their reaction with atmospheric OH radicals. Other potential degradation pathways involve reactions with the gaseous nitrate (NO_3) radical, ozone, and a variety of photooxidants. It is also noted that VOCs can form as the result of atmospheric reactions as is the case in the photochemical formation of formaldehyde and acetaldehyde in urban atmosphere (Grosjean 1991; and references therein). In the water environment reactions such a photolysis, hydrolysis, and biodegradation may occur. For some chemical classes, reactions with photooxidants in sunlit waters can be important. Generally, hydrolysis is not a significant reaction for nonpolar VOCs. Hydrolysis only applies to chemicals that have hydrolyzable functional groups such as esters, aliphatic halogens, amides, and carbamates. In the subsurface soil chemical transformations of organics include photodecomposition, chemical hydrolysis, chemical oxidation, and biotic reactions. In the aerobic zone (surface soil zone) bacteria may degrade many of the organic constituents, while under the usual anaerobic conditions in the deep soil region biodegradation of most hydrocarbons is significantly slower (or is virtually negligible). It should be noted that in some cases, as was shown in the early work of Wilson and McNabb (1983), VOCs can form as the result of biotransformations as is the case of vinyl chloride which can form in the soil environment due to biotransformations of halocarbons.

VOCs which are only sparingly soluble in water, are expected to reside mainly in the atmosphere which serves as the most significant reservoir. This behavior can be demonstrated

TABLE 4(*b*)—*Partition coefficients.*

- Air/water partition coefficient (or Henry's law constant)
- Bioconcentration and bioaccumulation coefficients
- Soil/water partition coefficient
- Soil/air partition coefficient
- Octanol/water partition coefficient (can be used in correlations to predict the soil/water partition coefficient)

TABLE 5(*a*)—*Intermedia transport pathways among the nonbiological compartments.*

Intermedia Transport Pathways	Transport Mechanism
Air/water	gas absorption dry deposition of particles and aerosol droplets precipitation scavenging of gases and particles
Water/air	volatilization aerosol formation
Air/soil	dry deposition of gases and aerosols precipitation scavenging of gases and aerosols
Soil/air	volatilization from the soil resuspension of particle-bound pollutants
Air/particulate phase	particle/gas conversion
Water/sediment	dissolved chemical diffusion sedimentation of particle-bound pollutants
Sediment/water	dissolved chemical diffusion resuspension of particles
Soil/water	runoff (transport of dissolved or particle-bound chemical)
Soil/ground water	diffusion and convection from the unsaturated soil to ground water
Ground water/soil	volatilization of contaminants from ground water, also convection due to rise of water table

by a multimedia partitioning analysis using the multimedia models developed by Cohen et al. (1990) and Cohen and Clay (1994). In this approach the environmental system under consideration is divided into a number of interacting compartments, as shown in Fig. 10, where chemical exchange among the different compartments takes place via the intermedia processes listed in Tables 5*a* and 5*b*. Using the above approach, the regional multimedia distributions were determined for benzene and MTBE in the Yorktown region, and for benzene, trichloroethylene (TCE), and methylene chloride in the Los Angeles region. The physicochemical properties of the above selected VOCs are given in Table 6 (Cohen et al. 1990; Clay 1992) with the resulting multimedia distribution displayed in Figs. 11*a* and 11*b*. It is noted that the combined contribution of the air, water, and soil compartments is less than 100% but this cannot be represented in the figures due to round-off of significant figures. Although the results displayed in Figs. 11*a* and 11*b* are region-specific, the trend that emerges for the above chemicals is that the VOCs with a high air/water partition coefficient (that is, TCE, benzene, methylene chloride) partition primarily to the air phase (with over 96% of the total mass of the VOC in the environment present in the atmosphere), less than 2% of the chemical present in the water compartment, and about 2% or less of the VOC mass in the soil. However, for MTBE, which has a significantly lower air/water partition coefficient than the sparingly soluble VOCs, a much higher fraction of MTBE (about 74%) partitions to the water compartment. For situations where the chemical is initially present in

TABLE 5(*b*)—*Intermedia transport pathways among the nonbiological and biological compartments.*

Compartment	Intermedia Transport Pathway	Transport Mechanism
Vegetation	air/vegetation	dry deposition of gases (uptake via the plant canopy)
		dry deposition of particle-bound pollutants (uptake via the plant canopy)
		precipitation scavenging of gases and particle-bound pollutants
	vegetation/air	emissions of volatile organics
	soil/vegetation	uptake via the roots
Fish and bivalves	water/fish and water/bivalve	uptake via diffusion through the body membranes (e.g., gills and via ingestion)
Cattle	feed/beef	ingestion
	air/beef	accumulation of contaminant in beef due to inhalation
	water/beef	accumulation of contaminant in beef due to water drinking
	feed/milk	ingestion
	air/milk	accumulation of contaminant in beef due to inhalation
	water/milk	accumulation of contaminant in beef due to water drinking

the soil environment (for example, subsurface contaminant plume), the driving force for VOC transport is also towards the distributions presented in Figs. 11*a* and 11*b*. However, since the volatilization rate from the soil subsurface is slow, the soil can act as a "source" for contaminants which are initially present in the soil subsurface (for example, due to disposal or leaky storage tanks) for a considerable time period.

Multipathway Exposure to VOCs

VOCs can readily disperse throughout the environment and thus the exposure of the human receptor can occur via multiple pathways (Fig. 12). Exposure to VOCs can occur via primary pathways (for example, inhalation of polluted air or drinking contaminated water) or via secondary pathways such as ingestion of contaminated beef in which contaminants have accumulated due to the exposure of beef cattle to contaminants via inhalation, water drinking, and contaminated feed. In order to illustrate the concept of multimedia exposure three hypothetical scenarios are discussed (Table 7). Scenario A pertains to multipathway exposure to trichloroethylene (TCE) in Los Angeles due to air emissions of TCE. The distribution of TCE in the environment was estimated using the SMCM model (Cohen et al. 1990) and the

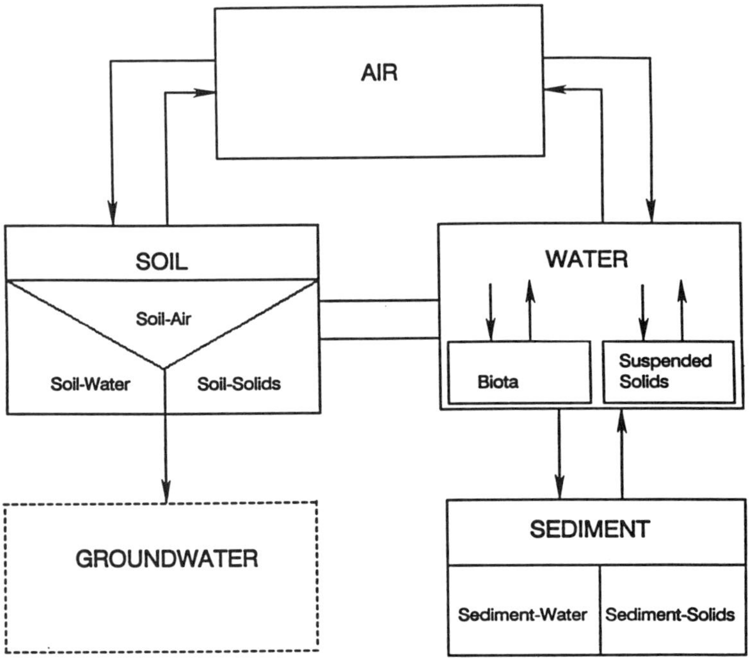

FIG. 10—*Schematic representation of the SMCM model (Cohen et al. 1990).*

exposure analysis was conducted using the standard and conservative EPA approach to estimating contaminant intake (USEPA 1988a, 1986). Scenario B is a hypothetical case in which the atmospheric concentration was selected to be about the average TCE concentrations reported in Los Angeles, the drinking water concentration was selected to be at the EPA action level, and the surface water concentration was selected the same as scenario A. In scenario C the air phase concentration was increased by a factor of two and the drinking

TABLE 6—*Physicochemical properties of selected VOCs.*[a]

Chemical	MW	Solubility, mg/L	Henry's Law Constant, Pa · m^3/mol	Octanol-Water Partition Coefficient	Atmospheric Degradation Half-Life (day)[b]
Benzene	78.1	1 780	550	80	2 to 20
Trichloroethylene	131.4	1 100	1 043	214	1 to 13
Methylene chloride	50.49	19 100	27	8	60 to 600
Perchloroethylene	165.8	149	2 900	2 511	16 to 160
MTBE	88.2	5.4×10^4	5	12	3[c]

[a] Physicochemical properties compiled from Howard, P. H., *Fate and Exposure Data,* Vols. 1–3, Lewis Publishers, Chelsea, MI, 1989–1991.

[b] Howard, O. H., Boethling, R. S., Jarvis, W. F., Meylan, W. M., and Michalenko, E. M., *Handbook of Environmental Degradation Rates,* Lewis Publishers, Chelsea, MI, 1991.

[c] Estimated from Atkinson, R. J., "Kinetic Mechanisms of the Gas-Phase Reactions of the Hydroxyl Radical with Organic Compounds," *Journal of Phys. and Chem. Ref. Data,* Monograph No. 1, 1989.

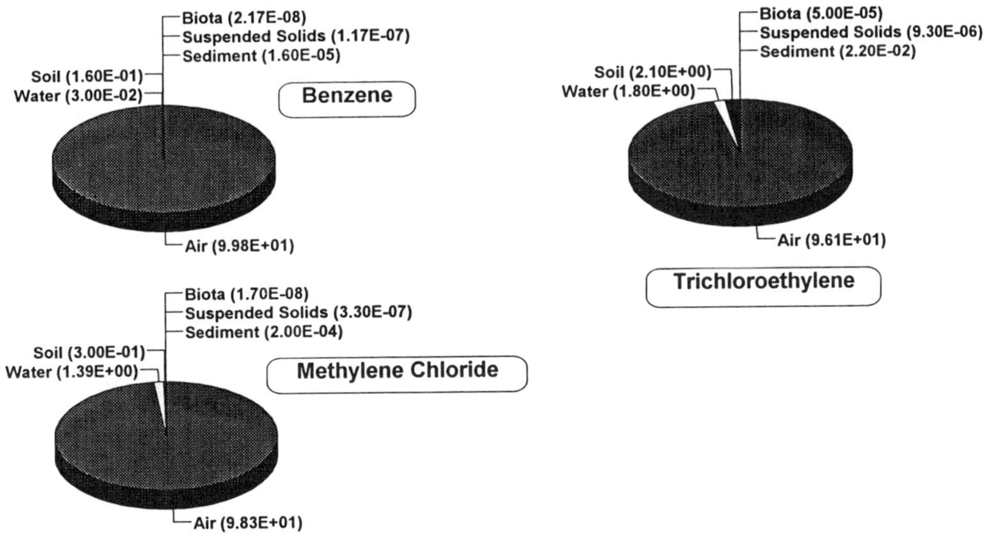

FIG. 11(*a*)—*Mass distribution (%) of benzene, trichloroethylene and methylene chloride in the Los Angeles region.*

water concentration was increased by a factor of four to simulate an exposure at high concentrations, relative to case A, as may occur near hazardous waste sites. Using standard conservative exposure calculations (USEPA 1988a,b, 1991a), the intake of TCE via the various pathways was calculated and the resulting individual lifetime risk for the major intake

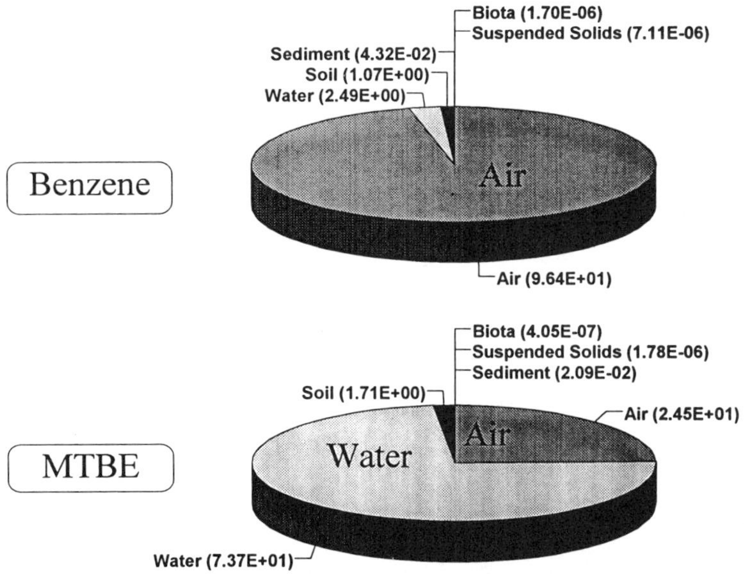

FIG. 11(*b*)—*Mass distribution (%) of selected VOCs in the Yorktown Region (Cohen et al. 1991).*

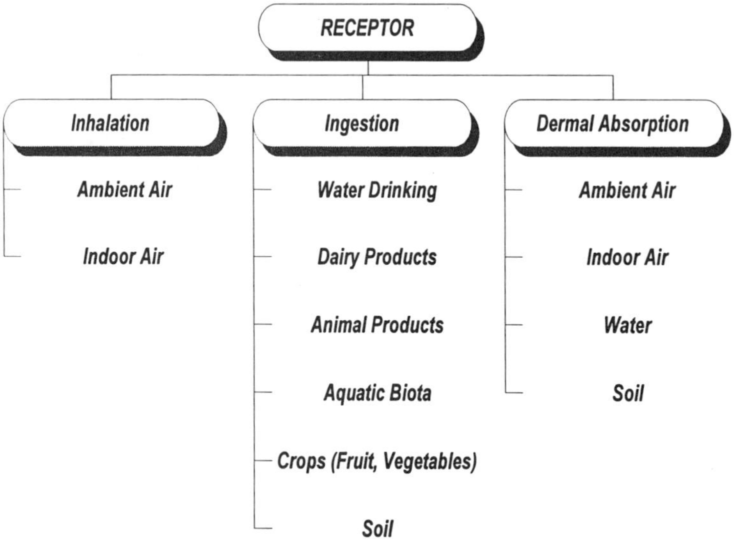

FIG. 12—*Exposure pathways.*

pathways are shown in Fig. 13. The major contributions to the individual lifetime risk are due to inhalation followed by drinking water. For the scenario where the individual relies on water from a highly contaminated ground-water source, exposure via both drinking and showering can be significant. Also, for receptors near or at a contaminated site, exposure to gas emissions from the soil may be nonnegligible. It is also noted that, in some situations, exposure can occur due to the consumption of vegetables grown at a contaminated site or consumption of meat and dairy products from cattle that are raised on such a site. Clearly, depending on the scenario played, even for VOCs, exposure pathways other than air can become important and thus may have to be considered. Over the years a significant effort has been devoted to describe the volatilization of chemicals from the soil environment as well as their migration to ground water, subsequent dispersion in aquifers, and human exposure to contaminants present in the unsaturated soil and ground-water media.

TABLE 7—*Environmental concentrations of TCE for hypothetical exposure scenarios.*

	TCE Concentration		
Environmental Medium	Scenario A	Scenario B	Scenario C
Air	0.2 μg/m^3	1.34 μg/m^3	2.68 μg/m^3
Drinking water	0.07 μg/L	5 μg/L	20 μg/L
Surface water	0.07 μg/L	0.05 μg/m^3	0.05 μg/m^3
Soil	0.7 ng/kg	0.75 ng/kg	0.75 ng/kg
Fish[a]	101 μg/kg	101 μg/kg	101 μg/kg
Meat[b]	1.3×10^{-4} μg/kg	9×10^{-4} μg/kg	1.7×10^{-3} μg/kg
Milk[b]	4.2×10^{-5} μg/kg	2.8×10^{-4} μg/kg	5.5×10^{-4} μg/kg
Vegetables[c]	1.9×10^{-6} μg/kg	1.3×10^{-5} μg/kg	5.4×10^{-5} μg/kg

[a] Calculated using bioconcentration factor for TCE (Howard et al. 1989).
[b] Calculated based on feed-to-meat and feed-to-milk biotransfer factors (Travis and Arms 1988).
[c] Calculated based on biococentration factor for vegetable uptake via the root and leaf pathways.

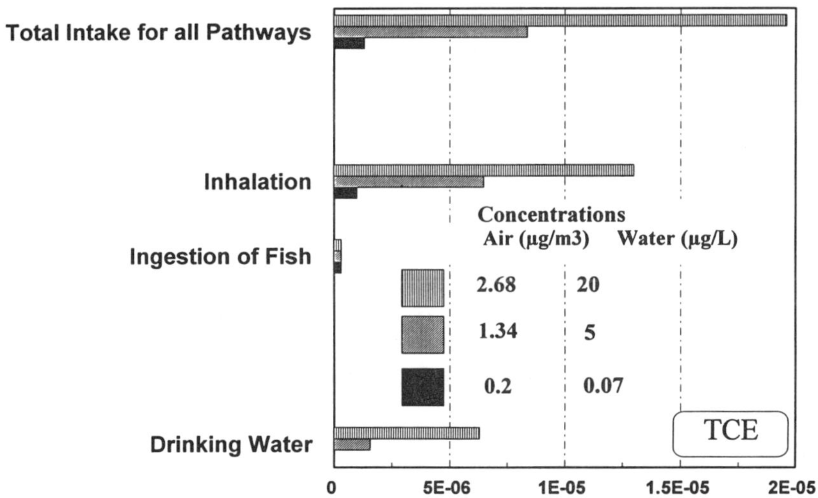

FIG. 13—*Individual lifetime cancer risk due to exposure to TCE: contribution of various pathways (hypothetical exposure scenarios corresponding to the concentration data given in Table 7).*

It is worth noting that in contrast to VOCs, the exposure profile for nonvolatile or semivolatile organics (for example, PAHs, PCBs, dioxins) can be very different since many of these compounds can bioaccumulate to a significant degree; thus, the intake of these contaminants via the food chain can be significant. In addition, unlike VOCs, a significant fraction of the nonvolatile and semivolatile organics present in the environment are transported in the particle-bound form; thus, these chemicals can accumulate in the terrestrial environment due to dry and wet deposition (Cohen and Clay 1994; Ryan and Cohen 1986).

VOCs in the Soil Environment

Overview

Direct exposure of humans to VOCs present in the soil environment may occur by the inhalation of the toxic solutes as they volatilize from the soil, dermal absorption due to contact with the soil, ingestion of soil particles, or from the drinking of contaminated ground water. The transport of organics in the soil matrix is a complex phenomenon which involves the various soil phases (soil-solids, soil-air, soil-water, and soil-organic matter) and the exchange of chemicals between the soil and ground water and the soil and the atmosphere (Fig. 14).

The exchange of pollutants between the soil and the atmosphere is determined largely by volatilization, an intermedia transport process which is governed by molecular diffusion, water and air movement, temperature variations in the soil, and aerodynamic factors operating at the air/soil interface (Cohen and Ryan 1989), sorption by the soil matrix, and chemical- and bio-transformations. The transport of pollutants to the atmosphere may also occur by the resuspension of particulate matter via wind erosion (Cowherd et al. 1988; Cohen and Clay 1994). Pollutants may also be transported from the soil to water bodies through runoff or through leaching to ground water (Novotny and Chesters 1981). VOC transport in the soil matrix is also affected by rain infiltration and subsequent moisture movement in the soil during nonrainy periods.

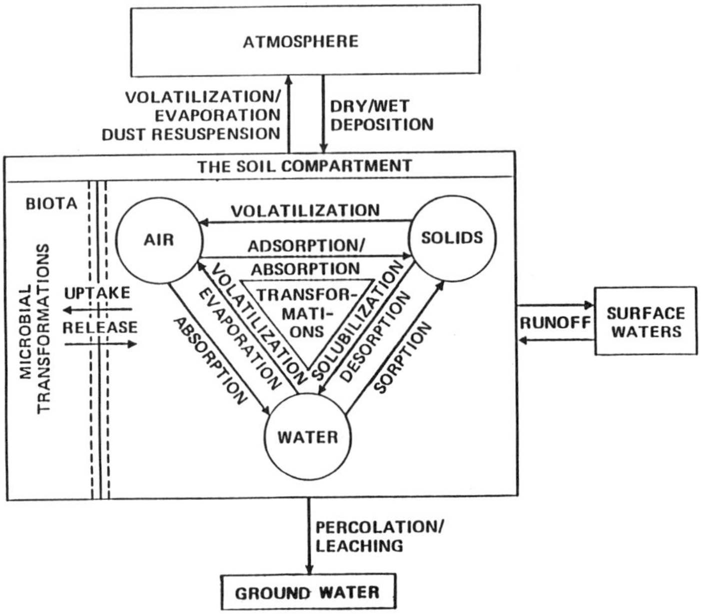

FIG. 14—*Intermedia transport pathways associated with the soil matrix (source: Cohen 1987).*

Volatilization from the Soil Matrix

The propensity of VOCs to volatilize through the soil matrix can be ranked according to a simple volatility as defined by Grifoll and Cohen (1994). Accordingly, the fraction of the VOC volatilization flux which is carried by the soil air phase is defined as a *volatility index.* For the special case, when VOC transport through the soil matrix is dominated by diffusional transport, and when the assumption of local equilibrium is reasonable, the *volatility index,* α_a, for a soil that consists of soil-air, soil-solids, and soil-water, can be written as

$$\alpha_a = \frac{1}{1 + q\,\dfrac{\theta_w \tau_a}{\theta_a \tau_w}} \tag{1}$$

$$q = \frac{1}{H_{aw}} \left[\frac{D_w}{D_a} \right] \tag{2}$$

where θ_i and τ_i are the volume fraction and tortuosity of phase *i*, respectively, D_i is the molecular mass diffusivity of the chemical in phase *i*, H_{ij} is the partition coefficient between phase *i* and phase *j* (that is, $H_{ij} = C_i/C_j$), and the subscripts *a* and *w* denote the air and water phases, respectively. If a specific model such as the Millington model (1959) is selected to evaluate the tortuosities (assumed to apply to both the soil-air and soil-water phases)

$$\tau_i = \frac{\epsilon^2}{\theta_i^{2.3}} \tag{3}$$

then α_a can be written as

$$\alpha_a = \frac{1}{1 + q\left(\frac{S_w}{1 - S_w}\right)^{3.3}} \tag{4}$$

where ϵ is the soil porosity (that is, $\epsilon = \theta_a + \theta_w$) and S_w is the water phase saturation (that is, $S_w = \theta_w/\epsilon$) which ranges in value from zero to unity. The volatility index as defined in Eq 4 is only a function of the dimensionless parameter q and the water saturation. The value of α_a ranges from a value of zero to unity. As the value of α_a approaches unity the chemical can be considered a highly volatile chemical for the soil conditions under consideration. Conversely, a chemical has little propensity to volatilize from the soil matrix when its *volatilization index* is small (that is, $\alpha_i \ll 1$) for the given soil conditions.

Typical values for q for a few different chemicals are given in Table 8. As noted by Cohen and Grifoll (1994) the ratio D_w/D_a appears to be a nearly chemical-independent constant with a value of about 10^{-4} for most VOCs; thus, the variability in the parameter q for the different chemicals is largely due to the variability in the Henry's law constant. As an example, if we focus on trichloroethylene for which $q = 2.5 \times 10^{-4}$, then for a soil-water phase saturation of 0.90 the soil volatility index attains a value of 0.74. This means that even for soils with a high water content, TCE will volatilize primarily through the soil air phase. In contrast, for 2,3,7,8-tetrachloro dibenzo-p-dioxin (2,3,7,8-TCDD), which has a q value of 0.105, α_a is about 0.007; this low value of α_a signifies that it is more difficult for 2,3,7,8-TCDD to volatilize from the soil at the same soil moisture condition. In other words, chemicals that have a low volatility index will be able to reach the soil/atmosphere interface more effectively due to bulk water movement in the soil rather than volatilization through the soil air phase.

VOC Migration in the Soil Under the Action of Rain and Water Evapotranspiration

Under the action of rainfall, contaminants can be transported downward through the soil column. Once the soil moisture content is increased, subsequent soil drying, after rainfall, results in a moisture flux upward (towards the soil/atmosphere interface) which can enhance the rate of volatilization (Cohen and Ryan 1989; Jury et al. 1990). The effect of rain infiltration can be illustrated using the model of Grifoll and Cohen (1995) which was developed to simultaneously track the coupled chemical and moisture transport in the soil matrix due to rain infiltration and soil drying. Benzene and methanol, with q values of 5.7×10^{-4} and 0.49, respectively, were selected to illustrate the effect of rain on infiltration rate. Methanol is completely water-miscible while benzene has a water solubility of 1.78 g/L. Simulations were carried out for these two compounds initially present in the soil at uniform concentra-

TABLE 8—*Values of* D_a, D_w, H_{aw} *and* q *for different chemicals.*

Compound	$D_a \times 10^5$, m^2/s	$D_w \times 10^9$, m^2/s	H_{aw}[a]	q
Benzene	0.88	0.9	0.18	4.6×10^{-4}
Methanol	1.6	1.5	1.9×10^{-4}	0.49
2,3,7,8 tetrachloro dibenzo-p-dioxin (TCDD)	0.51	0.36	6.7×10^{-4}	0.105
Trichloroethylene	0.81	0.835	0.41	2.5×10^{-4}

[a] Values of the dimensionless Henry's law constant were taken to be the values nearest to 20°C. When the temperature was not reported and various values were given, the value of the Henry's law constant reported in the table is the average of the reported values.

tions to a depth of 0.5 m, using a synthesized rainfall data set patterned after rainfall distribution in the Los Angeles region. Typical concentration profiles for January as the simulation start time are shown in Fig. 15. It is clear from the concentration profiles that the chemicals are migrating both up and down the soil column. The loss of methanol is less pronounced than benzene because methanol is less volatile and also because the loss of methanol from the soil is retarded, in part, due to its high solubility in the soil-water. Another important issue is the initial encounter between the rain infiltration and the chemical in the soil matrix. For example, a VOC which is introduced to the soil matrix during the summer months will volatilize to an appreciable extent over the summer, thus significantly depleting the total amount of VOC in the soil. In contrast, a chemical which is introduced to the soil matrix during the winter months will be driven deeper into the soil matrix thereby retarding its ability to volatilize. The result is a slower rate of volatilization loss of the chemical from the soil. The above situation is clearly illustrated in Fig. 16 where benzene, which is introduced into the soil during the summer, is lost at a greater rate than when it is introduced into the soil in the winter.

Another phenomenon which is often neglected in assessing the transport and fate of VOCs in the environment is that under some circumstances diurnal temperature cycles can lead to significant diurnal variations of the volatilization flux (Cohen et al. 1989; Cohen and Ryan 1989; Shonnard and Bell 1993). For VOCs which are not strongly sorbed onto the soil matrix (for example, benzene) this effect can result in up to a factor of two variations in the volatilization flux over the course of a day. In contrast, for strongly sorbed organics (which usually have a relatively low Henry's law constant) diurnal variations of the volatilization flux can reach up to two orders of magnitude (Cohen and Ryan 1989).

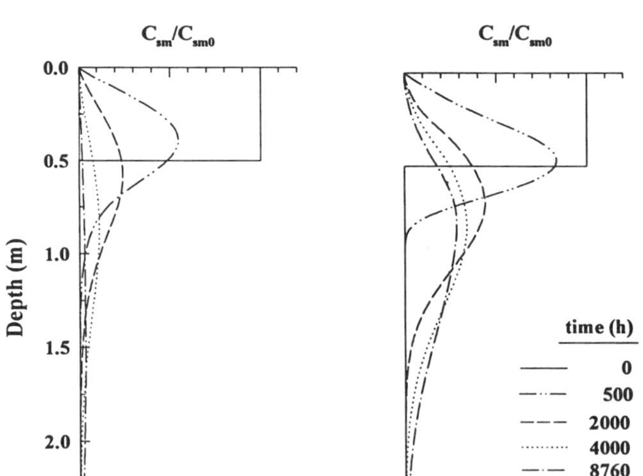

a) Benzene **b) Methanol**

FIG. 15—*Concentration profiles for benzene and methanol in the soil subsurface.* C_{sm} *and* C_{smo} *denote the concentration of benzene in the soil matrix at time* t *and at time* t = 0, *respectively. Soil composition (by volume): 10% clay, 30% silt, 60% sand; organic carbon content, 2% (mass basis); soil porosity = 0.35 (source: Grifoll and Cohen 1995).*

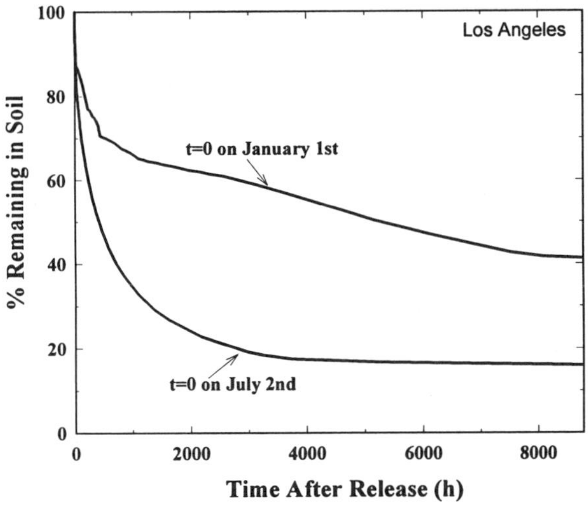

FIG. 16—*Percent of benzene remaining in the soil for two different scenarios. Scenario A—initial soil contamination in January; Scenario B—initial soil contamination in July (source: Grifoll and Cohen 1994).*

When the soil environment (both saturated and unsaturated soil zones) is contaminated with an organic free phase such as a water immiscible organic phase (also known as non-aqueous phase liquid, NAPL), VOCs contained in these phases can partition into the air, water, and solid phases of the soil. These VOCs can then migrate through the soil, air, and water ahead of the moving NAPL phases (Ryan and Cohen 1991). Hence, it is entirely possible that ground-water contamination can occur prior to the physical contact of the free NAPL phase with ground water. The fact that VOCs volatilize from NAPL phases is directly used in soil gas monitoring approaches designed to detect and map subsurface organic contaminant plumes. However, the ability to relate soil-gas monitoring data to the mapping of NAPL plumes depends on the availability of accurate information on VOC sorption isotherms for the specific area of concern, in addition to accurate characterization data of the soil subsurface.

Exchange of VOCs between the Atmosphere and Surface Water

The exchange of VOCs between the atmosphere and surface water bodies is governed by the degree of turbulence in the water body and the air above, temperature, diffusion coefficients (in the water and air phase) and the Henry's law constants for the VOC of concern. The interfacial flux, N, between air and water phases is generally expressed

$$N = K_G\,(C_g - HC_l) = -K_L(C_l - C_g/H) \tag{5}$$

where K_G and K_L are the overall mass transfer coefficients (cm/s) for the gas and liquid phase, respectively, H is the dimensionless Henry's law constant, and C_g and C_l are concentrations in gas and liquid phase, respectively. In Eq 5 it is assumed that local chemical equilibrium exists at the interface. The overall mass transfer coefficients for the gas and liquid phase can be defined as

$$1/K_G = 1/k_g + H/k_l \quad (6a)$$

$$1/K_L = 1/k_l + 1/Hk_g \quad (6b)$$

where k_g is a gas-phase mass transfer exchange coefficient, H is the dimensionless Henry's law constant (that is, air/water partition coefficient), and k_l is a liquid-phase mass transfer coefficient. Under typical environmental conditions the magnitude of k_l and k_g is approximately in the range of 10^{-4} to 10^{-6} m/s and 10^{-2} to 10^{-3} m/s, respectively. The individual-phase mass coefficients depend on several factors, including the prevailing turbulence level (as affected by water currents or wind), air and water temperature, solute mass diffusivity, and the geometry of the water body (Cohen 1986). Thus, one must ensure that k_g and k_l are determined for the appropriate environmental conditions using the appropriate estimation equations (Cohen 1986; Lyman et al. 1990; Shen et al. 1993; Thibodeaux 1970; Cohen et al. 1994).

Wet Deposition

Wet deposition (for example, pollutant scavenging by rain and snow) can increase the overall deposition rate of VOCs and thus lead to input of VOCs to the soil, water, and vegetation compartments. Since precipitation events are episodic, regional climatological conditions must be considered when evaluating the relative importance of wet deposition as a removal process.

The degree of wet deposition is usually estimated from knowledge of the washout ratio, W, defined as

$$W = C_{\text{rainwater}}/C_{\text{air}} \quad (7)$$

in which $C_{\text{rainwater}}$ and C_{air} are the VOC concentrations in the precipitation (at ground level) and in the gaseous phase (that is, in the atmosphere), respectively. Thus, the wet deposition flux, $\dot{M}_w$, is given by

$$\dot{M}_w = W \cdot R \cdot C_{\text{air}} \quad (8)$$

where R is the precipitation rate.

Rain scavenging of VOCs can be predicted using detailed mathematical models (Hales 1986; Tsai et al. 1991; Clay 1992). However, a simple upper-limit estimate can be obtained for VOCs which are not reactive in the aqueous phase by assuming that the raindrops are in equilibrium with the air phase when they reach ground level. This assumption is supported by the analysis of Cohen (1986) which implies that raindrops that have fallen through 10 m of polluted atmosphere will be nearly at equilibrium with the gaseous phase. Thus, the maximum concentration of the dissolved chemical in rain, $C_w^{(d)}$ (ng/m^3 water), is given by

$$C_w^{(d)} \cong \frac{C_a^{(g)}}{H_{aw}} \quad (9)$$

in which, $C_a^{(g)}$ is the concentration of the chemical in the atmosphere in the vapor phase (for example, ng/m^3 air), and H_{aw} is the dimensionless Henry's law constant (i.e., $H_{aw} = C_a^{(g)}/C_w^{(d)}$). The modeling of rain scavenging of chemicals that undergo significant chemical transformations in the aqueous phase requires detailed reaction kinetics information. Finally, it is noted that when VOCs are scavenged from contaminant plumes, a detailed analysis of the dynamic VOC uptake by raindrops may be needed (Hales 1986; Cohen 1986).

Summary and Conclusions

In closure, VOCs in the environment present a unique problem in that exposure to VOCs occurs in multiple environmental media due to multiple sources and the migration of VOCs across environmental phase boundaries. Emission inventories for VOCs are being developed and together with monitoring activities information necessary for the design of effective pollution prevention strategies is being accumulated. It must be recognized, however, that environmental monitoring of the large number of VOCs that are of concern is a formidable task. Thus, there is a need to couple monitoring and modeling activities in order to reduce the cost of monitoring and allow for future projections of potential impacts. Given the many scattered emission sources, an integrated approach that addresses the major source categories may be the logical approach to reducing the environmental impact of VOCs.

Acknowledgment

The completion of this paper was funded in part by the University of California Toxic Substances Research and Training Program.

References

Allen, D., Kaplan, I., and Cohen, Y., Eds., 1989, *Intermedia Pollutant Transport: Modeling and Field Measurements,* Plenum Press, New York.

Andelman, J. B., Giardino, N. J., Borrazzo, J. E., and Davidson, C. I., 1990, "Exposure of Volatile Chemicals from Indoor Use of Water" *Total Exposure Assessment Methodology: A New Horizon,* Air and Waste Management Association, Pittsburgh, PA.

Austin, S. B., Greenfield, S. M., Weir, B. R., Anderson, G. E., and Behar, J. V., 1992, "Modeling the Indoor Environment," *Environmental Science and Technology,* Vol. 26, pp. 851–858.

Baker, D. R. and Warren, J. L., 1992, "Generation of Hazardous Waste in the United States," *Journal of Hazardous Waste & Hazardous Materials,* Vol. 9, pp. 19–35.

Bowman, J. A., June 1989, "The Consequence of [SARA] Section 313," *Chemical Engineering Progress,* pp. 48–52.

Bull, R. J. and McCabe, L. J., 1985, "Risk Assessment Issues in Evaluating the Health Effects of Alternate Means of Drinking Water Disinfection" *Water Chlorination: Chemistry, Environmental Impact and Health Effects,* Vol. 5, R. L. Jolley, R. J. Bull, W. P. Davis, S. Katz, M. H. Roberts, Jr., V. A. Jacobs, Eds., Lewis Publishers, Inc., Chelsea, MI.

Clay, R. E., 1992, "Multimedia Environmental Distribution of Gaseous, Dissolved and Particle-Bound Pollutants," M.S. thesis, Department of Chemical Engineering, University of California, Los Angeles.

Cohen, Y. and Clay, R. E., 1994, "Multimedia Partitioning of Particle-Bound Organics," *Journal of Hazardous Materials,* Vol. 37, pp. 507–526.

Cohen, Y., Winer, A. M., Creelman, L., and Pesinova, V., May, 1994, "Development of Intermedia Transfer Factors for Benzene," Report prepared for the California Air Resources Board.

Cohen, Y., March 14, 1992, "Multimedia Fate and Effects of Airborne Petroleum Hydrocarbons in the Port Valdez Region," Report prepared for the Regional Citizens' Advisory Council, Anchorage, AK.

Cohen, Y. and Allen, D. T., 1992, "An Integrated Approach to Process Waste Minimization Research," *Journal of Hazardous Materials,* Vol. 29, pp. 237–253.

Cohen, Y., Anderson, G. E., Chinkin, L. R., Pascoe, G., Schmidt, C. E., and Winer, A., Aug. 25, 1992, "Review of the Valdez Air Health Study," Report prepared for the Regional Citizens' Advisory Council, Anchorage, AK.

Cohen, Y., Allen, D. T., Clay, R. E., Rosselot, K., Klee, H., Jr., and Blewitt, D. N., 1991, "Multimedia Assessment of Emissions (MAE) from the AMOCO Yorktown Refinery (AMOCO/EPA) Pollution

Prevention Project," Air and Waste Management Association, 84th Annual Meeting, Vancouver, BC, June 16–21.

Cohen, Y., Tsai, W., Chetty, S. L., and Mayer, G. J., 1990, "Dynamic Partitioning of Organic Chemicals in Regional Environments: A Multimedia Screening-Level Modeling Approach," *Environmental Science and Technology,* Vol. 24, pp. 1549–1558.

Cohen, Y. and Ryan, P. A., 1989, "Chemical Transport in the Top Soil Zone—The Role of Moisture and Temperature Gradients," *Journal of Hazardous Materials,* Vol. 22, pp. 283–304.

Cohen, Y., Taghavi, H., and Ryan, P. A., 1988, "Contaminant Diffusion Under Non-Isothermal Conditions in Nearly Dry Soils," *Journal of Environ. Quality,* Vol. 17, pp. 198–204.

Cohen, Y., Ed., 1986, "Intermedia Transport Modeling in Multimedia Systems," *Pollutants in a Multimedia Environment,* Plenum Press, New York.

Conway, R. A., Ed., 1982, "Environmental Risk Analysis for Chemicals," Van Nostrand Reinhold Co., New York.

Countess, R. J. and Browne, D., 1993, "Fugitive Hydrocarbon Emissions from Pacific Offshore Oil Platforms: Models, Emission Factors, and Platform Emissions," *Journal of Air Waste Manage. Assoc.,* Vol. 43, pp. 1455–1460.

Cowherd, C., Muleski, G. E., and Kinsey, J. S., Sept. 1988, "Control of Open Fugitive Dust Sources," NTIS Report Number PB89-103691.

Doa, M. J., 1992, "The Toxics Release Inventory," *Journal of Haz. Waste & Hazardous Materials,* Vol. 9, pp. 61–72.

Dunovant, V. S., Clark, C. S., Que Hee, S. S., Hertzberg, V. S., and Trapys, J. H., 1986, "Volatile Organics in the Waste Water and Airspaces of Three Wastewater Treatment Plants," *Journal of Water Pollut. Control. Fed.,* Vol. 58, pp. 886–895.

Federal Register, July 22, 1991, "The Proposed Rule: Organic Air Emission Standards for Hazardous Waste TSDFs," Vol. 56, No. 140.

Friedman, K. A., June 1989, "EPA and SARA Title III," *Chemical Engineering Progress,* pp. 18–25.

Goldstein, B. D., Tardiff, R. G., Baker, S. R., Hoffnagle, G. F., Murray, D. R., Catiozone, P. A., Kester, R. A., and Caniparoli, D. G., June 1992, "Valdez Air Health Study," prepared for the Alyeska Pipeline Service Company, Anchorage, AK.

Govind, R., Lai, L., and Dobbs, R., 1991, "Integrated Model for Predicting the Fate of Organics in Waste Water Treatment Plants," *Environmental Progress,* Vol. 10, pp. 13–23.

Grifoll, J. and Cohen, Y., 1994, "Chemical Volatilization from the Soil Matrix: Transport through the Air and Water Phases," *Journal of Hazardous Materials,* Vol. 37, pp. 445–457.

Grifoll, J. and Cohen, Y., 1995, "Contaminant Migration in the Unsaturated Soil Zone Under the Effect of Rainfall and Evapotranspiration," *Journal of Contaminant Hydrology,* in press (1996).

Grosjean, D., 1991, "Ambient Levels of Formaldehyde, Acetaldehyde, and Formic Acid in Southern California: Results of a One-Year Base-Line Study," *Environmental Science and Technology,* Vol. 25, pp. 710–715.

Hales, J. M., 1986, "Precipitation Chemistry: Its Behavior and Its Calculation," in *Air Pollutants and Their Effects on the Terrestrial Ecosystems,"* A. H. Legge and S. V. Krupa, Eds., J. Wiley & Sons, New York.

Howard, P., *Book of Environmental Fate and Exposure for Organic Chemicals,* 1989–1991, Vol. I–III, Lewis Publishers, Chelsea, MI.

Howard, P., Boethling, R., Jarvis, R., Meylan, W., and Michalenko, E., 1991, *Environmental Degradation Constants,* Lewis Publishers, Chelsea, MI.

Jury, W. A., Russo, D., Streile, G., and Abd, E., 1990. "Evaluation of Volatilization by Organic Chemicals Residing Below the Soil Surface," *Water Resources Research,* Vol. 26, pp. 13–20.

Klee, H. Jr., Schmitt, R. E., Podar, M. K., and Harrass, M. C., 1994, "Inventory on VOC Releases from AMOCO's Yorktown Refinery," Internal Symposium on Volatile Organic Compounds in the Environment, 11–13 April, Montreal, Canada.

Lagarias, J. S., Suer, C. B., Venturini, P. D., and Ames, D. J., 1993, "The California Experience in Toxic Air Pollutants Control," *Managing Hazardous Air Pollutants: State-of-the-Art,* W. Chow and K. K. Conor, Eds., Lewis Publishers, Chelsea, MI.

LaGrone, S., 1991, "Potential Community Exposure to Toxic Chemicals," *Environmental Science and Technology,* Vol. 25, pp. 366–368.

Lioy, P. J., 1990, "Assessing Total Human Exposure to Contaminants," *Environmental Science and Technology,* pp. 938–945.

Loehr, R. and Ward, C. H., 1987, "Waste Treatment and Cross-Media Transfer of Pollutants" *Cross-Media Approaches to Pollution Control,* National Academy Press, Washington, DC.

Los Angeles Times, 19 March 1994, "Superfund: Delays Raise Doubts About Program."

Loscutoff, W. V. and Poore, M. W., 1993, "Ambient Air Toxics Data from California's Toxic Air Contaminant Monitoring Program" *Managing Hazardous Air Pollutants: State-of-the-Art,* W. Chow and K. K. Conor, Eds., Lewis Publishers, Chelsea, MI.
Lyman, W. J., Reehl, W. F., and Rosenblatt, D. H., 1990, *Handbook of Chemical Property Estimation Methods,* American Chemical Society, Washington, DC.
McKone, T. E., 1987, "Human Exposure to Volatile Organic Compounds in Household Tap Water: The Indoor Inhalation Pathway," *Environmental Science and Technology,* Vol. 21, pp. 1194–1201.
Millington, R. J., 1959, "Gas Diffusion in Porous Media," *Science,* pp. 100–102.
National Academy Press, Washington, DC, 1987, *Multimedia Approaches to Pollution Control.*
Naugle, D. F. and Pierson, T. K., 1991, "A Framework for Risk Characterization of Environmental Pollutants," *Journal of Air and Waste Management Assoc.,* Vol. 41, pp. 1298–1307.
Novotny, V. and Chesters, G., 1981, *Handbook of Nonpoint Pollution,* Van Nostrand Reinhold, New York.
Onishi, Y., Shuyler, L., and Cohen, Y., 1990, "Multimedia Modeling of Toxic Chemicals" *Proceedings,* International Symposium on Water Quality Modeling of Agricultural and Non-Point Sources, Part II, Donn G. DeCoursey, Ed., published by the United States Department of Agriculture, ARS-81, pp. 479–509.
Ott, W. R., 1985, "Total Human Exposure," *Environmental Science and Technology,* pp. 880–886.
Ryan, P. A. and Cohen, Y., 1986, "Multimedia Transport of Particle Bound Organics: Benzo(a)Pyrene Test Case," *Chemosphere,* Vol. 15, pp. 21–47.
Ryan, P. A. and Cohen, Y., 1991, "One-Dimensional Transport of Non-Aqueous Phase Liquid Containing Sparingly Water Soluble Organics: A Front-Tracking Model," *Water Resources Research,* Vol. 27, pp. 1487–1500.
Scheff, P. A. and Wadden, R. A., 1993, "Receptor Modeling of Volatile Organic Compounds. 1. Emission Inventory and Validation," *Environmental Science and Technology,* Vol. 27, pp. 617–625.
Shen, T. T., Schmidt, C. E., and Card, T. R., 1993, *Assessment and Control of VOC Emissions from Waste Treatment and Disposal Facilities,* Van Nostrand Reinhold, New York.
Shonnard, D. R. and Bell, R. L., 1993, "Benzene Emissions from Contaminated Air-Dry Soil with Fluctuations of Soil Temperature or Relative Humidity," *Environmental Science and Technology,* Vol. 27, pp. 2909–2913.
Sleep, B. E. and Sykes, J. F., 1989, "Modeling of the Transport of Volatile Organics in Variably Saturated Media," *Water Resources Research,* Vol. 25, pp. 81–92.
South Coast Air Quality Management District, Sept. 1988, "Analysis of Ambient Data from Potential Toxics Hot Spots in the South Coast Air Basin."
Swann, R. L. and Eschenroeder, A., 1983, "Fate of Chemicals in the Environment," ACS Symposium Series, No. 225, American Chemical Society, Washington, DC.
Thibodeaux, L. J., 1970, *Chemodynamics: Environmental Movement of Chemicals in Air, Water, and Soil,* John Wiley & Sons, New York.
Travis, C. and Arms, A., 1988, "Bioconcentration of Organics in Beef, Milk, and Vegetation," *Environmental Science and Technology,* Vol. 22, pp. 271–274.
Tsai, W., Cohen, Y., Sakugawa, H., and Kaplan, I. R., 1991, "Dynamic Partitioning of Semi-Volatile Organics in Gas/Particle/Water Phases During Rain Scavenging," *Environmental Science and Technology,* Vol. 25, pp. 2012–2023.
USEPA, 1982, "Fugitive Emission Sources of Organic Compounds—Additional Information on Emissions, Emission Reductions, and Costs," Research Triangle Park, NC, Publication No. EPA-450/3-82-010.
USEPA, Sept. 24, 1986, "Guidelines for Carcinogen Risk Assessment," *Federal Register,* Vol. 51, p. 33992.
USEPA, 1988a, "Guidelines for Estimating Exposures," *Federal Register,* Vol. 51, pp. 34092–34054.
USEPA, 1988b, "Superfund Exposure Assessment Manual," EPA/540/1-88/001.
USEPA, 1988c, "National Ambient Volatile Organic Compounds (VOCs) Database Update," EPA ORD, Research Triangle Park, NC, EPA/600/3-88-010A.
USEPA, Aug. 1989, "Report to Congress on Indoor Air Quality."
USEPA, 1991a, "Health Assessment Summary Tables," OERR-9200.6-303 (91-1), NTIS#PB91-921199.
USEPA, 1991b, "Urban Air Toxics Monitoring Program," EPA OAQPS, Research Triangle Park, NC, EPA-450/4-91-001.
USEPA, March 1993, "Targeting Indoor Air Pollution," EPA 400-R-92-012.
Venturini, P. D., Denton, J. E., Howard, K., Huscroft, S. V., Shiroma, G. S., and Suer, C. B., 1991, "California Air Toxics Program: Review of Progress and Future Direction," presented at the Air and Waste Management Association, 84th Annual Meeting and Exhibition, Vancouver, BC, 16–21 June 1991.

Wallace, L. A., 1989, "Major Sources of Benzene Exposure," *Environ. Health. Perspectives,* Vol. 82, pp. 165–169.

Wallace, L. E., Pelizzari, E., Sheldon, L., Hartwell, T., Sparacino, C., and Zelon, H., 1986, "The Total Exposure Assessment Methodology (TEAM Study): Direct Measurements of Personal Exposures Through Air and Water for 600 Residents of Several U.S. Cities" in *Pollutants in a Multimedia Environment,* Y. Cohen, Ed., Plenum Press, New York.

Wallace, L., Nelson, W., Ziegenfus, R., Pelizzari, E., Michael, L., Whitmore, R., Zelon, H., Hartwell, T., Perritt, R., and Westordahl, D., 1991, "The Los Angeles TEAM Study: Personal Exposures, Indoor-Outdoor Air Concentrations, and Breath Concentrations of 25 Volatile Organic Compounds," *Journal of Exposure Analysis and Environmental Epidemiology,* Vol. 1, pp. 157–192.

Wilson, J. T., McNabb, J. F., and Eos, J. F., 1983, *Transactions,* American Geophysical Union, Vol. 64, pp. 505–506.

Zhang, J., He, Q., and Lioy, P. J., 1994, "Characteristics of Aldehydes: Concentrations, Sources, and Exposures for Indoor and Outdoor Residential Microenvironments," *Environmental Science and Technology,* Vol. 28, pp. 146–152.

Regulation and Assessment

Howard H. Klee, Jr.,[1] *Ronald E. Schmitt,*[1] *Mahesh K. Podar,*[2] *and Michael C. Harrass*[1]

Taking Inventory on VOC Releases from Amoco's Yorktown Refinery

REFERENCE: Klee, H. H., Jr., Schmitt, R. E., Podar, M. K., and Harrass, M. C., **"Taking Inventory on VOC Releases from Amoco's Yorktown Refinery,"** *Volatile Organic Compounds in the Environment, ASTM STP 1261,* W. Wang, J. Schnoor, and J. Doi, Eds., American Society for Testing and Materials, 1996, pp. 35–52.

ABSTRACT: Amoco's Yorktown, Virginia, refinery is a 35-year-old, 53 000 bbl/day facility that manufactures gasoline, heating oil, liquid petroleum gas, sulfur, and coke. In a cooperative and voluntary effort, Amoco Corporation and the U.S. Environmental Protection Agency conducted a joint project to study polluation prevention opportunities at an operating industrial facility.

Source reduction efforts—key to pollution prevention strategies—require knowledge of specific sources of releases. However, data on releases from individual process units are limited in favor of data to monitor existing end-of-pipe pollution control requirements. The study's sampling program sought to portray the distribution of releases within the refinery, their management within the refinery, and ultimate releases leaving the refinery. Subsequent tests of blowdown stack and fugitive emissions further improved total release estimates.

The initial study estimated that the refinery generates about 25 000 metric tons (t)/year of potential pollutants. Of these, about half are released from the refinery as airborne, waterborne, or land-disposed releases. Airborne releases comprise the majority of releases by mass, about 12 000 t/year. Most of the airborne releases are volatile organic compound hydrocarbons. The inventory sampling project and subsequent work identified differences with Toxic Release Inventory (TRI) values and standard emission factors (AP-42). The inventory and other data provided an opportunity to consider options for, and limitations of, specific pollution prevention or source reduction strategies.

KEYWORDS: pollution prevention, refinery, VOC release inventory

"Pollution prevention" represents a relatively recent term in the approaches advocated to improve the environment, although the concept is not at all new. The concept focuses on reducing or eliminating releases of materials into the environment rather than managing the releases later. Congress, in the Pollution Prevention Act of 1990, put great emphasis on source reduction as the primary, if not the exclusive, means to accomplish pollution prevention. However, the concept of pollution prevention can also involve recycling, treatment and environmentally sound disposal, as well as source reduction, among the potential choices in pollution management. All these actions effectively prevent pollution, that is, the release of materials into the environment.

In late 1989, Amoco Corporation (Amoco) and the United States Environmental Protection Agency (USEPA) began a voluntary, joint project to study pollution prevention opportunities

[1] Amoco Corporation, Chicago, IL, 60601.
[2] U.S. Environmental Protection Agency, Washington, DC, 20640.

at an operating industrial facility. A workgroup of Amoco, USEPA, and Commonwealth of Virginia staff agreed to use Amoco Oil Company's refinery at Yorktown, Virginia to conduct a multimedia assessment of releases to the environment, then to develop and evaluate options to reduce these releases. The purposes of the study were five-fold: (1) to inventory refinery releases to the environment, including their chemical type, quantity, source, and medium of release; (2) to develop options to reduce specific releases; (3) to rank and prioritize the options using a variety of criteria and perspectives; (4) to identify and evaluate factors that impede or invite pollution prevention; and (5) to enhance the participants' knowledge of refinery and regulatory systems. This paper focuses on the first objective. The project and executive summaries review findings about the other objectives (Amoco/USEPA 1992a,b, Industrial Economics, Inc. 1993).

Amoco's Yorktown refinery began operations in 1956 and the current capacity is 53 000 bbl/day (over 7000 metric tons (t) per day). The refinery manufactures gasoline, heating oil, liquefied petroleum gas (LPG), sulfur, and coke using processes that are reasonably typical of many U.S. refineries. It is located in an environmentally sensitive area along the York River in Virginia, near the Chesapeake Bay (Fig. 1). The relatively small size of the refinery enabled an environmental sampling and analysis program to be conducted within a reasonable timeframe. The proximity to Washington, DC encouraged over 100 Federal and state officials to visit and obtain firsthand information about refinery operations and practices.

The principal components of the refinery are listed in Table 1 along with their primary products and potential releases. As might be expected, there are a large number of potential sources of emissions. Sources of potential fugitive emissions—for example, valves, flanges, seals, and vents—number in the tens of thousands. One of the initial recognitions of the Amoco/USEPA workgroup was that any attempt to prevent pollution requires good information about what sources actually contribute to emissions, including quantitative information when available.

Current regulations require public reporting of certain emissions or releases. Title III of SARA, the "Emergency Planning and Community Right-to-Know Act," created the Toxic Release Inventory (TRI) in 1986. Title III requires regulated facilities in the Standard Industrial Category (SIC) Code 20-39 to submit annual release data on more than 300 chemicals manufactured, produced, or otherwise used in quantities exceeding certain thresholds. The TRI is one way of focusing corporate attention on opportunities to reduce releases.

TRI reports are based on either emission estimates, direct measurements, or a combination of methods. Each facility is responsible for the accuracy of the data reported. Industrial facilities frequently file amendments to TRI reports to reflect improvements in the accuracy of the estimation and measurement techniques. The emissions inventory provides an opportunity to compare the TRI-reported values with measured values or improved estimates.

Methods: Emissions Sampling Program

Monitoring resources are typically allocated to meet permit discharge requirements rather than at the point of generation. As a result, a comprehensive multimedia release inventory could not be assembled from information on individual process unit contributions. To overcome this deficiency, a multimedia sampling collection and analysis approach was used, including air, surface water, ground water, and solid wastes. Selected chemicals were targeted for measurement in order to make the number and types of analyses manageable.

Separate sampling protocols for air, surface water, and ground water were developed (Blewitt et al. 1992, Radian 1990). Solid waste generation data from 1988 to 1990 compiled by the refinery in response to an industry survey was used to evaluate solid waste, augmented by chemical analyses performed at selected locations (Kizior 1991). To evaluate surface

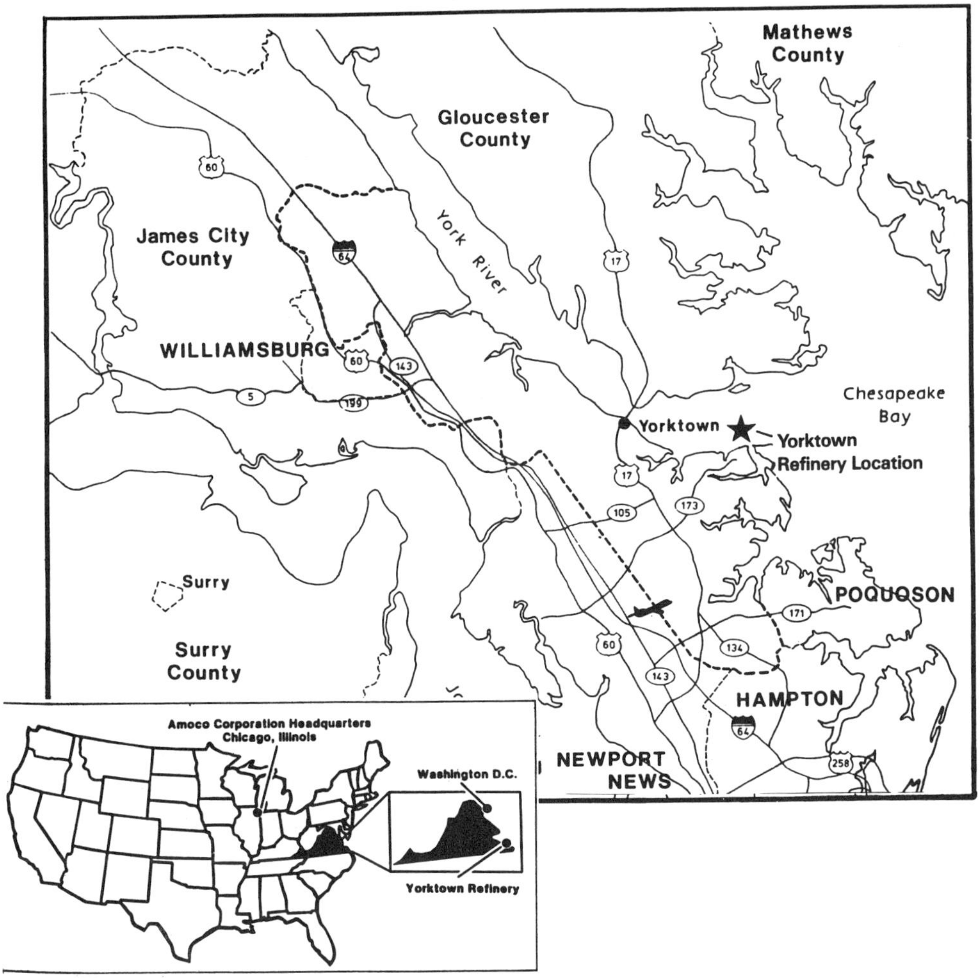

FIG. 1—*Yorktown refinery location.*

water, flows in sewers and ditches were measured and process flows were estimated indirectly, augmented by sample analyses from numerous locations and process streams (Amoco/USEPA 1991b). To evaluate ground water, samples from a network of 181 wells both on and off the facility proper were measured (Amoco/USEPA 1991a). All sampling programs were reviewed at some level by USEPA technical personnel at one or more regional laboratories in Ada, Oklahoma, Cincinnati, Ohio, and Research Triangle Park, North Carolina (Blewitt et al. 1992). The entire program was reviewed in general terms by an independent peer review group retained by USEPA (Glickman 1990).

Inventory Structure—The emissions inventory used combinations of direct measurement, default factors, mass balance, patterns of chemical speciation, and modeling. Table 2 details how these methods were used for the various inventory components. Emphasis in this discussion will be on the evaluation of air emissions. Evaluation of other emissions is detailed elsewhere (Amoco/USEPA 1991a,b, Kizior 1991).

TABLE 1—*Significant components of Yorktown refinery emissions.*

Unit	Process	Products	Type of Emission	Media
Crude unit	fractionates crude oil	intermediates	fugitive HC[a] process waste water	air sewer
Ultraformer	converts intermediates	gasoline	fugitive HC process water spent catalyst	air sewer solid waste
Fluid catalytic cracker unit	converts intermediates	distillate fuels and gasoline	fugitive HC particulates process water spent catalyst	air air sewer solid waste
Coker	treats heavier fractions, recycles sludge	intermediates, coke	fugitive HC particulates quench water	air air sewer
Distillate desulfurizer	removers sulfur	distillate fuels	fugitive HC spent catalyst	air solid waste
Gasoline treating	blends, stores final products	gasoline	fugitive HC	air
MTBE polymerization	makes additive	MTBE oxygenate	fugitive HC	air
Flare	burns trace gases	...	CO, NOx, SOx	air
Blowdown stacks	recover volatiles	...	fugitive HC	air
Waste water treatment	separation, treatment of water from processes, cooling, sewers	clean water	fugitive HC sludges treated water	air solid waste York River
Marine loading and unloading	loading products on barges, receiving crude	...	fugitive HC	air

[a] HC = hydrocarbons.

Air Emissions Sampling—Because air emissions contribute the major portion of releases, this review will focus on those methods and results. The chemicals of most interest in the air emissions studies were benzene, toluene, ethylbenzene, and xylene (total), (abbreviated collectively as BTEX), total nonmethane hydrocarbons (TNMHC). The air emissions were quantified by direct measurements and indirect analyses. Ambient measurements were made to provide verification of the BTEX emission inventory by comparing measured concentrations with concentrations predicted using the USEPA Industrial Source Complex Short-Term (ISCST) dispersion model (USEPA 1987).

Prior to field sampling, the potential sources, estimated emissions, and the quality of existing data used to develop emission estimates were reviewed. Previous emission estimates for the API separator, sewer vents, inactive landfarm, blowdown stacks, and the coker blowdown pond were judged to be the most uncertain of the values used to estimate emissions. For example, past emission estimates for the API separator used the AP-42 emission factor of 0.6 kg/1000 L of water from refinery wastewater treatment facilities (USEPA 1985). This factor apparently was based on a single mesurement from an API separator conducted in 1960 (Blewitt et al. 1992). Subsequently, additional tests of API separators have been published, but with apparently limited success (Hustvedt 1992). Similarly, little data existed on emission factors for speciated hydrocarbon emissions from an inactive landfarm or a coker blowdown pond. All of these emission sources are "fugitive" in nature; that is, they do not originate from a stack, do not have buoyancy or momentum, and cannot be easily sampled

TABLE 2—*Inventory components.*[a]

Air Emissions		
Type of Source	Basis for Emission Estimate	Method Used to Speciate Emissions
API Separator	direct measurement	direct measurement
Barge loading	AP-42	component fraction and ratios of total emissions
Blowdown stacks	AP-42 and speciation	AP-42 and direct measurement
Coker pond	direct measurement	direct measurement
Cooling tower	mass balance, water samples	direct measurement
Inactive landfarm	direct measurement	direct measurement
Loading rack	AP-42	ratio of total emissions using vapor or liquid fraction
Fugitive emissions from processes	AP-42	ratio of total emissions using vapor or liquid fraction
Sewer vent	direct measurement	direct measurement
Tanks	AP-42	ratio of total emissions using vapor or liquid fraction

Surface Water Emissions Type of Source
Sewer flow
Process flows
Drainage ditches

Subsurface Emissions Type of Source
Ground water

Solid Waste Emissions Type of Source
API separator sludge
biomass from activated sludge process
storm basin sludge
equalization basin sediment
process separator sludges
tank bottoms
area catch basins
spent catalysts
coke
soils
crude oil

[a] Sources investigated for each type of emissions are listed. For emissions other than air, direct measurements were used to estimate emissions and to determine hydrocarbon species. Modeling was used to assist in the determination of ground water flows and emissions.

to estimate mass emission rate. Further, these emission sources are likely to be very site specific.

API Separator Measurements—An isolation flux chamber, originally developed for USEPA by Radian Corporation (Keinbush et al. 1988, Radian 1985), was used to measure flux from a portion of the surface of the 2400 m^2 API oil-water separator. The chamber, made of inert material, was placed on the surface, thus isolating a known surface area. The chamber was purged with ultra-high purity air. The outlet was monitored for total hydrocarbons to determine when equilibrium conditions were achieved. A whole air sample was collected using an evacuated Suma canister and analyzed according to USEPA Method TO-14. Details of the procedure are reported separately (Williams 1991).

A flux rate (in g/s-m^2) was estimated from the concentration measurements, the purge gas rate, and the surface area of the chamber. The averge recovery of this technique has been reported to be 103%, ranging from 77 to 124% for 40 different compounds (Keinbush et al.

1988). Two replicate measurements were made within 2 h, taken at opposite ends of the separator box. The average of these measurements was used to calculate the emissions rate.

Several simplifying assumptions were made in order to estimate an annual emissions rate. The flux rate was assumed to be uniform over the entire area of the API separator, and was assumed to be constant over time. The flux chamber creates an artificial wind at the air-water interface that is likely to be greater than actual wind across the surface of the below-grade unit. Ambient temperature during sampling was about 27°C, which should result in high evaporation and flux rates. Data from the samples showed high variability between the replicates, suggesting that the assumption of uniform emissions over the entire separator surface is overly simplistic. One ancillary conclusion of the project was the recognition that measurement techniques are rudimentary and better sampling methods are needed (Amoco/USEPA 1992b).

Coker Pond Measurements—The isolation flux chamber was used to sample from both the quench section and the overflow sections of the coker pond. Samples in the quench section were obtained every 6 h during the 18 h coking cycle. Samples in the overflow section pond were taken as in the API separator. Replicate samples from both sections of the coker were taken. Total emissions were estimated by summing emissions from both sections of the coker pond.

Landfarm Measurements—A single measurement was made using the flux chamber in the center of the inactive landfarm. The surface of the soil had been tilled recently, which should have increased the emission rate, at least temporarily.

Sewer Vent Measurements—Emissions were measured at three different sewer vent locations along the main sewer trunkline (near the coking unit, the cooling tower, and the tank farm). Sewer grates were covered with a plastic tarp and made airtight. A cylindrical stack (diameter 18 cm, height 20 cm) was placed in the center of the tarp. Vent emissions were sampled by a charcoal sorbent tube connected to a battery-powered pump, set to a flow of 300 mL/min, for a 30 min period.

Cooling Tower and Water Storage Tanks—For the cooling tower and the waste water storage area, air emissions were estimated from hydrocarbon measurements of the water at the inlet and outlets. A mass emission rate was then determined using the water flow rate through the facility with the assumption that the differences between the inlet and outlet were air emissions.

Coker Blowdown Stacks—A sampling program was undertaken to measure emissions from blowdown stacks. However, the quality control program indicated probable errors in the data. The sources were retested, but sampling was not completed within time to be included in the inventory project. Consequently, AP-42 emission factors were used to estimate VOC emissions from the blowdown stacks. Schmitt (1994) presents the results of the later program.

Barge Loading, Loading Racks, Product Tanks—Emissions from barge loading were estimated using the emission factor from AP-42 (713 t/year). Estimates for loading rack and product tank emissions were also made using AP-42 emission factors. Estimates of hydrocarbon species were made from the ratio of total VOC emissions to specific chemicals, using information for the composition of various products.

Ambient Air Monitoring—Ambient air was sampled at ten sites near the perimeter of the refinery (Williams 1991, Amoco/USEPA 1992d–g). The locations were chosen to reflect the predominent seasonal wind pattern and the modeled benzene impacts using the SARA 313 emission inventory, seeking the locations with maximum predicted impact. Six of the sites were located along the north side of the refinery, to be upwind of the refinery during typical daytime sea breeze conditions (NE winds) and downwind during typical nighttime land breeze conditions (SW winds). Four sites were along the southwest side of the refinery, to be downwind during NE winds and upwind during SW winds. In addition, the sites were chosen relative to selected emission sources (coker unit, sewer vents, Ultraformer unit, combination unit, product storage area, wastewater treatment ponds, landfarm, and API separator) in order to associate specific emitted substances with specific sources.

Sequential samples were collected over 24 h periods using charcoal sorbent tubes with a calibrated flow rate of 1 L/min for a 2 h sample time (nominal 120 L air sample). A subset of the samples collected were chosen for analysis based on predominant wind direction during collection, that is, in order to obtain a suitable upwind-downwind orientation. Organic vapors sorbed to the activated charcoal were extracted and analyzed by gas chromatograph. Details of the procedure, including quality control, are provided elsewhere (Amoco/USEPA 1992d–g, Blewitt et al. 1992, Williams 1991).

In addition, several 24 h composite samples were collected at four perimeter sites using directional samplers in evacuated canisters. These provided data representative of fenceline conditions.

Fugitive Leak Detection and Repair—The Yorktown refinery did not have a routine inspection program for fugitive leaks at valves, flanges, and fittings. A portable gas chromatograph suitable for field use using a flame ionization detector was used to measure total hydrocarbon concentrations at the Ultraformer unit and polymer gasoline unit.

Ambient Air Dispersion Modeling—The USEPA ISCST dispersion model was used to predict BTEX concentrations in the area of the refinery (USEPA 1987). For those units where emissions were measured directly or estimated from the mass balance technique, the resultant emission values were entered into the model. Otherwise the 1989 TRI value was used. On-site meteorological data were used representing several thousand measurements. The modeled concentrations of BTEX were then compared with the ambient air monitoring data.

The dispersion model used (ISCST) is known to have an accuracy factor of 2 (Williams 1991). If the ambient BTEX measurements differed from the predicted by more than a factor of 2, then it was concluded that the source term in the model, that is, the inventory value, was incorrect and it was adjusted such that the predicted value fell within a factor of 2 of the measured value.

Results

Air Emissions—API Separator—Hydrocarbon emissions of BTEX and total VOCs (TNMHC) for the API separator are given in Table 3. Detailed data for other compounds are given elsewhere (Blewitt et al. 1992). Total VOC emissions were estimated to be about 55 t/year. For comparison, the total VOC emission estimate determined from the AP-42 factor was 1161 t/year. An attempt to corroborate this estimate using the CHEMDAT7 model

TABLE 3—*Hydrocarbon emissions from API Separator.*[a]

Compound	Sample A6	Sample A7	Average	Emission Rate	Annual
Benzene	11 300	27 300	19 300	743.65	0.96
Toluene	64 700	162 000	113 350	4 367.52	5.52
Ethylbenzene	16 900	56 300	36 600	1 410.24	1.78
Xylene (total)	77 200	244 300	160 750	7 800.90	7.45
Unidentified VOCs	197 000	604 000	400 500	15 431.78	19.50
Total VOCs	555 000	1 720 000	1 137 500	43 829.33	55.40

[a] Measurements of selected hydrocarbon emissions from the API oil-water separator are shown in $\mu g/m^3$. Emission rate is in $\mu g/min\text{-}m^3$. Annual emission estimate is in metric tons per year. Values are from Blewitt et al. (1992).

TABLE 4—*Hydrocarbon emissions from coker pond.*[a]

Compound	Sample A11	Sample A16	Average	Emission Rate	Annual
		QUENCH AREA			
Benzene	105 970	149 645	127 808	9 849	1.27
Toluene	169 980	360 769	265 375	20 450	2.63
Ethylbenzene	58 530	121 172	89 851	6 924	0.91
Xylene (total)	375 750	695 367	535 559	41 272	5.45
Unidentified VOCs	7 160 000	10 300 000	8 730 000	672 756	88
TNMHC	21 300 000	25 500 000	23 400 000	1 803 264	236
		OVERFLOW AREA			
Compound	Sample A20	Sample A21	Average	Emission Rate	Annual
Benzene	295	4.7	149.9	11.54	0.0009
Toluene	5 210	21.6	2 615.8	201.42	0.010
Ethylbenzene	2 810	2.7	1 406.35	108.29	0.0055
Xylene (total)	12 140	12.3	6 076.15	467.86	0.024
Unidentified VOCs	485 000	1 120	243 060	18 715.62	0.94
TNMHC	692 000	2 990	347 495	26 757.12	1.34

[a] Measurements of selected hydrocarbon emissions from the coker pond quench and overflow areas are shown in μg/m^3. Emission rate is in μg/min-m^3. Annual emission estimate is in metric tons per year. TNMHC: total nonmethane hydrocarbons. Values are from Blewitt et al. (1992).

(USEPA 1989) was unsuccessful because not all the data necessary to run the model were available.

Air Emissions—Coker Pond—Hydrocarbon emissions of BTEX and total VOCs (TNMHC) for the coker pond quench area and overflow area are given in Table 4. Approximately 37% of the total emissions were not identifiable. Total VOC emissions were estimated to be about 236 t/year.

Air Emissions—Inactive Landfarm—Hydrocarbon emissions of BTEX and total VOCs (TNMHC) for the landfarm are given in Table 5. Note that approximately 50% of the total emissions were not identifiable. Total VOC emissions were estimated to be about 48 t/year.

Air Emissions—Sewer Vents—Emissions examined from direct measurements are detailed in Table 6. Total VOC emissions along the main sewer trunkline were estimated at 106 t/year.

Air Emissions—Cooling Tower and Water Storage Tanks—For the cooling tower and the waste water storage area, air emissions were estimated from hydrocarbon measurements of the water at the inlet and outlets. A mass emission rate was then determined using the water

TABLE 5—*Hydrocarbon emissions from inactive landfarm.*[a]

Compound	Sample A8	Blank	Net	Emission Rate	Annual
Benzene	26.2	35.7	0.0	0.000	0.00
Toluene	505	195	310	0.199	0.42
Ethylene	477	63.3	413.7	0.265	0.56
Xylene (total)	2 576	276.6	2 299.4	1.474	3.10
Unidentified VOCs	18 900	801	18 099	11.602	24
TNMHC	37 900	2 420	35 480	22.744	48

[a] Measurements of selected hydrocarbon emissions from the inactive landfarm are shown in μg/m^3. Emission rate is in μg/sec-m^3. Blank values were subtracted to give net concentrations. Annual emission estimate is in metric tons per year. TNMHC: total nonmethane hydrocarbons. Values are from Blewitt et al. (1992).

TABLE 6—*Hydrocarbon emissions from sewer vents.*[a]

Compound	Concentration	Emission Rate	Annual
	Sewer Vent AF1 (Sample A-4)		
Benzene	530	48	0.42
Toluene	2100	189	1.65
Ethylbenzene	630	57	0.50
Xylene (total)	2300	207	1.82
C4 aliphatics	170	15	0.13
C5 aliphatics	350	32	0.28
C6 aliphatics	67	6	0.06
C7 aliphatics	710	64	0.57
C8+	550	50	0.45
	Sewer Vent AF2 (Sample A-3)		
Benzene	14	1.0	0.13
Toluene	19	1.4	1.00
Ethylbenzene	3.7	0.3	0.40
Xylene (total)	21	1.6	1.79
C4 aliphatics	9.4	0.7	0.03
C5 aliphatics	12	0.9	0.03
C6 aliphatics	40	2.9	0.03
C7 aliphatics	40	2.9	0.75
C8+	32	2.3	4.62
	Sewer Vent AF3 (Sample A-2)		
Benzene	130	15	0.27
Toluene	1000	112	1.32
Ethylbenzene	400	45	0.45
Xylene (total)	1600	173	1.71
C4 aliphatics	25	3	0.08
C5 aliphatics	30	3	0.15
C6 aliphatics	30	3	0.05
C7 aliphatics	770	88	0.66
C8+	4600	524	0.60

[a] Measurements of selected hydrocarbon emissions from the sewer vents are shown in μg/m^3. Emission rate is in g/h. Annual emission estimate is metric tons per year. C4, C5, C6, C7 are aliphatic hydrocarbons with 4, 5, 6, and 7 carbons. C8+ includes all C8 aliphatics and aromatics. Hydrocarbons lighter than C4 (butanes) are not collected by the technique used. Values from Williams (Radian Corp.) (1991) and Blewitt et al. (1992).

flow rate through the facility with the assumption that the differences between the inlet and outlet were air emissions.

Fugitive Leak Detection and Repair—Over 2500 valves, flanges, and fittings were surveyed. However, statistics about rates of leakage are not available because records were kept only for leaking fixtures. Technicians were able to make immediate repairs to 100% of the valves that were found to leak above 10 000 ppm (v/v) and 76% of the valves with leaks above the 500 ppm threshold, illustrating the effectiveness of a leak detection and repair (LDAR) program.

Overall Inventory of Emissions

A total of 987 samples were collected for use in the final emissions inventory, including 630 air samples, 181 ground-water samples, 100 water samples, 39 soil samples, and 37 solid waste samples. Using these data with supporting modeling work led to the overall inventory description shown in Fig. 2. A total of 25 004 t per year of pollutants were estimated to be generated by the refinery. Air pollutants comprised 48% of this amount, or 12 028 t, with hazardous and solid wastes comprising 29%, or 7367 t, waterborne material 14%, or 3408 t, and biosolids from waste water treatment 9%, or 2200 t. A fuller review of the inventory results is available in Amoco/USEPA (1992c).

Transfers to different media, recycling, and treatment reduced the generated pollutants, so that pollutant releases totaled 13 982 t/year, or 55.9% of the pollutants generated. Airborne pollutants comprised a total of 88.5% of the releases. Criteria pollutants (NOx, SO_2, CO,

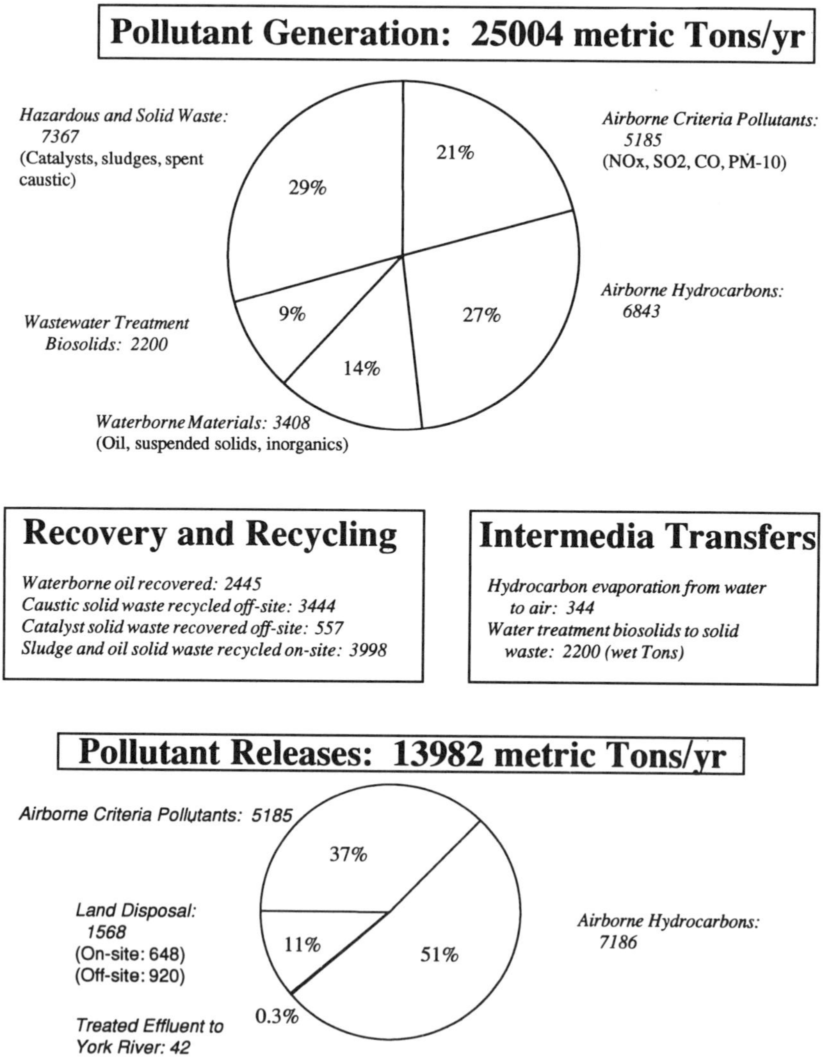

FIG. 2—*Pollutant generation, recovery/recycling, intermedia transfers and releases at the Yorktown refinery. Almost half of the pollutants generated at the refinery are recycled or removed by treatment. Some waterborne materials are transferred to air or solid waste. Values shown are in metric tons/year.*

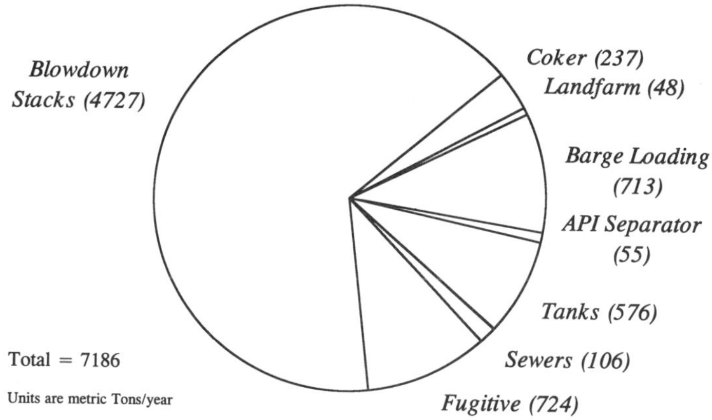

FIG. 3—*Sources of airborne hydrocarbons released from refinery. The contributions of various sources in the refinery are shown relative to the air emissions from the refinery, after recycling and treatment. The estimated contribution (value shown) of the blowdown stacks was measured after the inventory was completed (see text).*

PM-10) totaled 5185 t/year. Airborne hydrocarbons totaled 7186 t/year. Land disposal of solid wastes totaled 11.2% of the total releases, or 1568 t/year. Treated waste water releases to the York River totaled 42 t/year, or 0.3% of the total releases. Recycling processes included spent caustic to an off-site recycler, 3445 t/year, spent catalysts to an off-site recovery, 557 t/year; and sludges and oils to on-site recycling, 3998 t/year.

The hydrocarbon sources of airborne emissions are shown in Fig. 3. The airborne emissions were dominated by estimated blowdown stack emissions of 4727 t of hydrocarbons. Fugitive emissions and barge loading emissions each emitted about 10% of the total airborne hydrocarbons. Tanks contributed about 8% of total airborne hydrocarbons, with the coker, sewers, landfarm, and API separator contributing smaller fractions of the total.

A comparison of the TRI values (reported under SARA Title III) with the emissions inventory shows several distinct differences. Table 7 details the BTEX and VOC emissions and Fig. 4 graphically compares the VOC emissions. (As a result of the inventory completed for this project, Amoco submitted a revision to its 1989 TRI report.)

In work done since the joint Amoco/USEPA project was completed, Amoco measured the blowdown stack emission rate because this was such a major contribution to the total

TABLE 7—*Comparison of inventory with TRI Report.*[a]

	Chemical Emission in metric tons/year				
Emission Source	Benzene	Toluene	Ethylbenzene	Xylenes	Total
1989 TRI Report	37.3	83.2	22.3	97.3	337
Blowdown stack	29.5	51.5	41.5	110.5	391
Barge loading	13.6	48	12.7	62.8	150
Coker	1.6	2.8	0.9	5.7	11.8
Waste water	−0.09	−21.1	−6.6	−30.2	−80.9
Inventory project total	81.9	164.3	70.8	246.3	812

[a] The 1989 TRI values, as originally reported, are compared with values developed during the refinery inventory project. Wastewater entries show the net effects of reduced emissions from the API Separator and increased emissions from the sewer vents. Total VOC emissions are 0.3% of the total crude oil run. Source: Amoco/USEPA (1992c, Table 2–3).

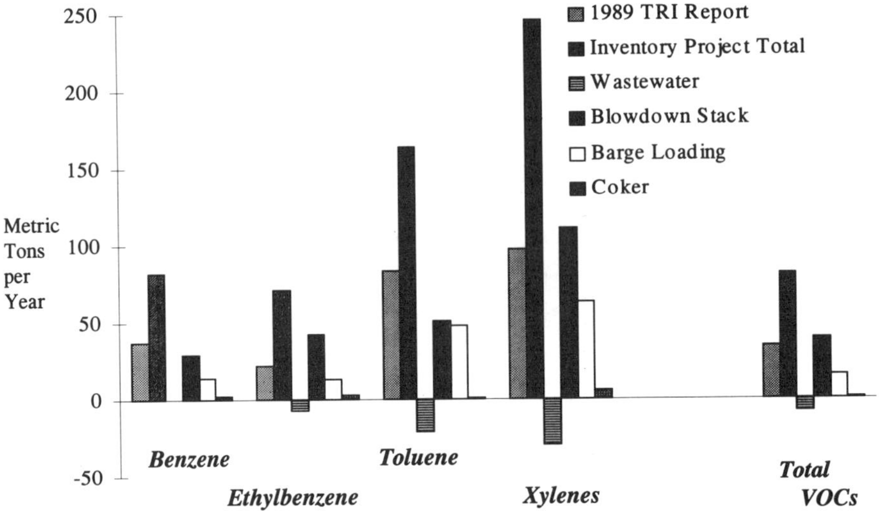

FIG. 4—*Impact of inventory project on TRI estimate. The changes in the 1989 TRI Report due to the refinery inventory are shown. Major increases were estimated from consideration of blowdown stacks and barge loading. Significant reductions were estimated from consideration of waste water. Minor increases were estimated for contributions from the coker and other sources. Values shown are in metric tons per year.*

air emissions estimate. Amoco's investigation used a sulfur hexafluoride (SF_6) tracer to measure stack flow rate and different analytical techniques to measure the gases sampled. The stack flow rate was extremely low, so it was difficult to quantify. Pitot tube measurements were used originally; however, the small pressure drop across the device led to overstating the velocity. The molecular weight of the stack gas was orginally assumed to be similar to air, but this, too, turned out to be a problematic assumption. Finally, the gas collection method appeared to have a significant positive bias when high levels of aliphatic hydrocarbons are present, resulting in an apparent overestimate (Schmitt 1994).

Using the measured flow rate and improved analytical method, the blowdown stack emissions were calculated to be 91 t/year, rather than the 4730 t/year obtained using the AP-42 estimate. Use of the SF_6 tracer allowed flow to be calcualted in three ways—tracer arrival, steady state, and decay—resulting in comparable estimates using several days' data. Analyses of TO-14 canisters also showed that dilution techniques appeared to introduce a positive bias in BTEX species by as much as a factor of 2 (Team Environmental Services 1992).

Discussion of the Refinery Release Inventory

Review of the inventory project led to several findings:

A. *Existing estimates of environmental releases were not adequate for making a chemical-specific, multimedia facility-wide assessment*

The Yorktown refinery had good information about the quantity of material released to the York River from Virginia Pollution Discharge Elimination System (VPDES) Permit monitoring requirements, and for solid wastes as a result of internal programs and participation in recent American Petroleum Institute surveys (American Petroleum Institute 1991). These releases, however, made up only 11% of the total releases from the facility. Available data did not include adequate chemical-specific characterization of the water discharge or solid waste streams.

The Yorktown refinery (and other refineries as well) could not easily identify specific airborne hydrocarbon compounds released or the quantity released because:

(1) Refineries typically do not manufacture products with specific chemical compositions and therefore do not routinely measure chemical compositions of their products or emissions. Rather, refinery products have specific properties such as octane rating, freeze point, and sulfur content. Crude oil, the raw material used to make these products, contains thousands of distinct chemicals that are never fully separated during the manufacturing processes. Airborne releases from this kind of facility are particularly complex.

(2) Most hydrocarbons are released through a large number of widely distributed sources (valves, flanges, pump seals, and tank vents). Even a small refinery may have more than 10 000 different potential sources. Direct measurement of each of these sources is not practical.

(3) The quantities released through any single source are extremely small—on the order of pounds per year—dilute, and difficult to measure. In addition, some large sources that emit pollutants in the amount of tons per year are difficult to measure and quantify. Total hydrocarbons released from the Yorktown refinery from all sources were approximately 0.3 weight percent of the total crude oil processed. Therefore, they would not be detected through normal mass balances and materials accounting (National Research Council 1990).

Thus, collecting detailed, chemical-specific release information used to characterize the refinery was expensive and time consuming. This project developed a sampling and monitoring program that included about 1000 samples. Each sample was analyzed for 15 to 20 chemicals. The sampling program took about 12 months to complete at a cost of about $1 million. Even with this time and dollar commitment, only selected sources were sampled. The final release inventory was assembled using a combination of sampling, measurements, dispersion modeling, and estimates based on emission factors.

Because this sampling program was a first-of-its-kind effort, its scope was intentionally broad. Subsequent analysis showed that not all of the information obtained was necessary to identify significant sources and potential reduction options. For the Yorktown refinery (and the petroleum refining industry overall), more general information, such as source specific VOC emissions, is adequate to identify many of the pollution prevention projects developed in this study. Total VOC emissions are a good indicator of overall emissions and can be used for tracking emissions reduction programs.

B. *A substantial portion of the pollution generated at this refinery is not released to the environment*

The release inventory process allowed a comparison of pollutant generation, on-site management and ultimate releases to the environment. The refinery generates about 25 000 t/year of pollutants. As a result of site hydrogeology, on-site waste water treatment, and solid waste recycling practices, about 10 900 t are recovered, treated, or recycled and do not leave the refinery site. Of the remaining 14 000 t about 90% are released to the air.

The site-wide analysis allowed closer focus on the largest releases—airborne emissions. Modeling studies indicated relatively little naturally occurring transfer of hydrocarbon emissions from air into other media (Cohen et al. 1991). Most hydrocarbons are not very water soluble, so they are not easily removed from the air by rainfall. Measurements and modeling results showed small transfers from some surface water ponds to ground water. Ground water enters the waste water treatment system through the underground sewers, resulting in a net ground-water inflow, that is, ground water is not leaving the facility.

Transfers of pollutants between media were observed, particularly as a result of pollution management activities. Over 335 t/year of hydrocarbons initially present in waste water are volatilized into air from the water collection system. More than 1800 t/year of biosolids are produced by treating waste water in the refinery's activated sludge system.

C. *The TRI database does not adequately characterize releases from this refinery*

The TRI database has become the *de facto* national release inventory even though the quality and utility of data reported can vary widely. For example, at a plant that uses a single solvent to wash manufactured parts and purchases enough extra solvent to make up for evaporative losses, the quantity of solvent emissions is well known and tracked through purchasing records. The TRI report for this solvent at this manufacturing plant should be quite accurate. However, at the refinery, the TRI cannot report total facility emissions with the same accuracy. This is because:

1. The TRI is based on estimates rather than measurements. Estimating accuracy varies widely. Emissions from waste water sources had been overestimated at the refinery. Amoco has amended past TRI reports for Yorktown to reflect this new data. However, estimates will continue to be used because of the large number of potential sources and will constitute a limitation on the accuracy possible in a TRI report.
2. The TRI does not include all activities at the facility. Some activities and emissions are excluded by USEPA from record-keeping requirements, such as emissions from barge loading. At Yorktown, barge loading operations accounted for about 20% of total benzene emissions.
3. The TRI does not include all chemicals. The TRI focuses on specific chemicals that account for only a portion of the total emissions. In the refinery's case, the TRI report covers only 9% of the total hydrocarbons released and only 2.4% of the total releases to all media. Pollutants that are criterion pollutants for other USEPA programs, such as CO, NOx, SO_2, and PM-10 are not reportable in the TRI.
4. The TRI does not consider the significance of the reported chemicals to nearby populations or ecosystems. The TRI includes estimates and measurements of selected materials released from the site. TRI data by themselves do *not* allow for meaningful risk evaluation or comparisons between facilities because they do not describe the facility's relationship to its environment, for example, people who live nearby or possibly sensitive ecological regions. An airborne emission may cause significant risk if there are people living nearby, but the same emission may have minimal risk to an aquatic community.

D. *Site-specific features determined during the facility-wide assessment affect releases and release management options*

National programs, by design, address overall problems in specific media. But these programs seldom consider site-specific differences when developing standards. Other refineries, and indeed other industrial facilities, can use the general sampling approach developed at Yorktown to obtain a facility-wide release inventory. However, each site will exhibit unique geophysical and process characteristics. Each assessment plan must include these site-specific characteristics in its design and focus. As an example, the Yorktown refinery does not have a hydrofluoric acid (HF) alkylation unit, so HF was not measured. HF is used by other refineries and can pose a significant health risk if managed improperly, so HF may need to be tracked at facilities that use it.

The project identified features specific to the Yorktown site that significantly affected the releases and options for their management:

1. *Ground Water*—As a result of a clay soil layer, unique hydrogeology, the placement of the underground drainage system relative to the water table, and local climate, ground-water movement at the Yorktown site is minimal. In fact, the ground-water drainage system is acting as a ground-water collection unit, sending ground water to the refinery's waste water treatment plant. Thus, ground water at this site is not leaving the property. Furthermore, sampling showed surprisingly low levels of ground-water contamination compared to other refineries.
2. *Marine Loading Emissions*—The Yorktown refinery uses marine transportation for receiving all crude oil and shipping more than 80% of its products. Estimated releases from product loading operations are 713 t/year of VOCs. Computer modeling analysis showed this source had the greatest impact on exposure of nearby residences to refinery hydrocarbon emissions. Therefore, marine loading emissions would be a useful factor to include in this facility's environmental management plans. Other refineries rely more on pipeline, rail, and truck shipments to handle crude products and marine loading would not be a significant factor in their environmental management options.
3. *Airshed Status*—The refinery is located in an airshed classified as an "attainment area" for all criteria pollutants including ozone. Therefore, relatively few hydrocarbon emission controls have been required or installed at the Yorktown facility. The sampling program and release reduction options focused on hydrocarbon releases. Many other refineries in ozone nonattainment areas have already installed extensive hydrocarbon emission controls. Consequently, other facilities may have a significantly lower percentage of hydrocarbon emissions. Similarly, NOx, CO, PM-10, and SO_2 emissions have been more tightly controlled in some other airsheds (such as the Los Angeles Basin) that do not meet National Ambient Air Quality Standards (NAAQS) for these pollutants.

Conclusions and Recommendations

The working group reported several recommendations based not only on the results of the VOC inventory, but also on the regulatory framework, costs, and other relevant factors. The recommendations are listed in Table 8, but are described more fully in the project documentation (Amoco/USEPA 1992a,b, Industrial Economics, Inc. 1993).

The second and third recommendations are particularly relevant to this discussion of the VOC inventory. Data from this study show that an emissions inventory could be improved by measuring releases and developing new emission factors. For example, the emissions inventory at the beginning of the project did not account for all potential releases to the environment. Some releases were excluded because USEPA had excluded them from reporting (for example, barge loading operations); some releases were not included because the sources and the amount were thought by Amoco to be insignificant (such as blowdown stacks); some emissions were overestimated (such as the API separator); and some releases were underestimated (such as the coker pond). Jointly established sampling and analysis protocols could help improve data quality, so that reported values more accurately portray facility releases.

TABLE 8—*Yorktown project recommendations (Amoco/USEPA 1992a).*

1. Explore opportunities to produce better environmental results more cost-effectively.
2. Improve environmental release data collection, analysis and management.
3. Provide incentives for conducting facility-wide assessments and for developing multimedia release reduction strategies. Such strategies should consider multimedia consequences of environmental management decisions.
4. Encourage additional public/private partnerships on environmental management.
5. Conduct research on the potential health and ecological effects of VOCs.

Data currently collected in response to regulatory or permitting requirements could be evaluated to determine how its utility and quality might be improved. For example, TRI data quality and utility could be improved by:

1. Providing more inclusive estimates of facility-wide releases to all media. The project found the exclusion of marine loading operations from TRI reporting requirements conveyed an inaccurate picture of total facility releases.
2. Reporting groups of chemicals, rather than individual species, especially if these chemicals have similar structure, physical, and toxicological properties. Requiring reporting of all VOCs for refineries, rather than specific compounds like xylene (and its individual isomers), would provide a meaningful measure of refinery releases. That is because xylene poses approximately the same risks and has physical characteristics similar to the hundreds of undifferentiated VOC compounds not covered in TRI. For a refinery, where a complex mixture of chemicals are releases from most sources, tracking many separate chemicals does not make good use of technical, laboratory, and environmental management resources.
3. Reporting other selected chemicals of concern for demonstrated human health or ecological impact separately. At a refinery, chemicals such as butadiene, benzene, and nickel may be good indicators of risk/release potential and management practices. Other industrial sectors would need to track different specific chemicals.
4. Improving emission factors for estimating releases based upon information developed in this project, and additional work by USEPA/industry task groups that could focus on the different data collection needs of discrete industry sectors.

The project had great difficulty collecting and verifying environmental release data from the site. Emissions from these sources are complex and measurement techniques are rudimentary. Many emission measurements varied with time. For example, due to the cyclic nature of the coking process, the coker pond emissions varied by a factor of three within a few hours. Better sampling and analysis methods and statistical tools are needed to analyze variability. Research is also needed to develop methods that can verify release inventories within reasonable confidence limits, accounting for specific differences in emission factors.

The project demonstrated that more cost-effective environmental protection programs can be designed by allowing companies to consider site specific factors and focus on results.

A detailed facility-wide, multimedia assessment identified the most significant medium (air) and release sources, both in terms of quantity and potential impact on the surrounding area. Specific technology options could then be developed to deal with these sources. The significance of sources identified in this project were not initially known or apparent to the participants. Proposed solutions could not have been developed in the absences of data which identified their importance.

For example, hydrocarbon emissions estimates from barge loading operations (713 metric tons annually) and blowdown stacks (4700 t) appeared significant. However, the refinery did not know this prior to this project, nor did the existing regulations require the collection of these data. Thus the refinery had not carefully evaluated these sources and had not developed control options to reduce these emissions.

Several technologies considered for reducing releases transfer pollutants from one medium to another or convert pollutants to different forms. Since human health and environmental consequences vary from one medium to another, viewing a release problem in the context of *net* environmental effects is essential to developing more sound solutions.

The current institutional framework and procedures for developing regulations do not include multimedia assessments and analysis. Current practices should be reviewed to determine how they could be modified to use information from such assessments. An integrated pollution prevention and management strategy would facilitate development of release management options that produce better environmental results (Science Advisory Board 1990a,b, Office of Management and Budget 1991). At present, industry has little incentive to conduct such assessments because it does not have an opportunity to implement their findings (Raffle and Mitchell 1993).

References

American Petroleum Institute, 1991, "The Generation and Management of Wastes and Secondary Materials in the Petroleum Refining Industry, 1987–1988," American Petroleum Institute, Washington, DC.

Amoco/USEPA (Amoco Corporation, U.S. Environmental Protection Agency), 1992a, "Amoco–U.S. EPA Pollution Prevention Project: Project Summary," NTIS No. PB 92-228527, National Technical Information Service, Springfield, VA.

Amoco/USEPA, 1992b, "Amoco–U.S. EPA Pollution Prevention Project: Executive Summary," NTIS No. PB 92-228519, National Technical Information Service, Springfield, VA.

Amoco/USEPA, 1992c, "Amoco–U.S. EPA Pollution Prevention Project, Yorktown, Virginia: Refinery Release Inventory," NTIS No. PB-92 228550, National Technical Information Service, Springfield, VA.

Amoco/USEPA, 1992d, "Amoco–U.S. EPA Pollution Prevention Project, Yorktown, Virginia: Air Quality Data, Volume II—Appendix B. Benzene and Toluene," NTIS No. PB-92 228691, National Technical Information Service, Springfield, VA.

Amoco/USEPA, 1992e, "Amoco–U.S. EPA Pollution Prevention Project, Yorktown, Virginia: Air Quality Data, Volume II—Appendix B. Ethylbenzene and Xylene," NTIS No. PB-92 228709, National Technical Information Service, Springfield, VA.

Amoco/USEPA, 1992f, "Amoco–U.S. EPA Pollution Prevention Project, Yorktown, Virginia: Air Quality Data, Volume II—Appendices C, D & E," NTIS No. PB-92 228717, National Technical Information Service, Springfield, VA.

Amoco/USEPA, 1992g, "Amoco–U.S. EPA Pollution Prevention Project, Yorktown, Virginia: Air Quality Data, Volume II—Appendix F," NTIS No. PB-92 228725, National Technical Information Service, Springfield, VA.

Amoco/USEPA, 1991a, "Amoco/USEPA Amoco–U.S. EPA Pollution Prevention Project: Groundwater and Soil Data," NTIS No. PB 92-228576, National Technical Information Service, Springfield, VA.

Amoco/USEPA, 1991b, "Amoco/USEPA Amoco–U.S. EPA Pollution Prevention Project: Surface Water Data," NTIS No. PB 92-228584, National Technical Information Service, Springfield, VA.

Blewitt, D. N., Keating J. W., and Yohn, J. F., 1992, "An Air Quality Evaluation of the Amoco Yorktown Refinery as Part of the Amoco–USEPA Pollution Prevention Project," *Air Quality Data,* Vol. II, with Appendices A, I, and J, NTIS No. PB 92-228600, National Technical Information Service, Springfield, VA.

Cohen, Y., Allen, D. T., Blewitt, D. N., and Klee, H., 1991, "A Multimedia Emissions Assessment of the Amoco Yorktown Refinery," presented at the 84th Annual Air and Waste Management Association, Vancouver, British Columbia, 17 June.

Glickman, T. S., 1990, "Minutes of the First Meeting of the Peer Review Committee for the Amoco/USEPA Pollution Prevention Project," Resources for the Future, Washington DC, 13 Dec.

Hustvedt, K. C., 1992, Memorandum to Ann Bailey. Appendix J in Blewitt et al. 1992, pp. 5-1 to 5-11, with Discussion pp. 5-12–5-23.

Industrial Economics, Incorporated, 1993, "Projects, Evaluations, and Rankings (Amoco/EPA Pollution Prevention Project) Prepared for U.S. Environmental Protection Agency," NTIS No. PB-92 228626, National Technical Information Service, Springfield, VA.

Keinbush, M. R., Runum, D., and Ekland, B., 1988, "Evaluation of Flux Chamber Method for Measure Air Emissions from Surface Impoundments," Radian Corp., Austin, TX.

Kizior, G. J., 1991, "Amoco/USEPA Pollution Prevention Project Solid Waste Sampling," NTIS No. PB 92-228568, National Technical Information Service, Springfield, VA.

National Research Council, 1990, "Tracking Toxic Substances at Industrial Facilities: Engineering Mass Balance versus Materials Accounting," National Academy Press, Washington, DC.

Office of Management and Budget, 1991, "Budget of the United States Government, Fiscal Year 1992, Section IX.C. Reforming Regulation and Managing Risk-Reduction Sensibly," Feb. 4, Washington, DC.

Radian Corporation, 1985, "Measurements of Gaseous Emission Rates from Land Surfaces using an Emission Flux Chamber—User's Guide," Environmental Monitoring Standards Laboratory, U.S. Environmental Protection Agency.

Radian Corporation, 1990, "The Monitoring Plan for a Program of Emission and Ambient Air Sampling at the Amoco Yorktown, Virginia Refinery," 17 Sept. E. L. Hockman, "Groundwater Sampling and Monitoring Plan for Amoco Oil Company's Yorktown, Virginia Refinery," June 1990, Amoco Groundwater Management Services, Tulsa, OK.

Raffle, B. I. and Mitchell, D. F. 1993, "Effective Environmental Strategies: Opportunities for Innovation and Flexibility Under Federal Environmental Law. Draft—for discussion purposes," Amoco Corporation, Chicago, IL.

Schmitt, R. E., 1994, "Amoco–U.S. EPA Pollution Prevention Project, Yorktown Virginia: Supplemental Report, December 1994," Amoco Corporation, Chicago, IL.

Science Advisory Board, 1990, "Reducing Risk: Setting Priorities and Strategies for Environmental Protection," U.S. Environmental Protection Agency, Washington, DC.

Science Advisory Board, 1990, "The Report of the Strategic Subcommittee, Relative Risk Reduction Project, Reducing Risk, Appendix C," U.S. Environmental Protection Agency, EPA SAB-EC-90-021C, Washington, DC.

Team Environmental Services, Inc., 1992, "Flow and Emissions Estimates from Three Blowdown Stacks at the Amoco Yorktown Refinery," Report to Amoco Oil Company, San Marcos, CA.

U.S. Environmental Protection Agency, 1989, "Hazardous Waste Treatment, Storage, and Disposal Facilities (TSDF)—Air Emissions Models," Office of Air Quality Planning and Standards, Research Triangle Park, NC, (Review draft).

U.S. Environmental Protection Agency, 1987, "Industrial Source Term Complex (ISC) Dispersion Model User's Guide—Second Edition (Revised)," Office of Air Quality Planning and Standards, Research Triangle Park, NC.

U.S. Environmental Protection Agency, 1985, "Compilation of Air Pollutant Emission Factors," Office of Air Quality Planning and Standards, Research Triangle Park, NC.

Williams, C. H. (Radian Corporation), 1991, "Final Report: Measurements of Hydrocarbon Emissions and Ambient Air Concentrations at the Amoco Yorktown Refinery," Amoco–U.S. EPA Pollution Prevention Project, Yorktown, Virginia: Air Quality Data Volume 1, NTIS No. PB 92-228592, National Technical Information Service, Springfield, VA.

Air Quality

Claude S. Davis[1] *and Rein Otson*[2]

Estimation of Emissions of Volatile Organic Compounds (VOCs) from Canadian Residences

REFERENCE: Davis, C. S. and Otson, R., **"Estimation of Emissions of Volatile Organic Compounds (VOCs) from Canadian Residences,"** *Volatile Organic Compounds in the Environment, ASTM STP 1261,* W. Wang, J. Schnoor, and J. Doi, Eds., American Society for Testing and Materials, 1996, pp. 55–65.

ABSTRACT: Estimates of emissions of ten VOCs from Canadian residences based on Eq 1, where C_i and C_o, respectively, are the indoor and outdoor concentrations, ach is the number of air changes per hour, V the average house volume, and N the number of houses, ranged from 14 tonnes/y (benzene) to 227 tonnes/y (toluene).

$$E = (C_i - C_o) \times \text{ach} \times V \times N$$

Measured ambient benzene concentrations in Metropolitan Toronto (annual mean of 2 to 3 $\mu g/m^3$) were within a factor of 10 of box model predictions based on a benzene emission inventory. Predicted ambient concentrations for the VOCs with the three highest residential emissions (toluene, decane, and 1,4-dichlorobenzene) were a negligible fraction of measured ambient concentrations in Metropolitan Toronto. Canadian indoor emissions for selected VOCs found in household solvents were a small fraction (0.2 to 2.3%) of the residential solvent usage. Indoor and outdoor VOC concentration data in conjunction with air exchange rates are useful in estimating VOC emissions from residences.

KEYWORDS: indoor VOC sources, indoor emissions, residential emissions

Full assessment of the human health risks associated with exposure to volatile organic compounds (VOCs) must consider exposures in both indoor and outdoor environments. Information on exposure of the Canadian population to VOCs is receiving increased attention as indicated by the expansion of routine monitoring of ambient VOCs in Canada (River Road Environmental Technology Centre 1991) and by the compilation of indoor VOC concentration data for Canadian residences (Davis and Otson 1993; Otson et al. 1992). The National Air Pollutant Surveillance (NAPS) network, which was designed to monitor routinely the so-called criteria pollutants at selected urban centers in Canada, includes stations at which 24 h average ambient concentrations of VOCs are measured. Ambient VOC monitoring in the NAPS network has been underway since 1984 (River Road Environmental Technology Centre 1991). In 1992 there were 24 NAPS stations in ten urban centers and two rural locations at which ambient VOC concentrations were measured (Dann 1994). The Ontario

[1] Principal consultant, BOVAR-CONCORD Environmental, 2 Tippett Road, Downsview, ON M3H 2V2, Canada.

[2] Senior scientist, Bureau of Chemical Hazards, Environmental Health Directorate, HPB, Health Canada, Rm. B-19, EHC, Tunneys Pasture, Ottawa, ON K1A OL2, Canada.

Ministry of the Environment and Energy (OMEE) also routinely monitors VOCs at 18 stations throughout Ontario (OMEE 1992). A database containing indoor and simultaneously measured outdoor concentrations of VOCs and other pollutants has been compiled (Davis and Otson 1993). A national pilot survey in which 26 VOCs were measured in over 700 Canadian homes has been completed recently (Fellin and Otson 1993).

The major portion of human exposure to VOCs is likely to occur indoors since people spend over 90% of their time indoors (U.S. EPA 1989a). In a recent study of indoor and outdoor concentrations of VOCs and other pollutants in Ontario, it was found that indoor concentrations of some VOCs were higher than outdoor concentrations (Bell et al. 1991). Higher indoor than outdoor concentrations (Andelman 1985a; Wallace et al. 1985) are invariably due to the presence of indoor sources such as building materials, consumer and commercial product usage, and human activities such as residential water usage (Andelman 1985b) and cooking. VOCs present in outdoor air also contribute to indoor airborne VOC concentrations as a result of air exchange between indoor and outdoor environments. Indoor emissions of VOCs and other pollutants eventually end up in ambient air as a result of air exchange, but little is known of the contribution of indoor sources to ambient VOC emissions inventories and thus to ambient air VOC concentrations.

Ambient emissions of pollutants, compiled in inventories, are estimated by multiplying an emission factor by an activity level for each process that leads to emission of the pollutants of interest (U.S. EPA 1985). Residential fuel combustion sources are included in the 1985 Canadian VOC emissions inventory but other sources such as building materials and consumer products which are associated with residences, are not directly included (Kosteltz and Deslauriers 1990). For the 1985 Canadian VOC emissions inventory, the VOCs were not speciated. Benzene is the only VOC for which Canadian emissions inventories have been compiled for the years 1985 (Jacques 1990) and 1991 (Solomon et al. 1993). In the United States, emission factors for VOC emissions from consumer and commercial products and product usage information have been compiled to support emissions inventories of toxic air pollutants in urban areas (U.S. EPA 1989b; U.S. EPA 1990; Rogozen et al. 1985).

Knowledge of indoor and outdoor air VOC concentrations together with air exchange rates between indoor and outdoor air can be used to estimate the emissions from buildings. Such estimates will allow evaluation of the contribution of indoor sources to VOC emissions inventories and to ambient air VOC concentrations.

In this study we explored the use of indoor VOC concentration data obtained from a national survey of over 700 Canadian homes together with estimated air exchange rates to provide preliminary estimates of VOC emissions from residences. The validity of the estimates of indoor emissions for selected VOCs was examined by comparison with available emissions inventory estimates (for benzene) and with residential solvent use data (benzene, dichloromethane, toluene, and xylenes). A simple model was used to estimate typical ambient concentrations that would occur in a selected urban region (Metropolitan Toronto) if VOC emissions from residences in the region were mixed in the boundary layer of the atmosphere. The predicted concentrations were compared with measured mean ambient VOC concentrations. If the estimates of indoor emissions for the VOCs found in residential solvents and for benzene are reasonable, then the approach could be useful to indicate the contribution of VOC emissions from residences to ambient concentrations and to indicate the relative importance of indoor and outdoor sources to human exposure to VOCs.

Methodology

The emission rate, E (μg/h) to the outdoors of an individual indoor VOC (x), for a number of houses was calculated by use of Eq 1

$$E = (C_{i(x)} - C_{o(x)}) \times \text{ach} \times V \times N \tag{1}$$

where

$C_{i(x)}$ = indoor concentration for species x, $\mu g\ m^{-3}$,
$C_{o(x)}$ = outdoor concentration for species x, $\mu g\ m^{-3}$,
ach = air exchange rate between indoor and outdoor air, ach,
V = average house volume, m^3, and
N = number of houses.

The daily average ambient concentration, $C_{o(pred)}$, of a VOC was predicted from a box model (Weber 1982) which is described by Eq 2

$$C_{pred} = \frac{Q}{F} \tag{2}$$

where

Q = the daily emission rate of VOC into the region of interest, ton d^{-1}, and
F = the ventilation factor, that is, the horizontal wind velocity multiplied by the surface area perpendicular to the wind direction, $m^2\ h^{-1}$.

The box model assumes that pollutants emitted are well mixed. The area perpendicular to the wind direction was the annual mean of the maximum afternoon mixing height multiplied by the width of the region of interest.

Estimates of solvent supply and usage in Canada for seven categories of solvents are available (Camford Information Services Inc. 1991). Data for two of the solvent categories, namely, hydrocarbons and chlorinated solvents likely to be associated with residential use (paints and coatings, adhesives and sealants, household products, paint strippers, and aerosol formulations), were of relevance to this study.

Results and Discussion

Annual emissions from Canadian residences for those VOCs considered to have indoor sources were estimated using Eq 1. The residential emissions, as defined here, occur first by emission indoors and then to the outdoor atmosphere through air exchange between indoor and outdoor air. Residential VOC emissions are not specifically included in the Canadian VOC emission inventory except for residential fuel combustion sources where the emissions are assumed to occur directly to the atmosphere through chimneys and not through air exchange between indoor and outdoor air. Independent estimation of the residential emissions of individual VOCs from fuel combustion is not possible since the VOC emission inventory is not speciated. Most of the discussion is concerned with benzene since it is the only VOC for which both detailed emissions inventory and ambient concentration data are available. The validity of the estimates of residential VOC emissions was assessed by two approaches. In the absence of independent comparable estimates for residential emissions of benzene from combustion sources and for VOCs used in residential solvents, the reasonableness of the estimates was assessed based on gross assumptions on likely residential emissions of benzene and by comparing residential VOC emissions with residential solvent use data. In the second approach, model estimates of the ambient benzene concentrations that would result from the residential emissions are compared with ambient concentration measurements.

The reliability of the model was first checked by comparison of measured ambient benzene concentrations with predicted concentrations based on emissions inventory data for benzene.

Estimation of VOC Emissions from Indoor Sources

Equation 1 is appropriate for VOCs which have predominantly indoor sources as indicated by C_i/C_o ratios that are greater than one. Indoor VOC concentrations obtained in the national survey were based on a three-stage probability-based sampling design to obtain measurements that were representative of Canadian residences. Twenty-four hour average indoor VOC concentrations were measured with the 3M passive monitoring device (OVM 3500). Details of the national survey methods are described elsewhere (Otson et al. 1992).

Since simultaneous measurements of indoor and outdoor VOC concentrations were not made in the national survey, ambient measurements obtained at NAPS network stations were used to provide values of the outdoor VOC concentration, C_o. The mean and maximum concentrations for the 21 VOCs common to the NAPS network and the national survey of indoor VOC concentrations are listed in Table 1. Mean VOC concentrations reported for the national survey data were calculated by substituting zeros for values below the detection

TABLE 1—*Summary of ambient VOC concentrations at urban sites in Canada and indoor VOC concentrations in Canadian residences.*

Compound	Ambient Canadian[a] (24 sites in 1990–1991) ($\mu g\ m^{-3}$)		National Survey[b] ($\mu g\ m^{-3}$)		Likelihood for indoor sources[c]
	Mean[d]	Maximum	Mean[c]	Maximum	
Toluene	9.02	389	40.8	5730	high
Decane	1.3	74.4	31.4	6450	high
m- and p-xylene	5.02	389	20.7	1470[e] 280[f]	high
1,4-dichlorobenzene	0.64	16.0	18.9	1390	high
Dichloromethane	1.55	32.3	16.1	1690	high
1,2,4-trimethylbenzene	1.52	11.0	11.5	1920	high
Ethylbenzene	1.3	142	8.2	540	high
o-xylene	1.59	143	5.6	320	high
Benzene	3.84	61.1	5.4	67.9	medium
1,3,5-trimethylbenzene	0.49	6.52	2.7	640	high
Tetrachloroethylene	0.98	45.7	2.7	313	low
Chloroform	0.21	3.78	1.5	68.6	low
Hexane	3.91	417	1.2	124	N
p-cymene	0.15	9.57	1.2	115	N
Trichloroethylene	0.28	20.0	0.5	165	N
Styrene	0.60	41.8	0.3	130	N
1,2,4-trichlorobenzene	0.05	1.13	<0.1	5.8	N
1,1,2,2-tetrachloroethane	0.09	0.48	<0.1	5.2	N
1,2-dichloroethane	0.16	8.71	<0.1	1.7	N
1,3-dichlorobenzene	0.05	0.56	<0.1	<0.1	N

[a] From National Air Pollution Surveillance Network data, (Dann 1994).
[b] Fellin and Otson 1993.
[c] See text.
[d] Means were calculated using zero for measurements below the detection limits and are considered unreliable—see text.
[e] Maximum for m-xylene.
[f] Maximum for p-xylene.

limits. The detection limits for the compounds listed in Table 1 were 0.1 $\mu g\ m^{-3}$ for the ambient measurements (Dann 1994) and were in the range 1.6 to 5.9 $\mu g\ m^{-3}$ for indoor measurements (Fellin and Otson 1993). Toluene had the highest annual mean outdoor concentration (9.02 $\mu g\ m^{-3}$) and also had the highest mean indoor concentration (40.8 $\mu g\ m^{-3}$). The range of toluene measurements outdoors and indoors was considerable since the maximum concentrations outdoors and indoors were, respectively, about 40 and 140 times greater than the corresponding mean values (Table 1). The maximum concentrations found indoors were generally higher than the maximum ambient concentrations except for hexane and 1,2-dichloroethane. This suggests that individual houses can have sources that can contribute to high indoor levels. Other compounds listed in Table 1 that had high annual mean ambient concentrations relative to the detection limit were m- and p-xylene, hexane, and benzene. Compounds with annual mean ambient concentrations below (1,3-dichlorobenzene, 1,1,2,2-tetrachloroethane, 1,2,4-trichlorobenzene) or close to the detection limit (p-cymene, chloroform, 1,2-dichloroethane, trichloroethylene) also were found indoors with mean concentrations that were below the detection limits. Mean indoor concentration values below the detection limits may be unreliable when used with ambient data to estimate indoor/outdoor ratios to infer the presence of indoor sources.

Six of the compounds in Table 1 (toluene, m- and p-xylene, o-xylene, tetrachloroethylene, and 1,4-dichlorobenzene) had indoor mean concentrations that were at least two times greater than the annual ambient mean concentrations and were designated as having a high likelihood of being emitted from indoor sources. The maximum ambient and indoor concentrations for benzene were similar, while the mean indoor benzene concentration was 40% higher than the annual mean ambient concentration. Benzene was designated as having medium likelihood of having significant indoor sources. Compounds with ratios less than one or with indoor concentrations below the detection limit were not likely to have significant indoor sources and were designated as "N." Ten compounds were designated as having high or medium likelihood of having predominant indoor sources (Table 1), but the fact that the indoor and ambient concentrations were not obtained simultaneously at the same locations and that the sampling and analytical methods were different must be borne in mind. The estimates of the emissions based on Eq 1 must be considered as order-of-magnitude estimates.

Indications of the presence of indoor sources for some of the 26 compounds were obtained using factor analysis of the national survey data which included indoor VOC concentrations as well as house characteristics. The national survey included the administration of a questionnaire that solicited information on the type of heating, the type of ventilation, renovations in the house in the one-year period prior to sampling, and the amount of time spent indoors by occupants during the 24-h sampling period. The factor analysis indicated that five compounds (o-, m-, and p-xylene, toluene, and ethylbenzene) were frequently associated (occurred in the same factor) when all data or various subsets of the data were analyzed. This suggested a common indoor source for these compounds. These five compounds were associated with 1,3,5- and 1,2,4-trimethylbenzenes and decane for some subsets of the data which suggested association of these eight compounds with one or more indoor source. Indoor concentrations of all eight compounds were higher than annual mean ambient concentrations (Table 1). This supports the notion that the factors associated with these compounds correspond to indoor source(s). The other compounds with medium or high likelihood of having significant indoor sources based on the indoor/outdoor concentration ratio (benzene, dichloromethane, and 1,4-dichlorobenzene) did not occur consistently in the factors with the highest loadings but this does not rule out the presence of any indoor sources for these three compounds.

For compounds believed to be associated with indoor sources, their emissions to the outdoors can be estimated from Eq 1. An average air exchange rate of 0.34 ach was used in Eq 1 for calculating VOC emissions from residences in all regions of Canada. Measurements of air exchange rates obtained by a perfluorocarbon tracer technique in 24 homes in the national survey (Fellin and Otson 1993) and in 44 houses in Saskatchewan and Ontario during the winter (mid-January to mid-February of 1991) (Dumont and Snodgrass 1992), both averaged 0.34 ach. A value of 0.3 ach previously used as an average for Canadian houses (Hamlin et al. 1990) is similar to the value selected for use in this study. The measurements were made primarily in the winter and fall months when air exchange rates were expected to be lowest since windows and doors are generally closed. Air exchange rate measurements using the fan pressurization method at 50 Pa (AC/h_{50}) are available for a larger sample of Canadian homes. For example, in 200 Canadian houses built between 1980 and 1982, the average air exchange rate was 4.89 AC/h_{50} (Sulatisky 1984). In a 1989 Canada Mortgage and Housing Corporation (CMHC) survey of 194 houses across Canada, the mean was 3.44 AC/h_{50} with a ratio of almost 4:1 between the highest and the lowest air exchange values (Hamlin et al. 1990). Homes conforming to the R-2000 insulation standard had air exchange rates of about 1.0 AC/h_{50} (Riley 1990), and more than 70% of the new merchant built homes (not R-2000) had air exchange rates less than 0.3 AC/h_{50} during the 1989 heating season (Haysom et al. 1990). Although the fan pressurization method does not provide reliable air exchange measurements under normal house operating conditions, the measurements indicate that widely varying air exchange rates occur in Canadian houses. Browser (1991), in a study of 28 homes, translated fan pressurization measurements into air exchange rates equivalent to the tracer technique measurements by dividing the AC/h_{50} by 17 for bungalows and 145 for two-storey houses. These conversion factors yield air exchange rates in the range of 0.35 ach (4.89 $AC/h_{50}/14$) to 0.20 ach (3.44 $AC/h_{50}/17$) which are similar to the air exchange rates measured by the tracer technique (Browser 1991).

The final terms in Eq 1, the average house volume and the number of households, were calculated as follows. The average house volume was assumed to be the average volume per person in residential housing (including volumes for living and dining areas, etc.), namely, 100 m^3, multiplied by the average number of persons per household. The most recent Statistics Canada survey of Household Facilities and Equipment (Statistics Canada 1992a) includes the number of households, the average number of persons per household, and the average number of rooms per dwelling by province in 1992. The national average for the number of persons per household was 2.63 (Statistics Canada 1992b) which yields an average house volume, V, of 263 m^3. Table 2 lists the annual emissions from Canadian residences as calculated from Eq 1 for those VOCs with high or medium likelihood of having predominant indoor sources contributing to indoor VOC concentrations. The validity and reasonableness of the estimates are assessed by comparisons with emissions inventory data for benzene, residential solvent use data, and by comparing measured and predicted ambient VOC concentrations that are based on emissions estimates in a simple box model.

Comparison with Benzene Emissions Inventory Data and Solvent Use

The annual residential emissions range from 227 tonnes (t) for toluene to 14 t for benzene (Table 2). The ranking of the emissions is similar to the ranking for the mean indoor concentrations found in the national survey. The indoor benzene emissions estimates can be compared with the benzene emissions inventory data for residential fuel combustion sources (Solomon et al. 1993). A summary of the Canadian benzene emissions inventory for 1991 is given in Table 3. Also included in Table 3 are estimates of the benzene emissions for Ontario (Solomon et al. 1993) and Metropolitan Toronto. The estimates of benzene emissions

TABLE 2—*Estimates of annual VOC emissions (tonnes) from residences in Canada.*

Compound	Canadian Emissions (tonnes)
Toluene	227
Decane	179
1,4-dichlorobenzene	128
m- and p-xylene	115
Dichloromethane	105
1,2,4-trimethylbenzene	68.9
Ethylbenzene	58.2
o-xylene	28.1
1,3,5-trimethylbenzene	15.0
Benzene	14.0

for Metropolitan Toronto were derived in this study from the Ontario emissions and are discussed later.

The Canadian benzene emissions for 1991 were determined (Table 3). The total emissions were 22 142 t of which transportation sector sources were the majority, 19 140 t or 86.4% of the total, followed by industrial sector sources 1664 t (7.5%), and miscellaneous sources, 1347 t (6.1%). Benzene emissions from residential fuel combustion sources, which were included in the miscellaneous sector of the emissions inventory, were 991 t (4.5% of the total benzene emissions). The benzene emissions from residential fuel combustion were from fuelwood combustion (Solomon et al. 1993). The majority of the emissions from fuelwood combustion will be emitted directly to the outside through the chimney, but small amounts are expected to be entrained indoors by combustion spillage.

Recent analysis of ventilation and airtightness in new detached Canadian houses suggested that more than 18% of new detached homes can be depressurized by exhaust bathroom or laundry fans. This could lead to an increased potential for spillage of combustion products from naturally aspirating, fuel-fired heating equipment (Hamlin et al. 1990). Combustion

TABLE 3—*Benzene emissions in Canada, Ontario, and Metropolitan Toronto for 1991.*

Sector/Source	Benzene Emissions (tonnes)		
	Canada[a]	Ontario[a]	Metropolitan Toronto[b]
Transportation sources	19 140	6 141	1 386
Industrial sources			
petroleum, chemical, iron, and steel	1 238	721.6	0
solvent use	409.0	151.8	34.2
fuel combustion	15.4	1.6	0.4
Miscellaneous sources			
municipal refuse	357.0	13.5	3.0
residential fuel combustion	991.0	286.3	64.6
Total	22 142	7 316	1 488

[a] Solomon et al. 1993.

[b] Based on prorating transportation sources, solvent use and residential fuel combustion emissions by population.

spillage from residential wood heating systems in Canadian homes was found to be a common occurrence in almost all houses with woodburning systems. The frequency, duration, and severity of combustion spillage varied widely and it is not possible to estimate what percentage of residential wood combustion emissions are emitted indoors as a result of combustion spillage. Approximately 20% of Canadian single family dwellings use wood for either the primary or supplementary heating (Performance Woodburning Inc. and I.R.T.A. 1991). The benzene emissions from residential sources estimated in this study (14 t) are 1.4% of the benzene emissions from residential fuelwood combustion in the benzene emissions inventory. If all the indoor air benzene concentrations in residences originated from combustion spillage during residential wood combustion, then combustion spillage would account for a maximum of about 1.4% of the fuelwood emissions entering houses. Other than combustion spillage during residential fuelwood combustion, cigarette smoke is expected to be an important source of indoor benzene exposure (Wallace 1989).

Other indoor sources of benzene include emissions from consumer products and building materials. Although benzene was not reported in the VOC species profiles of consumer and commercial solvent use (U.S. EPA 1989b), about 400 of the 5000 materials tested by the National Air and Space Administration (Ozkaynak et al. 1987) and latex paints (Wallace et al. 1987; Sheldon et al. 1988) emitted benzene. The contributions of the various indoor sources to indoor benzene levels are unknown.

The residential emissions for those VOCs found in solvents (from Table 2) and estimates of the residential solvent use in Canada (Camford Information Services 1991) are listed in Table 4. The residential use of solvents includes paints, coatings, adhesives, sealants in residential constructions, and household products. These solvents are used both indoors and outdoors and the fraction of indoor use and thus indoor emissions is difficult to estimate. The data in Table 4 show that the indoor emissions from residences of the VOCs found in residential solvents range from approximately 0.2 to 2.3% of the usage of residential solvents. The emissions estimates must be considered as crude but appear to be the right order of magnitude.

Estimation of Ambient Concentrations Based on Emissions Inventory Estimates

In order to assess the validity of emissions estimates, model predictions of ambient concentrations can be compared with ambient measurements. Such a comparison was possible for benzene for which emissions inventory and ambient concentration data were available for Metropolitan Toronto. If these model predictions are consistent with ambient measurements, then similar model predictions can be used to estimate the contributions that VOCs emitted only from residences make to ambient VOC concentrations.

TABLE 4—*Summary of estimates of residential emissions and solvent use for selected VOCs in Canada.*

Compound	Residential Solvent Use in Canada[a] (ktonnes)	Indoor Emissions (ktonnes)	Indoor Emissions as % of Solvent Use
Xylenes	53	0.143	0.3
Toluene	49	0.227	0.5
Dichloromethane	4.5	0.105	2.3

[a] Data from Camford Information Services Inc. (1991).

Estimates of the benzene emissions in the Metropolitan Toronto area were calculated from the provincial benzene emissions inventory (Solomon et al. 1993) by assuming that emissions for transportation, solvent use, and residential fuel combustion (Table 3) are proportional to population. The industrial source benzene emissions in Metropolitan Toronto were not similarly prorated and were assumed to be negligible since most of the industrial sources in Ontario are not located in Metropolitan Toronto. In 1991 the Ontario and Metropolitan Toronto populations were, respectively, 10 084 885 and 2 192 721 (Statistics Canada 1992b). The annual benzene emissions in Metropolitan Toronto were estimated at 1488 t.

The ambient 24 h average concentration, which would result from daily benzene emissions in Metropolitan Toronto, were calculated using the box model described by Eq 2. Metropolitan Toronto is approximately 42 km in the east-west direction and 22 km in the north-south direction. The equivalent square with the same area is 30.4 by 30.4 km and the surface area perpendicular to the wind direction as required to calculate the ventilation factor (Eq 2) for Metropolitan Toronto was taken as the annual mean of the daily maximum afternoon mixing height multiplied by 30 km. The annual mean of the daily afternoon maximum mixing height for Buffalo Airport, the upper air meteorological station nearest to Metropolitan Toronto (Portelli 1981), was 1026 m. Annual mean wind speed for the Toronto International Airport is 15.2 km/h (Environment Canada 1982).

The predicted benzene concentration was 0.4 $\mu g\ m^{-3}$ as calculated from Eq 2. Benzene concentrations at remote sites in Ontario were about 1 $\mu g\ m^{-3}$ (Dann 1994) and represent a background benzene concentration level that should be added to the model prediction of 0.4 $\mu g\ m^{-3}$. The model prediction plus the background is within a factor of two of the measured mean ambient benzene concentrations at two Metropolitan Toronto stations in the NAPS network where the annual mean benzene concentrations in 1991 were 2.8 and 2.5 $\mu g\ m^{-3}$ (Dann 1994). The model predictions of the ambient benzene concentrations are considered reasonable since they are within an order of magnitude of the measured benzene concentrations. Model predictions can therefore be credibly applied to estimate the contributions of residential emissions of benzene and other VOCs to ambient concentrations.

The benzene emissions from indoor sources in Metropolitan Toronto residences will be approximately 1.2 t or 8.4% of the total emissions from all Canadian residences (14 t) since Metropolitan Toronto has approximately 8.4% of the Canadian population and 8.6% of the dwellings in Canada (Statistics Canada 1992b). The benzene emissions from Metropolitan Toronto residences were predicted to contribute 0.003 $\mu g\ m^{-3}$ to the ambient benzene concentration. This contribution is negligible (~ 0.01% of the total benzene emissions in Metropolitan Toronto). Model predictions of the ambient concentrations that would occur from the VOCs with the three highest emissions from residences (toluene, decane, and 1,4-dichlorobenzene) were, respectively, 0.047, 0.037, and 0.027 $\mu g\ m^{-3}$. These predicted concentrations were, respectively, 0.5, 2.8, and 4.2% of the annual mean for the measured concentrations of toluene, decane, and 1,4-dichlorobenzene in Metropolitan Toronto (Dann 1994).

Summary

The estimates of emissions of VOCs from Canadian residences based on mean indoor concentrations obtained in a national survey and on ambient measurements obtained in a routine monitoring network and typical air exchange rates for Canadian houses are considered crude. Ideally, simultaneous indoor and outdoor measurements should be made along with suitable air exchange rate measurements for a representative sample of Canadian residences. The estimates of emissions for those VOCs found in solvents that are used in residential applications (xylenes, toluene, tetrachloroethylene, and dichloromethane) were small fractions (less than 2%) of the residential solvent use data. Residential benzene emissions estimated in this study were a small percentage (1.4%) of the benzene emissions from residential fuelwood combustion and represent a reasonable level of the fuelwood combustion emissions that could enter a house instead of being emitted directly to the atmosphere.

Model predictions of the ambient benzene concentrations in Metropolitan Toronto, based on benzene emissions from all sources derived from an emissions inventory, were within a factor of two of the measured annual mean benzene concentrations. Similar model predictions of the contribution of only residential emissions to ambient concentrations suggest that 1,4-dichlorobenzene was predicted to make the greatest contribution (about 4%) to ambient concentrations of 1,4-dichlorobenzene in Metropolitan Toronto. This contribution is considered significant and indicates that residential sources for 1,4-dichlorobenzene are important. The predicted contributions of residential emissions of other VOCs to ambient concentrations are much smaller. The use of indoor VOC measurements together with air exchange rate information provides a method for estimating residential emissions of VOCs. The method appears to be useful for providing order-of-magnitude estimates of the contributions of indoor sources to ambient emissions and ambient concentrations of VOCs.

References

Andelman, J. B., 1985a, "Human Exposure to Volatile Halogenated Organic Chemicals in Indoor Air and Outdoor Air," *Environmental Health Perspectives,* Vol. 62, p. 313.

Andelman, J. B., 1985b, "Inhalation Exposure in the Home to Volatile Organic Contaminants in Drinking Water," *Science of the Total Environment,* Vol. 47, pp. 4433–4460.

Bell, R. W., Chapman, R. E., Kruschel, B. D., Spencer, M. J., Smith, K. V., and Lusis, M. A., 1991, "The 1990 Toronto Personal Exposure Pilot (PEP) Study," Report ARB 207-90, prepared for the Atmospheric Research and Special Programs Section, Air Resources Branch, Ontario Ministry of the Environment.

Browser Technical Inc., 1991, "Testing of Older Homes for Microbiological Pollutants," Report prepared for Canada Mortgage and Housing Corporation, Ottawa, ON, Canada.

Camford Information Services Inc., 1991, "Study on the Identification of Solvent Users and Their VOC Emissions on Canada," Report prepared for Conservation and Protection, Environment Canada.

Dann, T., 1994, "Personal communication, Environment Canada, Conservation and Protection, Pollution Measurement Division," River Road Environmental Technology Centre, Ottawa, ON K1A 0H3, Canada.

Davis, C. S. and Otson, R., 1993, "Canadian Indoor Air Quality (CIAQ) and Canadian Indoor Source Profile (CISP) Databases," *Indoor Environment,* Vol. 3, pp. 39–47.

Dumont, R. and Snodgrass, L., 1992, "Volatile Organic Compounds Survey and Summarization of Results," Saskatchewan Research Council Publication No. I-4800-11-C92, Report prepared for Canada Mortgage and Housing Corporation, Ottawa, ON, Canada.

Environment Canada, 1982, "Canadian Climate Normals, Volume 5, Wind, 1951–1980. A Publication of the Canadian Climate Program," Atmospheric Environment Service, Environment Canada.

Fellin, P. and Otson, R., 1993, "Seasonal Trends in Volatile Organic Compounds (VOCs) in Canadian Homes," *Proceedings,* Indoor Air '93, 4–8 July, Helsinki, Finland, Vol. 2, p. 117.

Hamlin, T., Forman, J., and Lubun, M., 1990, "Ventilation and Airtightness in New and Detached Canadian Housing," Research Division, Canada Mortgage and Housing Corporation, Ottawa, ON, Canada.

Haysom, J. C., Reardon, J. T., and Monsour, R., 1990, "1989 Survey of Airtightness of New Merchant Builder Houses," *Proceedings,* Indoor Air 90, The 5th International Conference on Indoor Air and Climate, Toronto, ON, Canada, 29 July to 3 Aug., Vol. 4, p. 263.

Jaques, A. P., 1990, "National Inventory of Sources and Emissions of Benzene (1985)," prepared by Regulatory Affairs and Program Development Branch, Environmental Protection Division, Environment Canada, Report EPS 5/AP/1.

Kosteltz, A. and Deslauriers, M., 1990, "Canadian Emissions Inventory of Common Air Contaminants (1985)," Regulatory Affairs and Program Integration Branch, Environmental Protection, Conservation and Protection, Environment Canada.

OMEE, 1992, "Volatile Organic Compounds Monitoring Network: Ambient Air Concentration Data Listing, 1991," Ontario Ministry of Environment Report ARB-106-92, prepared by Atmospheric Research and Special Projects Branch, Air Resources Branch, Ontario Ministry of the Environment, ISBN 0-7778-0426-3.

Otson, R., Fellin, P., and Whitmore, R., 1992, "A National Pilot Study on Occurrence of Airborne VOCs in Residences—Design and Progress," *VIP-25 Proceedings,* 1992 EPA/AWMA Int. Symposium on Measurement of Toxic and Related Air Pollutants, Air and Waste Management Association, Pittsburgh, PA.

Ozkaynak, H., Ryan, R. B., Wallace, L. A., Nelson, W., and Behard, A. W., 1987, "Sources and Emission Rates of Organic Chemical Vapours in Homes and Buildings, *Indoor Air '87: Proceedings,* Fourth International Conference on Indoor Air Quality and Climate, Institute for Water, Soil, Air and Hygiene, Berlin, August 17–21, pp. 17–21.

Performance Woodburning Inc. and I.R.T.A., 1991, "That Nice 'Woodsy' Smell: Combustion Spillage from Residential Wood Heating Systems," Report prepared for the Research Division, Policy Development Research Sector, Canada Mortgage and Housing Corporation.

Portelli, R. V., 1981, "Canadian Mixing Depth Data: July 1965 to June 1969," Atmospheric Environment Service, Environment Canada.

Riley, M., 1990, "Indoor Air Quality and Energy Conservation: The R-2000 Home Program Experience," *Proceedings, Indoor Air 90,* The 5th International Conference on Indoor Air and Climate, Toronto, Canada, July 29 to August 3, 1990, Vol. 5, p. 143.

River Road Environmental Technology Centre, 1991, "Update and Summary Report, Measurement Program for Toxic Contaminants in Canadian Urban Air," River Road Environmental Technology Centre, unpublished report, PMD 91-2.

Rogozen, M. B., Rapaport, R. D., and Shochet, A., 1985, "Development and Improvement of Organic Compound Emissions Inventories for California," prepared by Science Application International Corporation for the California Air Resources Board, SAIC-85/1531.

Sheldon, L. S., Zelon, H., Hartwell, T. D., Perritt, R., Michael, L., and Pellizari, E. D., 1988, "Indoor Air Quality and Public Buildings Volume II," EPA 600/6-88/009b, U.S. Environmental Protection Agency, Research Triangle Park, NC.

Solomon, P., Davis, C. S., and Tharby, R., 1993, "Programs and Strategies to Minimize and Prevent Benzene Releases From Industrial, Commercial and Transportation Sources in Canada," Concord Environmental Corporation Report CEC J3131, prepared for Environment Canada, Industrial Sectors Branch, Hull, Quebec.

Statistics Canada, 1992a, "Household Facilities and Equipment Catalogue 64-202 Annual," ISBN 0318-5273.

Statistics Canada, 1992b, "Population Catalogue 64-202 Annual," ISBN 0318-5273.

Sulatisky, M., 1984, "Airtightness Tests on 200 New Houses Across Canada: Summary of Results," BERR Publication No. 84.02, Energy, Mines and Resources, Canada.

U.S. EPA, 1985, "Stationary Point and Area Sources, AP-42 Compilation of Air Pollution Emission Factors," U.S. Environmental Protection Agency, Research Triangle Park, NC.

U.S. EPA, 1989a, "Report to Congress on Indoor Air Quality, Volume II: Assessment and Control of Indoor Air Pollution," Report EPA/400/1-89/001C, U.S. Environmental Protection Agency, Washington, DC.

U.S. EPA, 1989b, "Compilation and Speciation of National Emissions Factors for Consumer/Commercial Solvent Use," Information compiled to support Urban Air Toxics Assessment Studies, Report No. EPA/450/2-89/008.

U.S. EPA, 1990, "VOC/PM Speciation Data System (SDS) (Version 1.32a)," Environmental Protection Agency, Research Triangle Park, NC.

Wallace, L. A., 1989, "Major Sources of Benzene Exposure," *Environmental Health Perspectives,* Vol. 82, pp. 165–169.

Wallace, L. A., Pellizari, E. D., Hartwell, T. Y., Sparacino, C. M., Sheldon, L. S., and Zelon, H., 1985, "Personal Exposures, Indoor-Outdoor Relationships, and Breath Levels of Toxic Air Pollutants Measured for 355 Persons in New Jersey," *Atmospheric Environment,* Vol. 19, pp. 1651–1661.

Wallace, L. A., Pellizari, E. D., Leaderer, B., Hartwell, T. Y., Perritt, R., and Sheldon, L., 1987, "Emissions of Volatile Organic Compounds from Building Materials and Consumer Products," *Atmospheric Environment,* Vol. 21, pp. 385–393.

Weber, E., 1982, "Air Pollution: Assessment Methodology and Modeling," Report of the NATO/CCMS Pilot Study on Air Pollution Assessment Methodology and Modeling, ISBN 0-306-40997-6, Plenum Press, New York.

Rein Otson[1]

Vapor-Phase Organic Compounds in Canadian Residences

REFERENCE: Otson, R., "**Vapor-Phase Organic Compounds in Canadian Residences,**" *Volatile Organic Compounds in the Environment, ASTM STP 1261,* W. Wang, J. Schnoor, and J. Doi, Eds., American Society for Testing and Materials, 1996, pp. 66–76.

ABSTRACT: A national average of 0.464 mg/m^3 was calculated for the total concentration of 40 organic compounds detected previously in 757 randomly selected Canadian homes. Reanalysis of a composite sample representative of all the homes by gas chromatography/mass spectrometry allowed quantitation of 52 additional compounds. Overall, the total of concentrations for the 92 vapor-phase organic compounds (VPOCs) averaged 0.829 mg/m^3 in the homes. The average total VPOCs burden was estimated to be at least 1.4 mg/m^3, and probably >2 mg/m^3, when extraction recoveries and unidentified organics in the composite sample were considered. Factor analysis of data for 26 VOCs and other variables collected in the national pilot study indicated associations between some variables (for example, hydrocarbons, season, temperature, humidity). Possible sources (fuels, carpets, paints, etc.) of the 26 VOCs were identified by comparison of the factor analysis results with information in the Canadian Indoor Source Profile database and the scientific literature.

KEYWORDS: organics, occurrence, Canadian, residential, indoor air, sources, factor analysis, exposure

Concerns about the hazards of exposure to indoor airborne pollutants have been widely expressed in news media reports and in scientific and other publications. Adverse health effects from exposure to some pollutants have been demonstrated (WHO 1989), and guidelines for acceptable indoor air quality have been developed (Health and Welfare Canada 1989). Assessments for the 44 substances on the Canadian Priority Substances List (Government of Canada 1989) have shown that exposure through indoor air is the major medium of exposure for most of the volatile organics on the list (Government of Canada 1993). It was concluded that many of these substances, when considered alone, may not pose a risk to the health of the general population of Canada. However, the effects of exposure to mixtures of such organics found in residential and other environments and media are not known.

Information on the exposure of a whole population to airborne pollutants is important for health risk assessment, but is difficult to obtain. In addition, there is little information on the total organics load in residential indoor air. Usually indoor air quality (IAQ) surveys focus only on a small group of compounds, a small geographic area, and a small population. Only a few IAQ studies have been national in scope (for example, Krause et al. 1987). Because of variations in indoor sources, climatic conditions, and building construction and operation in different locations, data collected in one country or location are not likely to be repre-

[1] Research scientist, Environmental Health Directorate, Health Protection Branch, Health Canada, Environmental Health Center, Tunney's Pasture, Ottawa, ON, Canada K1A OL2.

sentative of circumstances in other locations. In a large country such as Canada, there are likely to be significant variations, and perhaps regional trends, in residential pollution profiles. Before a study conducted in 1992 (Otson et al. 1992, 1993), there was no information on typical levels of indoor airborne organics across the nation.

The Canadian survey (Otson et al. 1993) provided the means to investigate the potential exposure of the population to a wide range of indoor airborne organics and the factors affecting their occurrence. It was expected to provide new information which could be used for risk assessments, design of future surveys and investigations, and identification of potential problems in indoor air quality.

Procedures

The collection of VOCs in 757 randomly selected Canadian homes by suspending passive samplers (OVM 3500, 3M Company) in living areas of the residences has been described elsewhere (Otson et al. 1992, Whitmore et al. 1992). The OVM 3500 badge elements were extracted *in situ* with carbon disulphide containing 1,2-dichloroethane-D4 and the extracts were transferred to 3.5 mL amber glass vials and stored in the dark at 4°C. A composite of air sample extracts (Comp A) was prepared (Otson et al. 1993) by transferring 20 μL of each extract to a precleaned 20-mL-capacity vial. A composite of blank extracts (Comp B) was similarly prepared from single extracts obtained from 47 unexposed OVM 3500 badges which had been transported to and from the residences. Comp A (about 15 mL) was divided into three equal portions stored in sealed amber glass vials: one for long-term storage, another for analysis (Comp A), and the third for preparation of a concentrate for determination of 53 target compounds (Otson et al. 1993). Portions of Comp A and Comp B were also concentrated for determination of additional organics on the Finnigan gas chromatograph/ mass spectrometer (GC-MS).

The determination of 26 VOCs on a Hewlett-Packard (HP) 5890 GC-5971A mass selective detector (MSD) operated in the selected ion monitoring (SIM) mode and equipped with a J & W DB-Wax fused silica capillary column (30 m × 0.25 mm, 0.25 μm film thickness) and a HP7673 automatic injector has been previously described (Otson et al. 1993). For determination of the 27 other target compounds, the GC-MSD system was equipped with a Hewlett-Packard Ultra-2 (30 m × 0.20 mm, 0.33 μm film thickness) column.

For determination of the additional organics in concentrates of Comps A and B, analyses were done on a Finnigan model 4021 GC-MS upgraded to a model 4500 equipped with a Super Incos data system. Final analyses were done on a J & W DB-5 (30 m long, 0.25 mm inside diameter, 1 μM film thickness) fused silica capillary column. The GC-MS was operated in both the electron impact (EI) and chemical ionization (CI) modes. Operating conditions were: carrier helium gas flow rate 2 cm^2/min; injector, transfer oven, and analyzer temperatures, respectively, 250, 285, and 110°C; column oven temperature program −30°C for 5 min, 4°C/min to 300°C; 50 eV electron energy; 0.3 mA emission current; methane CI gas; scan from 45 to 600 amu each second; resolution better than 1 amu.

Compounds corresponding to the major peaks in the EI chromatogram were identified by comparing mass spectra with those found in the NBS library of mass spectral responses using the INCOS data system. The spectra of the five best library fits for the sample spectrum were then examined for the molecular and fragmentation ions to ensure validity of the computer identification. If standard compounds were available, they were analyzed and confirmation of the identification was achieved by comparing both the mass spectra and the gas chromatographic retention times of those of the peak in question with those of the standard compound. The spectra for the major peaks which could not be adequately identified by computerized spectral searches were manually compared with published data. Information on molecular ions, M + 1 and M − 1, determined by the CI method was used to confirm the identity of compounds. Quantitation was based on the mass spectrometer response (one major ion) to a chemical standard whenever available, and on response factors of similar compounds when a standard was not available. To avoid a loss of accuracy by rounding off in the estimation of total VOC levels, some data are reported with three significant digits although two significant digits are more representative of the analytical precision.

SPSS/PC$^+$, version 5.0 software (Norusis 1992) was used in the factor analysis (FA) of the data for the 757 homes. The principal component method was used for factor extraction and the Varimax method was used for factor rotation. The suitability of the data for factor analysis was determined by comparison of observed correlation coefficients with the magnitude of the partial coefficients as specified in the Kaiser-Meyer-Olkin (KMO) measure, with KMO < 0.5 deemed as the criterion for unacceptability for factor analysis (Norusis 1992). Variables used in FA included each of the 26 VOCs, building age, building type, building renovations, heating fuel, main heating/cooling system, fresh-air source, indoor and outdoor temperature, indoor and outdoor relative humidity (% RH), and season.

Initially, the entire data set was used in factor analysis, with the exception of the variable 1,3-dichlorobenzene, which was not detected in any of the homes. Then the data were segregated into subsets according to geographic regions, renovations in the homes during the past year, and the use of wood as a means of heating in the homes. Factor analysis was performed on each subset (Table 1). The four regions of Canada were defined by provincial boundaries, that is, region 1—British Columbia; region 2—Alberta, Saskatchewan, Manitoba; region 3—Ontario, Quebec; and region 4—New Brunswick, Nova Scotia, Newfoundland (the random selection process yielded no homes in Prince Edward Island). The variable for renovation activity was segregated, based on the frequency distribution of responses, into homes with no reported renovations, homes where there was only painting, and only carpeting, and homes where there was any other type or combination of changes. The type of fuel was segregated according to homes with wood as the sole means of heating, and those with a means of heating that included wood burning.

Results and Discussion

Measurements of organics which would be representative for the whole nation were obtained by carefully designing and implementing statistical procedures for random selection of residences (Whitmore et al. 1992). Selective sampling of only vapor-phase organic compounds (VPOCs) was ensured by use of the OVM3500 passive samplers. After completion of the survey and initial analyses (Otson et al. 1993), stored sampler extracts were reanalyzed to obtain information on additional organics in Canadian residences. The integrity of the extracts was monitored during their storage. In addition, previously obtained survey results (Otson et al. 1993) were examined by statistical procedures to identify patterns or associations between the occurrence of 26 VOCs and other variables.

VPOCs Occurrence

The occurrence of 53 organics in the 757 Canadian homes has been previously reported (Otson et al. 1993). A mean total airborne organics load of 464 $\mu g/m^3$ was subsequently calculated from the average concentrations of the 40 compounds detected in the homes. Since it was evident that these VPOCs constituted only a small portion of the organics in the sample extracts, the composite blank and sample were reanalyzed when funds became available.

TABLE 1—*Summary of some factor analysis results.*

		Regions				Renovations			Fuel	
	All	1	2	3	4	Paint	Carpet	Any	Wood	Any
No. of homes	754	90	113	457	94	178	27	548	22	162
KMO	0.71	0.58	0.68	0.71	0.52	0.72	0.41	0.64	0.72	0.77
No. of factors	14	10	12	12	11	10	3	14	2	3
% variance	67.6	71.8	73.3	67.4	73.3	65.3	80.3	70.6	78.1	77.1
VARIABLES						FACTORS				
n-hexane	9	10	6	11	4	8	...	8	...	...
dichloromethane	4	2	7	6	8	3	3	5	...	...
benzene	10	4	1	9	5	9	1	10	...	...
n-decane	2	4	...	4	...	1	1	1	1	2
trichloroethylene	5	...	8	...	7	10	1	4	...	...
chloroform	10	7	8	12	8	9	...	10	...	...
a-pinene	7	...	5	5	6	3	3	7	...	...
tetrachloroethylene	11	7	5	10	4	7	2	12	2	3
toluene	1	4	9	9	...	6	1	14	...	2
1,2-dichloroethane	12	...	4	...	...	...	...	12	...	...
ethylbenzene	1	1	...	1	1	1	...	1	...	1
p-xylene	1	1	2	1	1	1	...	1	1	1
m-xylene	1	1	...	1	1	...	...	1	1	1

TABLE 1—*Continued.*

		Regions				Renovations			Fuel	
	All	1	2	3	4	Paint	Carpet	Any	Wood	Any
o-xylene	1	1	2	1	...	1	1	1	...	2
d-limonene	6	...	3	1	6	4	...	3	...	...
1,3,5-trimethylbenzene	2	3	...	2	1	1	1	1	...	2
styrene	4	...	12	11	3	10	...	14	...	...
p-cymene	6	8	3	5	6	4	...	9	...	...
1,2,4-trimethylbenzene	2	1	...	2	...	1	...	1	1	1
p-dichlorobenzene	4	9	7	6	7	3	...	5	...	...
1,2,4-trichlorobenzene	5	...	10	4	9	7	...	4	...	...
naphthalene	11	8	9	10	3	1	...	11	1	2
indoor °C	3	3	2	3	5	2	...	2	...	...
outdoor °C	3	2	2	3	2	2	2	2	...	...
indoor %RH	3	2	2	3	2	2	...	2	...	...
outdoor %RH	9	3	2	8	5	5	...	9	...	...
season	3	2	2	3	2	2	...	2	...	...
building age	7	5	6	5	3	5	2	6	...	...
building type	...	...	...	...	...	...	...	...	2	3
renovation	7	9	6	11	9	...	...	...	...	...
fuel type	8	2	4	7	3	6	...	6	...	...
main system	8	6	9	7	3	5	...	6	...	3

NOTE: See text for description of headings and variables.

Upon reanalysis of the composite sample obtained by combining aliquots of passive sampler extracts for the 757 homes, 134 major peaks were defined in the EI chromatogram. By examination of EI and CI mass spectral data, it was possible to identify 67 compounds in the composite sample. Comparisons with analytical data for available standard compounds allowed confirmation of the identity and quantitation of 32 of the 67 VPOCs. Quantitation of the remaining 33 VPOCs was based on the MS response factors for similar compounds. Since 15 of the more abundant of the 67 VPOCs had been previously reported for the 757 homes, a total of 52 new compounds was identified.

The GC-MS chromatogram for the blanks composite showed seven major peaks. One of these corresponded to bis(2-ethylhexyl) phthalate, but no mass spectral matches could be found for scans of the remaining six peaks. The mass spectrometer responses for contaminants in the blanks composite were considered when estimates of concentrations of VPOCs in the sample composite were calculated.

Based on the nature of the 52 new VPOCs and information on the nature and sampling rates of the previously characterized 26 VOCs, the sampling rates for the new target compounds were estimated to be near 30 cm^3/min on average. Calculations based on the average sampling duration of 1420 min (about 24 h) and the estimated passive sampling rate indicated that the sum of the average concentrations of the new VPOCs in the homes was about 0.365 mg/m^3. The other 40 VPOCs identified previously (Otson et al. 1993) contributed an additional 0.464 mg/m^3 to the airborne organics load. Thus, the total identified VPOCs concentration in Canadian homes was estimated to be about 0.829 mg/m^3 on average. The number and concentrations in chemical classes represented by the 92 VPOCs detected in Canadian homes are listed in Table 2.

The 52 new VPOCs accounted for only 39% of the total peak area for the sample chromatogram. If an estimated 0.571 mg/m^3 of unidentified VPOCs in the composite sample are included, the total concentration of VPOCs in the average Canadian home is 1.4 mg/m^3. The unidentified VPOCs estimate was based on the total peak areas for the chromatograms from previous (Otson et al. 1993) and current analyses, and from assuming that the average GC-MS response factor for identified compounds was the same as for unidentified compounds. It was also assumed that all peaks represented organic compounds. Since total peak area values were based only on areas of the major peaks in the chromatogram, the calculated concentration of unidentified VPOCs based on the composite sample analysis is probably higher than the estimated 0.571 mg/m^3. In addition, except for the value of 0.464 mg/m^3 determined previously for 40 VPOCs, the estimates do not account for extraction recoveries from the carbonaceous sorbent used in the passive samplers. Based on previous results (Otson et al. 1993), some of the polar and large molecules which were detected among the 52 new

TABLE 2—*Classes of VPOCs in Canadian homes and other studies.*

	Concentration ($\mu g/m^3$)		Number of VPOCs detected	
	Canada	WHO[a]	Canada	WHO[a]
Aliphatics	162	140	26	21
Aromatics	287	156	35	16
Chlorinated aliphatics	64	10	9	6
Chlorinated aromatics	24	1	7	6
Oxygenated organics	116	48	9	9
Amines	176	0	6	0
Total	829	355	92	58

[a] World Health Organization (WHO) 1989.

VPOCs are likely to show recovery values well below 100%. Also, the sampling and analysis methods were not suitable for a segment of VPOCs, such as vinyl chloride and other very volatile gases, in the homes. Consequently, it is probable that the average total concentration of identified and unidentified VPOCs in Canadian residences is >2 mg/m^3.

These are important findings, in that they clearly demonstrate that total concentrations of vapor-phase organic compounds to which occupants are exposed in their homes are typically well above 1 mg/m^3 in air. It is sometimes assumed, particularly by those unfamiliar with air sampling and analysis, that identified compounds constitute a major component of airborne organics, and occur at total levels below 1 mg/m^3 since such values are usually reported. Summation of data compiled for other large-scale studies (Lebret et al. 1986, Krause et al. 1987, Shah and Singh 1988, WHO 1989) shows that levels of identified organics in indoor air are typically <0.5 mg/m^3. Calculations based on a summary of such indoor data (WHO 1989) yield a total of 355 μg/m^3 and a distribution of these substances among chemical classes which is somewhat different from that observed in this Canadian study (Table 2).

Also TVOC values, obtained by GC-FID (flame ionization detector) or speciation techniques, are on average <1 mg/m^3 in indoor air (Molhave 1990, Otson and Fellin 1993). Note that TVOC analytical methods are often calibrated with a hydrocarbon reference material, and that aliphatic and aromatic compounds constitute only about 50% of the identified VPOCs in this Canadian study (Table 2). Amines, chlorinated hydrocarbons, and oxygenated organics, which were also abundant, are likely to give lower GC-FID-TVOC readings than the hydrocarbons. Due to the variability of indoor air composition, TVOC measurements obtained by the traditional GC-FID technique, or for a limited number of compounds, are not likely to accurately reflect the airborne organics load.

Factor Analysis

There was interest in information on factors influencing the occurrence of airborne organics due to the variety and abundance of VPOCs found in the 757 homes. The availability of data for individual residences (Otson et al. 1993) made it possible to investigate the association between the occurrence of 26 VOCs and other variables which may influence their occurrence, and to identify patterns which may indicate important VOCs sources.

Factor analysis (FA) results for the variables are summarized in Table 1. The number of homes used in the analyses, the KMO index values, the number of factors, the cumulative percentage variance values for the factors, the factors in the solutions for all the data (second column), and the data subsets are listed. The variables *m*-dichlorobenzene and fresh air source did not occur in any factors. Pentachloroethane, hexachloroethane, and 1,1,2,2-tetrachloroethane are not listed in Table 1 since they were typically found in relatively unimportant factors, that is, factor 6 or higher. Since the Varimax method could not provide a solution for the no-renovations variable, this category is not shown in Table 1.

Factor analyses of the whole data set showed that the two most dominant factors (factors 1 and 2) consisted only of hydrocarbons. The third factor showed an association of only season, indoor temperature and percent relative humidity (%RH), and outdoor temperature (Table 1). There were regional differences in factor analysis results, particularly as demonstrated by the relatively unimportant factors. Groups of hydrocarbons in the two most dominant factors, and season, temperature, and humidity in the second and third factors, were found for most of the data subsets. In a wintertime survey of 134 postwar homes in Ede, the Netherlands, Lebret et al. (1986) also found that dominant factors consisted of hydrocarbons.

Pairs or groups of compounds which were frequently associated in several data subsets were ethylbenzene/xylene isomers (six times), 1,3,5- and 1,2,4-trimemthylbenzenes (three times), limonene/*p*-cymene (three times), dichloromethane/*p*-dichlorobenzene (four times), and benzene/toluene (three times). The last three pairs were found mostly in the less important factors. Of the chlorinated hydrocarbons, only trichloroethylene, tetrachloroethylene, and dichloromethane for carpet renovations, tetrachloroethylene in the fuels subsets, and dichloromethane and *p*-dichlorobenzene in paint renovation subsets were found in the three most important factors.

Published information on indoor sources (Otson and Fellin 1992) and a computerized database (Davis and Otson 1993) for Canadian indoor source profiles (CISP) were used to attempt identification of sources of the VOCs. Compounds which were listed with the three dominant factors and a few possible sources for these compounds are summarized in Table 3. In fact there are many more possible sources, and the VOCs listed in Tables 1 and 3 may each have originated from a large number of sources in the 757 homes.

The results for each data subset were examined for relationships between components of dominant factors and the possible sources listed in Table 1. Four C2-benzenes in the first factor of the all-regions data set also occurred consistently in five of nine subsets, and may originate from tobacco smoke or automobiles. Two trimethylbenzenes, which were sometimes found together in the dominant factors, may have originated from carpets, woodburning, and paints. As compared to the other subsets, the dominance of factors containing hydrocarbons in the fuel subsets is noteworthy. The differences in emissions from the woodburning, and heating sources which included wood, presumably account for the slight differences in hydrocarbons listed in the dominant factors for the corresponding data subsets.

The variables listed in the dominant factor in the painting subset included the hydrocarbons, 1,3,5- and 1,2,4-trimethylbenzenes, which are identified as components of latex paint in the CISP. Dichloromethane and *p*-dichlorobenzene, which occurred together in several factors, have been reported in foam board emissions (Table 3). The third factor contained

TABLE 3—*Possible sources and the components associated with dominant factors.*

Source	Components
Latex paints[a]	benzene, 1,2,4- & 1,3,5-trimethylbenzenes, toluene
Alkyd paints[a]	1,1,2,2-tetrachloroethane, 1,2,4-trichlorobenzene
Carpets[a]	1,2,4- & 1,3,5-trimethylbenzenes, benzene, toluene, styrene
Glued carpets[a]	*p*-dichlorobenzene, n-decane
Wood burning[a]	xylenes, 1,2,4- & 1,3,5-trimethylbenzenes, naphthalene, toluene, styrene
Foam board[b]	dichloromethane, dichlorobenzenes
Paint removers[b]	dichloromethane, toluene
Spray products[b]	dichloromethane, xylenes
Adhesives/tape[a]	trichloroethylene
Adhesives/tape[b]	n-decane, toluene, styrene
Room deodorizers[b]	*p*-dichlorobenzene, limonene, decane
Detergents[b]	limonene, pinene
Waxes[b]	pinene
Tobacco smoke[b]	benzene, toluene, *m*- and *p*-xylenes, styrene, ethylbenzene
Gasoline/driving[b]	benzene, toluene, xylenes, styrene, trimethylbenzene
Solvents[b]	toluene, ethylbenzene
Dry cleaning[b]	tetrachloroethylene
Tap water[b]	chloroform

[a] CISP (Davis and Otson 1993).
[b] Otson and Fellin 1992.

dichloromethane, a component of paint removers, which would be expected to be associated with painting. Chlorinated compounds in the first three factors in the carpet and other subsets may be associated with various household products and drycleaned clothing. Limonene is present as an odorizer in detergents and air fresheners, but *p*-cymene has not been reported for such products. Other possible sources of the VOCs in dominant factors are alkyd paints, glued carpets, adhesives/tape, waxes, and solvents (Table 3). These assignments are only tentative since the VOCs may also originate from a number of other sources. However, these FA results have allowed identification of sources and compounds for future investigation.

Concluding Remarks

The national survey provided a unique opportunity to examine the residential indoor air quality across a country with a population of about 28 million people occupying about nine million residences, and with the second largest land area, 10×10^6 km^2, among nations. In fact, random sampling was limited to an area of about 6×10^6 km^2 since virtually the whole population resides in the ten provinces.

Information collected on the occurrence of VPOCs in Canada was important for several reasons. First, this provided data which could be stored in a Canadian Indoor Air Quality database, or CIAQ (Davis and Otson 1993), and then used for comparison with results of more limited and focused studies aimed at determination of problem areas or the significance of exposures of subpopulations with particular characteristics. The VPOCs data in the CIAQ will also be useful for investigations such as those aimed at prediction of emission rates within (Hamlin and Cooper 1992) or from homes (Davis and Otson 1994). A particularly important application has been the use of the data in assessment of the risk from exposure of Canadians to priority substances (Government of Canada 1989, 1993). The information will also be important in the establishment and assessment of substances on a new Priority Substances List.

The conclusion that the average total VPOCs concentration in residences is >1.4 mg/m^3, and is not necessarily accurately reflected by traditional TVOC measurements, suggests that the exposures of Canadians, and other populations, have been underestimated. The total organics exposure load on occupants in Canadian homes is likely significantly larger than indicated by the data reported here since all VPOCs were not effectively recovered from the samplers, and any organics bound to airborne particles were not collected. The abundance and variety of VPOCs in indoor air (Table 2) are often overlooked because attention is usually focused on a few organics of immediate interest. The assessment of the health risk from exposure to the complex mixtures is very difficult due to the variability in composition of VPOCs in different indoor environments, and differences in the biological effects induced by the various compounds, acting alone or in combinations. However, such assessments will be eventually required for better understanding of the impact of the environment on humans.

Factor analysis of some of the national survey data led to identification of associations between some variables (Table 1) and topics for future investigations. A disadvantage in interpretation of the results was the number of possible VOC sources and the lack of appropriate emissions data available for storage in the CISP (Davis and Otson 1993) and use in evaluation of the results. Many of the available source profiles are incomplete and do not allow for variations in source emissions according to use patterns. Nevertheless, patterns in the occurrence of certain hydrocarbons and chlorinated hydrocarbons were observed, and there was an evident dependence of VOC occurrences on temperature, humidity, and season of the year. More detailed statistical analyses of the data may reveal more significant correlations. When significant and commonly occurring sources can be clearly identified, then mitigation of hazardous pollutants can be initiated.

Due to the high costs required for such extensive and elaborate studies, this type of national survey is likely to remain unique for some time. Its uniqueness demands that the collected data be utilized to the fullest potential. Therefore, examination of the survey data from different perspectives, and reanalyses of the air sample extracts, will continue as resources for conducting such work become available.

Acknowledgments

The contributions of Claude Davis and Philip Fellin of Concord Environmental Corp., and Mel Strosher and Eric Peake of the Alberta Research Council, who conducted some of this work under contracts funded by the Canadian Program on Energy Research and Development, are gratefully acknowledged.

References

Davis, C. S. and Otson, R., 1993, "The Canadian Indoor Air Quality (CIAQ) and the Canadian Indoor Source Profile (CISP) Databases," *Indoor Environment,* Vol. 3, pp. 39–47.

Davis, C. S. and Otson, R., 1994, "Estimation of the Contribution of Residential Sources to Emissions Inventories of Selected Volatile Organic Compounds (VOCs)," presentation at the International Symposium on Volatile Organic Compounds (VOCs) in the Environment, 11–13 April, Montreal.

Government of Canada, 1989, *Canada Gazette, Part I, February 11, 1989, Priority Substances List,* Supply and Services Canada, Ottawa.

Government of Canada, 1993, *Priority Substance List Assessment Reports,* Canadian Environmental Protection Act, Minister of Supply and Services Canada, Ottawa.

Hamlin T. and Cooper, K., 1992, "CMHC Residential Indoor Air Quality—Parametric Study," *Proceedings,* Ventilation for Energy Efficiency and Optimum Indoor Air Quality, 13th AIVC Conference, Nice, France, pp. 207–216.

Health and Welfare Canada, 1989, *Exposure Guidelines for Residential Air Quality,* Federal-Provincial Working Group on Indoor Air Quality, Supply and Services Canada, Ottawa.

Krause, C., Mailahn, W., Nagel, R., Schulz, C., Seifert, B., and Ullrich, D., 1987, "Occurrence of Volatile Organic Compounds in the Air of 500 Homes in the Federal Republic of Germany," *Proceedings,* Indoor Air '87, 4th International Conference on Indoor Air Quality and Climate, Berlin, Vol. 1, pp. 102–106.

Lebret, E., van de Wiel, H. J., Bos, H. P., Noij, D., and Boleij, J. S. M., 1986, "Volatile Organic Compounds in Dutch Homes," *Environment International,* Vol. 12, pp. 323–332.

Molhave, L., 1990, "Volatile Organic Compounds, Indoor Air Quality and Health," *Proceedings,* Indoor Air '90, International Conference on Indoor Air Quality and Climate, Ottawa, Vol. 5, pp. 15–33.

Norusis, M. J., 1992, *SPSS/PC+ Professional Statistics Version 5.0,* SPSS Inc., Chicago.

Otson R. and Fellin, P., 1992, "Volatile Organics in the Indoor Environment: Sources and Occurrence," *Gaseous Pollutants—Characterization and Cycling,* J. O. Nriagu, Ed., Wiley, New York, pp. 335–421.

Otson, R., Fellin, P., and Whitmore, R., 1992, "A National Pilot Study on Occurrence of Airborne VOCs in Residences—Design and Progress," *Proceedings,* 1992 EPA/AWMA International Symposium, Measurement of Toxic and Related Air Pollutants, VIP-25, Air & Waste Management Association, Pittsburgh, PA, pp. 176–181.

Otson, R. and Fellin, P., 1993, "TVOC Measurement: Relevance and Limitations," *Proceedings,* Indoor Air '93, 6th International Conference on Indoor Air Quality and Climate, Chemicals in Indoor Air, Materials Emissions, Helsinki, Vol. 2, pp. 281–286.

Otson, R., Fellin, P., and Tran, Q., 1993, "Investigation of VOCs in Canadian Residences," *Proceedings,* Indoor Air '93, 6th International Conference on Indoor Air Quality and Climate, Chemicals in Indoor Air, Material Emissions, Helsinki, Vol. 2, pp. 141–146.

Shah, J. J. and Singh, H. B., 1988, "Distribution of Volatile Organic Chemicals in Indoor and Outdoor Air. A National VOCs Data Base," *Environmental Science & Technology,* Vol. 22, pp. 1381–1388.

Whitmore, R. W., Williams, S. R., Fellin, P., and Otson, R., 1992, "Design of a National Study of Residential Air Quality in Canada," *Proceedings,* Section on Statistics and Environment of the 1991 Joint Statistical Meetings of the American Statistical Association (151st Annual Meeting), and the Biometric Society, Institute of Mathematical Statistics (54th Annual Meeting), Alexandria, VA, pp. 207–212.

WHO, 1989, *Indoor Air Quality: Organic Pollutants Report. Report on a WHO Meeting. Euro Reports & Studies 111,* World Health Organization Regional Office for Europe, Copenhagen.

W. J. Muller[1] *and M. S. Black*[1]

Sensory Irritation in Mice Exposed to VOC Emissions from Indoor Products

REFERENCE: Muller, W. J. and Black, M. S., **"Sensory Irritation in Mice Exposed to VOC Emissions from Indoor Products,"** *Volatile Organic Compounds in the Environment, ASTM STP 1261,* W. Wang, J. Schnoor, and J. Doi, Eds., American Society for Testing and Materials, 1996, pp. 77–92.

ABSTRACT: An ASTM standard method for estimation of sensory irritation (SI) in mice of airborne chemicals (E 981-84) was modified and applied to the estimation of irritation caused by volatile organic compound (VOC) emissions of products used in indoor environments. This method has the potential ability to relate the results from animal studies to expected human responses. VOC emissions of indoor products are commonly tested using small environmental chambers, according to ASTM standard guide D 5116-90. This standard guidance was used as the basis for generating airborne VOCs emitted from samples of carpet, ceiling tile, wall-covering, resilient flooring, and veneer. Four mice in each test were exposed to these emissions. The changes in the respiratory waveform and respiratory frequency of exposed mice were used to determine the relative irritancy during each exposure. Experimental variables included the product chamber air temperature and the construction material of the environmental chambers (glass or stainless steel) that were used to generate the exposure atmosphere. Increased temperature was found to induce the irritation response; however, chemical analysis indicated that the relative concentrations of major VOCs in product emissions also change in response to elevated temperature. The type of construction material of the environmental chamber was not found to affect the response of the animals above biological variability, although a trend toward increased irritation was observed using glass chambers. The chemical atmospheres were comparable for the two chamber types.

KEYWORDS: sensory irritation, VOC emissions, indoor air, bioassay

Sensory irritation in mice is a reflex response to chemical exposure upon stimulation of the trigeminal nerve endings in the eye and upper respiratory tract (Alarie 1973). The degree of irritation caused by a particular chemical atmosphere may be quantified in mice by a standardized bioassay [ASTM Standard Test Method for Estimating Sensory Irritancy of Airborne Chemicals (E 981-84)]. Subjective irritation in humans on exposure to these chemicals has been found to correlate with the response in mice (Schaper 1993). Human physiological responses to sensory irritants are similar to complaints associated with sick building syndrome, including eye and upper respiratory irritation, as well as nonspecific symptoms such as headache, nausea, and memory loss. As a result, some recent work has been undertaken to determine the degree of sensory irritation associated with emissions from common indoor products (Hansen et al 1991).

Measurement of chemicals emitted by commonly used indoor products under conditions approximating those found under "realistic" conditions is commonly done using dynamic

[1] Laboratory director and chief scientist, respectively, Air Quality Sciences, Inc., 1337 Capital Circle, Suite D, Marietta, GA 30067.

environmental chamber technology [ASTM Standard Guide for Small-scale Environmental Chamber Determinations of Organic Emissions From Indoor Materials/Products (D 5116-90)]. Controlled experimental conditions include the product loading, chamber airflow, chamber air temperature, and relative humidity. Results allow estimation of indoor levels of chemicals from the test results. The air passing over a test product is pulled through appropriate sampling devices to determine the amount of formaldehyde or other volatile organic chemicals being emitted by the test product over a measured time period.

This paper discusses studies evaluating sensory irritation from products commonly found in indoor environments. Some previous product testing has involved heating products to evaluated temperatures which would not commonly occur in indoor situations (Anderson 1993, Costa 1993, Kennedy 1993). The effects of varying the product chamber air temperature from ambient to 70°C on both the chemical emissions and biological response are elevated. Since most sensory irritation testing is done using glass test systems, and most environmental chamber testing is done using stainless steel test symptoms, it has been hypothesized[2] that some differences in the animal response to chemical atmospheres may occur when there is stainless steel in the test system. This hypothesis was based on conflicting results from different laboratories exposing mice to individual chemicals and to carpet emissions, using methodology (Anderson 1993, Stott 1993). The variability between the glass and stainless steel chambers used for generation of chemical emissions in the sensory irritation response of the animals, and the chemicals detected during the exposures, is also discussed.

Materials and Methods

Exposure System

The animal exposure system developed for this study attaches a test product chamber to a 2.7 L glass animal exposure chamber, identical to that described by ASTM E 981-84. Two product chambers were used: one manufactured from borosilicate glass (volume = 54.5 L), and one from stainless steel (50 L). Experiments using the glass system have been previously described (Muller and Black 1995). Four animals were exposed simultaneously for each test condition.

Purified (contaminant-free) air is supplied to the product chamber from a proprietary air purification system where it mixes with product emissions. The outlet from the product chamber opens to an exhaust port or to the animal exposure chamber. A valve system [all-glass and trifluoroethylene or stainless steel] placed after the product chamber allows purified air or product chamber air as the supply to the animal chamber. The temperature of the air in this valve system, and in the connections between the product and animal chambers, could be controlled to avoid temperature changes between the product chamber airspace and the entrance of the animal exposure chamber. Experimental test conditions of air flow rate, temperature, and relative humidity were monitored and controlled during the tests.

Test Materials

All test samples were representative of freshly manufactured products, and were handled, stored, and wrapped to avoid any loss of volatile emissions prior to exposure testing. Products tested included carpet, vinyl resilient flooring, ceiling tile, wallcovering, and oak veneer (used

[2] M. Goldman, personal communication, April 21, 1993.

in the manufacture of products such as cabinets). Product characteristics are described in Table 1. Products were shipped wrapped in nonemitting foil and handled with nonemitting gloves. All samples were freshly trimmed prior to testing and placed within the product chamber on a glass rack or stainless steel support with both sides of the product exposed to the airflow.

Animals

Specific pathogen-free (SPF) male Swiss-Webster outbred mice [Hsd:ND4] were obtained at between 16 and 26 g and between 30 and 50 days of age and acclimated to laboratory conditions for at least six days prior to study. All animals were uniquely identified and weighed upon arrival and provided with feed (Purina 5002 diet, Purina Mills, Inc., St. Louis, Missouri), and deionized water *ad libitum* except during exposures. Animals were housed in pairs in polycarbonate cages equipped with micro-barrier tops (Ancare, N., Bellmore, New York), with heat-treated α-cellulose chips (ALPHA-dri, Shepherd Specialty Papers, Kalamazoo, Michigan) as bedding. Each cage remained within a Bio-Bubble clean room (Bubble Systems/CSA, Inc., Ft. Collins, Colorado) from arrival until experimental use. The animal facility was maintained at a temperature between 18 and 30°C, a relative humidity of about 50%, and a 12-h light/dark cycle.

Animal weights were monitored prior to experimental use to assure health maintenance of the animals. Four animals were used for each study group, each between 20 and 33 g body weight. Once removed from the Bio-Bubble, animals were housed in micro-isolator cages in a separate area of the animal facility outside of the Bio-Bubble.

Product Chamber Parameters

Parameters controlled during product testing included air exchange rate through the chamber, product loading (m^2 of exposed product surface per m^3 test chamber volume), air temperature, and relative humidity of the supply air. A flow rate of 6.7 L/min was used in these experiments. The product surface area, with both sides of the product exposed to flow, was 0.32 m^2.

Temperature and relative humidity within the product chamber were maintained by passing purified, dehumidified supply air through a custom-built humidifier adjoining the chamber. Vaisala temperature and relative humidity sensors (Model HMP 1334, Helsinki, Finland) were used to monitor these parameters at regular intervals during a test. Two test conditions

TABLE 1—*Characteristics of products used in study.*

Product	Size	Description	Loading Conditions
Carpet	26.4 × 61 cm	slate blue fiber, hardback	placed on glass rack, both sides exposed
Ceiling tile	26.4 × 61 cm	white	placed on glass rack, both sides exposed
Flooring	26.4 × 61 cm	resilient, green-blue-gray patterned	placed on glass rack, both sides exposed
Wallcovering	26.4 × 61 cm	Spanished, white, embossed and filled with beige ink	placed on glass rack, both sides exposed
Veneer	cut into 10 12-in. × 2-in. pieces	oak, about 0.75-in. thickness	pieces placed onto glass rack

of temperature and relative humidity within the airspace in the product chamber were studied. The first test situation, designed to simulate normal ambient conditions, involved a temperature of 23 ±3°C and relative humidity of 50 ±5%. The second test condition involved maintaining the product chamber airspace temperature at 70 ±5°C. Product surface temperature was also monitored and was found to be maintained at the same temperature as the airspace above the product. Maintaining the same absolute water content of the air decreased the relative humidity to 4.5 ±1%, although in some experiments this was increased when water was driven off the product during heating.

When products in the current study were tested with product maintained at 70°C, the elevated air temperature was maintained constant between the product chamber outlet and the animal chamber inlet, to avoid chemical condensation in the connecting lines. This resulted in an elevated air temperature in the breathing zone of the animal of about 42°C (107°F) and a relative humidity of about 4.5%.

To avoid potential condensation of chemicals emitted into the product chamber airspace in the lines connecting the product and animal chambers, heat tape (Omega Engineering, Stamford, Connecticut) was used to maintain the air temperature between the product chamber outlet and the animal chamber inlet at the same temperature as the product chamber airspace. Animal chamber inlet air temperature was monitored in the airspace near the diffuser region of the chamber with an RTD sensor (Omega Engineering) and maintained at either 23 ±1°C or 70 ±5°C.

Exposure Time Course

Mice were exposed to product emissions for a total of four one-hour periods over the course of two days. The exposure periods corresponded with 1, 4, 24, and 27 h following the beginning of dynamic ventilation of product, allowing comparison of sensory irritation effects with the emission factor decay of a product.

Animals were loaded into the animal chamber and exposed to purified air. The method for measuring respiratory activity is described below. Baseline respiratory activity of each animal was determined during a 60-min acclimation period on the first day of exposure. Individual animal respiratory activity was monitored throughout a 5.5 h period on each day.

Product was loaded into the product chamber during the acclimation period on the first day of exposure, and dynamic ventilation of the product chamber began 30 min prior to the first period of animal exposure. From this point, product remained dynamically ventilated for the entire test period, with 6.7 L/min of purified air supplied to the product chamber.

During the 60-min animal acclimation, product chamber outlet air was directed to an exhaust line. After this period, inlet flow to the animal exposure chamber was changed from purified air (at 6.7 L/min) to the outlet air from the product chamber. Exposure to the product chamber air continued for an additional 60 min. The animal exposure chamber was then flushed with purified air for a two-hour control breathing period for stabilizing and recovery, followed by a second 60-min exposure to the product chamber air. A 30-min recovery period ended the exposure train.

Respiratory Activity During Exposure

Each animal was placed in a glass plethysmograph sleeve attached to a trifluoroethylene collar using tape, as described previously (Muller and Black 1995). Proper respiratory waveforms were verified by inspection of the pressure tracings according to ASTM E 981-84. Further validation of this method, as comparable with the use of rubber dams for determination of respiratory frequency, may be made by comparing baseline respiratory frequencies.

For all current experiments, baselines averaged 253 breaths/min, while ASTM E 981-84 cites a normal baseline breathing frequency for mice of 240 to 275 breaths/min.

Each plethysmograph was connected to an individual differential pressure transducer (Validyne Engineering, Model DP45-14, Northridge, California). The transducers were connected to a computer-based data acquisition system which displayed in real time the respiratory activity of all four animals and digitally recorded the individual respiratory waveforms. Respiratory frequencies were continuously calculated and digitally recorded during the course of the exposure.

Analysis of Respiratory Response

Respiratory frequencies were determined digitally from the pressure traces recorded from the plethysmograph. Frequencies were determined from the average of 15-s periods throughout the course of the exposure. The baseline respiratory frequencies were determined from the final 30 min of the 60-min acclimation period of the first day. This procedure was done to eliminate waveforms associated with the initial stress of animal confinement. The effects of body and head movements were removed from the pressure traces by a time-averaging algorithm, which calculated a running time-averaged frequency of an individual mouse over a three-minute interval.

Average frequencies for a group of four animals were determined from the individual frequency data for each time interval. Sensory irritation is indicated when the characteristic change in respiratory waveform is accompanied by a decrease in frequency of 12% or greater (ASTM E 981-84).

Chemical Sampling

VOCs and formaldehyde were sampled for 60-min periods from the animal chamber outlet air at times corresponding to 1, 4, 24, and 27 h of product dynamic ventilation. Background measurements were performed each day prior to testing to ensure nondetectable background levels of chemicals.

VOCs were collected on triple bed sorbent tubes and analyzed by gas chromatography/mass spectrometry, using a NuTech 8533 universal sample concentrator (Durham, North Carolina), a Hewlett-Packard 5890 Series II gas chromatograph, and a Hewlett-Packard 5971 mass selective detector (Houston, Texas). Formaldehyde was collected and analyzed by a spectrophotometric technique similar to NIOSH Method 3500 (NIOSH 1990).

Gas Monitoring

Carbon dioxide (CO_2) levels in the animal chamber were monitored with a Gas Tech CO_2 sensor (Model R1-411A, Newark, California), and recorded at regular intervals. CO_2 levels were maintained under 2100 ppm at the exposure flow rates studied. Oxygen, carbon monoxide, and ammonia (O_2, CO, and NH_3) concentrations in animal chamber outlet air were monitored and electronically recorded continuously during testing with Sensor Stik flow-through gas sensors (EIT, Exton, Pennsylvania).

Statistical Methods

All statistical analysis referred to in the text was performed using SYSTAT® for Windows (Systat, Inc. Evanston, Illinois) analytical software. Mean respiratory depression for experiments using the different chamber types was compared for given conditions of temperature using a *t*-test. Levels of significance were taken at 95% confidence ($P < 0.05$). The hypothesis that there is no difference between mean values of a given parameter is rejected when the *t*-test produces a *P*-value of less than 0.05. A standard least-squares regression analysis on concentration levels for chemical groups measured using a stainless steel product chamber and a glass product chamber was also done.

Results

Respiratory Response to Exposure

The mean (and standard deviation) of the maximum respiratory depression of four mice during all exposure periods from each test, as a percent of baseline mean respiratory rate, is shown in Fig. 1. The descriptive degrees of sensory irritation, based on the degree of respiratory depression according to ASTM E 981-84, are also shown. For products maintained at close to ambient temperatures (≈23°C), little evidence of sensory irritation was noted during any exposure period. Significant respiratory depression was observed during four exposure periods; however, the characteristic sensory irritant waveform was not found in conjunction with this waveform change. A single mouse was observed to have the characteristic waveform change during exposure to carpet emissions; however, this was not accompanied by respiratory depression.

For products heated to 70°C, respiratory depression consistent with slight to extreme sensory irritation was observed during exposures to all products, though not at all exposure time

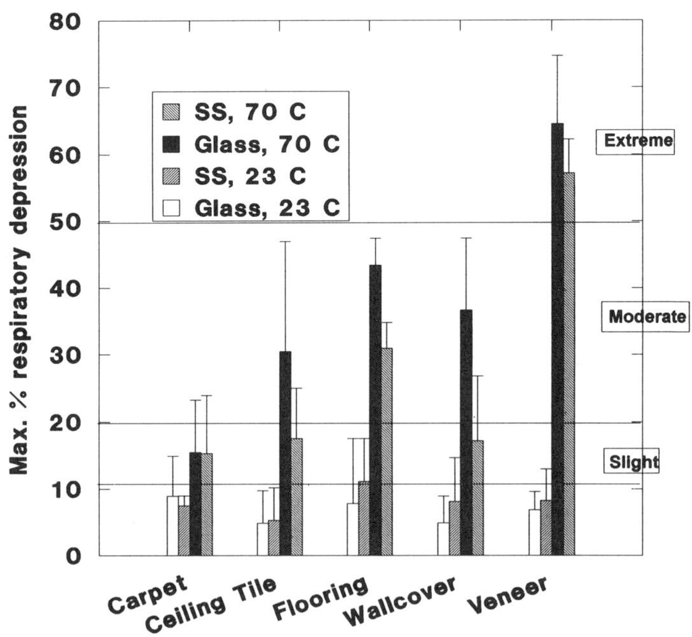

FIG. 1—*Mean (and standard deviation) of the maximum respiratory depression of four mice during all exposure periods from each test, as a percent of baseline mean respiratory rate. Sensory irritation was not found from any product at 23°C; however, at 70°C all products except the carpet sample produced some degree of sensory irritation. Asterisks indicate significant differences (P < 0.05) between data from stainless steel and glass product chambers. A general trend of increased respiratory depression using a glass product chamber is observed at elevated temperature.*

points. The presence of the characteristic sensory irritant waveform was confirmed for all products except carpet during all of these exposures. The sample of carpet was not found to result in irritation during any exposure periods. Table 2 shows the descriptive waveform responses of the animals during any test. The number of animals demonstrating sensory irritation waveforms using the stainless steel system is nearly identical to the number using the all-glass system. The response using the all-glass system appeared to persist longer than in the stainless steel system for ceiling tile and wallcovering at elevated temperature.

Figure 1 also compares the response to exposure when a stainless steel product chamber is used with the response when a glass chamber is used. At 23°C, no difference in response is observed. At 70°C, a significant difference ($P < 0.05$) in mean respiratory depression during exposure was found only for resilient flooring; however, a trend toward increased response when using the glass chamber is observed for all products except carpet.

Figure 2 shows the mean respiratory rate from four mice (as a percent of baseline frequency) for the first control period, first exposure period, and 30 min of the subsequent control period for each product under each condition. The respiratory depression associated with sensory irritation is observed in the 70°C tests, as is the trend towards increased response with the glass system. Comparison of the nature of the respiratory depression for different products suggests different responses; for example, the respiratory depression from veneer emissions is rapid and sustained during exposure, with rapid recovery to baseline, while that from resilient flooring emissions is delayed, with a slow recovery. These different responses have been observed for different individual chemicals, and this difference has been discussed previously (Bos et al. 1992).

Summed Chemical Emissions from Products

Figure 3 shows the average total concentration of volatile organic compounds (VOCs) and formaldehyde for the four exposure periods for emissions from each product in experiments using a stainless steel chamber as compared with those using a glass chamber. For all prod-

TABLE 2—*Summary of observed waveform characteristics during exposure to product emissions.*

Product	Chamber	23°C	70°C
Carpet	stainless steel	...	+ (E2)
	glass	+ (E2)	+ (E1)
Ceiling tile	stainless steel	...	++++ (E1)
	glass	...	++++ (E1) +++ (E2)
Resilient flooring[a]	stainless steel	...	++++ (E1, E2, E3, E4)
	glass	...	++++ (E1, E3) +++ (E2, E4)
Spanished wallcovering	stainless steel	...	++++ (E1, E2)
	glass	...	++++ (E1, E2, E3, E4)
Oak veneer[b]	stainless steel	...	++++ (E1, E2, E3, E4)
	glass	...	++++ (E1, E2, E3, E4)

[a] Recovery from sensory irritation during exposure to flooring was not as rapid as for other products.
[b] Some recovery from sensory irritation during exposure to veneer was observed using the SS system, but sensory irritation persisted throughout while using the all-glass system.

NOTE: "+" indicates the presence of the characteristic sensory irritant waveform; the number of "+" marks indicates the number of animals (out of four) which were observed to demonstrate the sensory irritant waveform. The exposure periods during which the observations were made are indicated in parentheses (numbered 1–4). "−" implies that sensory irritant waveforms were not observed.

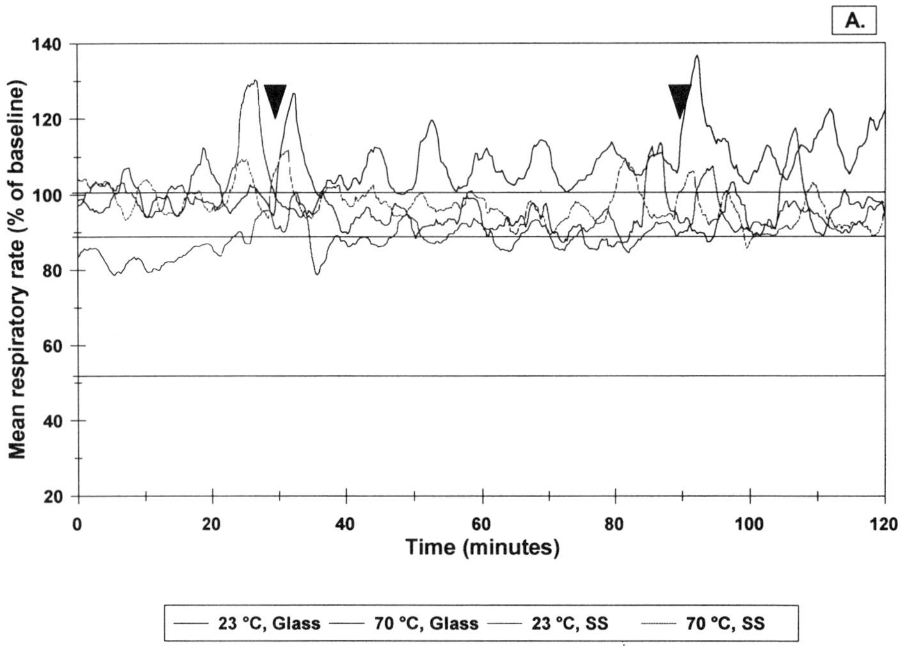

FIG. 2—*Mean respiratory rate of four mice (as a percent of baseline respiratory rate) during the initial control period, first exposure, and 30 min of the first recovery:* (A) *carpet.*

ucts, increasing the temperature results in an increased emission of total VOC and formaldehyde. Product emissions show a decay with time for all products at each temperature, which is typical for product emissions (Table 3). Formaldehyde and total VOC emissions are generally comparable for each chamber type at similar conditions; however, some differences in formaldehyde emissions were observed from the wallcovering, and the VOC levels measured from flooring differed between the chamber types at the same temperature.

Composition of Chemical Emissions at Different Temperatures

The historical basis for increasing the product temperature above ambient for sensory irritation testing included increasing the total exposure to product emissions of the animals. As shown in Fig. 3 and Table 3, the total chemical emissions from each product were increased when the product was heated. However, a realistic consideration of increased exposure to product emissions would also require that individual chemical concentrations in the test atmospheres increase in relative proportion to the total increase in dose, that is, that the relative amounts of each individual chemical be identical. Table 4 compares the top five identified VOCs from the first exposure period (1 h of product dynamic ventilation), ranked by concentration, for each product at each temperature. Although the chemical composition of the exposure atmosphere is often similar at 23°C and 70°C, there are significant differences in individual compounds. These differences are both in the absolute and relative amounts of

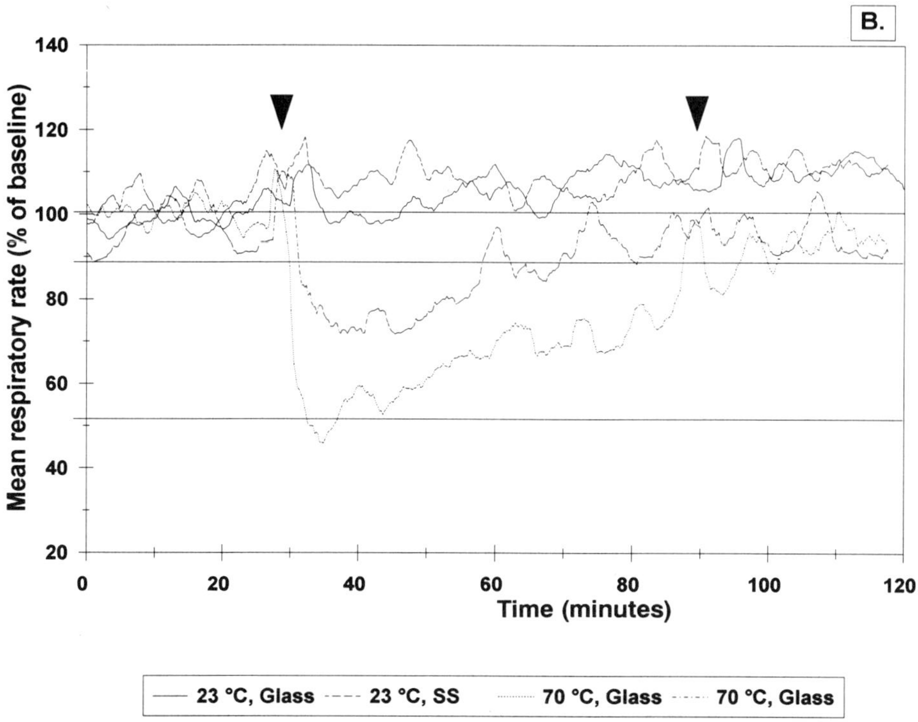

FIG. 2—*Mean respiratory rate of four mice (as a percent of baseline respiratory rate) during the initial control period, first exposure, and 30 min of the first recovery:* (B) *ceiling tile.*

many chemicals. Identical results are found during the subsequent exposure periods (data not shown).

The chemicals identified from each product during each exposure period may be grouped to allow an easier comparison of the emission characteristics using the different types of chambers. The chemical groups chosen were intended to reflect those groups associated with different mechanisms of activation of the sensory irritant receptor (Nielsen 1991), as well as other major chemical groups observed in the chemical atmospheres. These groups included alkanes, benzene-related compounds, other aromatics, alcohols, ketones and ethers, allyl compounds, aldehydes (including formaldehyde), isocyanates, amines, carboxylic acids and esters, and halogenated aliphatics.

Figure 4 compares the measured chemical concentrations from each group for product testing using the stainless steel product chamber with the glass product chamber at each exposure period. The least-squares regression line for these data is also shown and is comparable with the identity line. The slope of the regression line (0.84) is also comparable.

Discussion

Sensory irritation from product emissions can be induced by heating of the products. However, as shown in Table 4, the change in the exposure atmosphere caused by heating changes the characteristics of the exposure. Therefore, a meaningful comparison with the

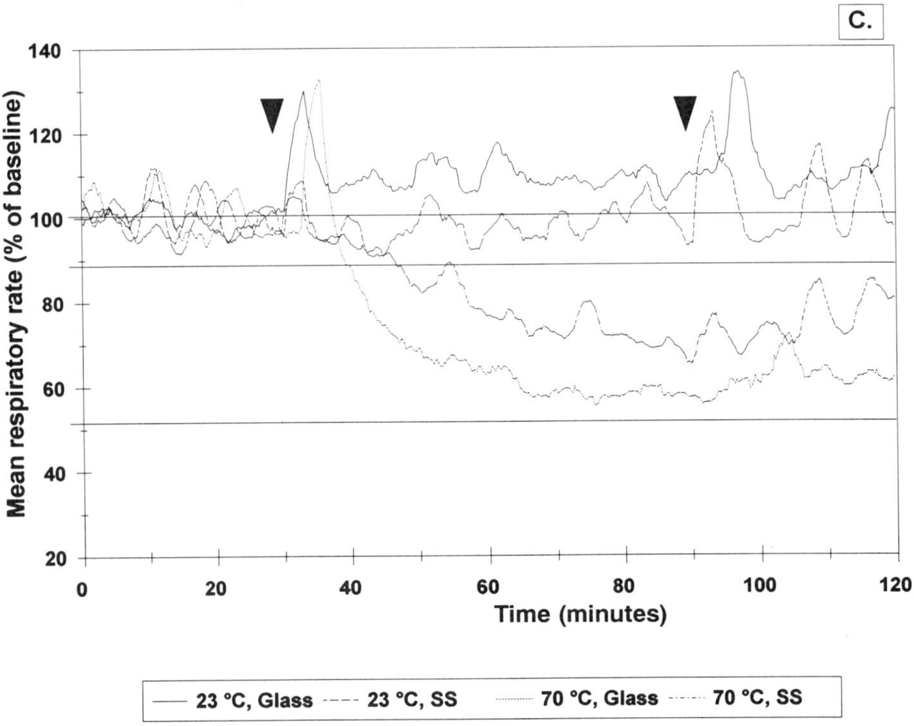

FIG. 2—*Mean respiratory rate of four mice (as a percent of baseline respiratory rate) during the initial control period, first exposure, and 30 min of the first recovery:* (C) *resilient flooring.*

exposure at lower temperatures is impossible to make and an exposure-response curve for sensory irritation to product emissions cannot be determined by manipulating exposure levels through temperature.

The fact that little evidence of sensory irritation was observed in the current tests at 23°C does not mean that these chemical atmospheres are incapable of producing irritation in humans. For comparison of animal responses with human responses, an RD_{50} value (the exposure which results in a 50% decrease from baseline in the mean respiratory rate of four animals) for product exposures would need to be generated by exposing animals to test atmospheres of identical relative chemical compositions but varied concentrations. A more useful strategy for assessing the sensory irritation caused by exposure to product emissions would utilize changes in product loading and airflow through the product chamber to vary the exposure atmospheres. The effect of increased loading and decreased airflow on the chemical emissions from products, including characterization of the exposure atmosphere, is necessary before a meaningful dose-response study may be undertaken. Animal exposures may then be done on a range of product doses to properly interpret the respiratory response and relate to anticipated human irritation as done previously for individual chemicals (Kane et al 1979). The exposure conditions can also be related to those for "typical" product use.

The use of multiple exposures was originated in other product tests (Anderson 1993, Costa 1993, Kennedy 1993), presumably to investigate any sensitization effects analogous to the development of a sick building syndrome-related irritant response in humans over time. However, the changes in respiratory response as the emission characteristics of the products

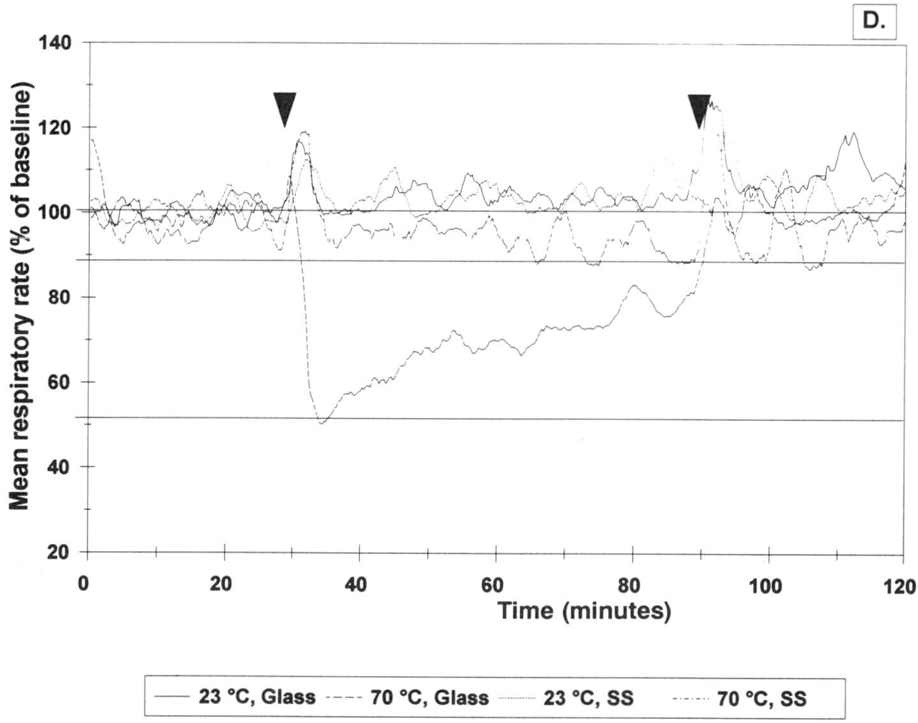

FIG. 2—*Mean respiratory rate of four mice (as a percent of baseline respiratory rate) during the initial control period, first exposure, and 30 min of the first recovery:* (D) *wallcovering.*

change over time are also observed. Also, the differences in respiratory depression observed for different products (Fig. 2) may indicate that different mechanisms are responsible for at least part of the response observed. Longer-term (sub-chronic or chronic or both) exposure protocols, including whole body exposures, would be a recommended way for investigation of mechanisms of toxicity if irritant effects are observed in "dose"-response studies discussed above.

Comparison of the respiratory responses observed using glass and stainless steel chambers for the same products at identical conditions showed no differences in response at ambient conditions (Figs. 1 and 2). At the elevated temperature, there was a trend for an increased response when the glass chamber was used, which was statistically significant for the resilient flooring. The experimental significance of this trend, however, may reflect biological variability and sample-to-sample variability in product emissions rather than any intrinsic differences (including adsorption and surface reactions) caused by product chamber composition.

The possibility that the differences in respiratory response are within biological variability is supported by comparing respiratory response to exposures to individual chemicals in published data. For example, concentration-response relationships for formaldehyde and acrolein were shown by Kane and Alarie (1977) to have a 10% or greater variability in response to virtually identical chemical concentrations. Since the minimum respiratory depression for sensory irritation to be considered significant is 12% (ASTM E 981-84), one would expect biological variability during exposure to be at least ±12%.

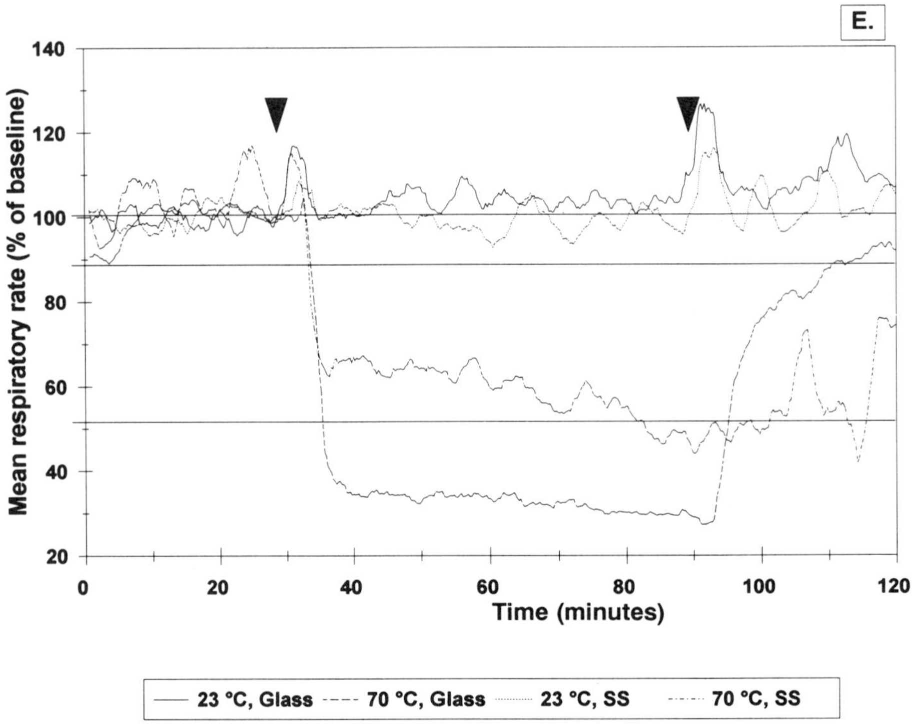

FIG. 2—*Mean respiratory rate of four mice (as a percent of baseline respiratory rate) during the initial control period, first exposure, and 30 min of the first recovery:* (E) *veneer.*

Comparison of the chemical emissions (TVOC and formaldehyde) using the two types of chambers (Figs. 3 and 4) shows good agreement, particularly considering the variability which may be exhibited by samples from individual products. These data would also tend to support the hypothesis that sensory irritation testing of products is limited as much by biological variability as by any differences in test chamber composition.

Differences in emissions from products in tests at identical temperatures using the different chambers were principally evident in the formaldehyde measurement from wallcovering. It is not clear why there is a dramatically higher increase in formaldehyde measurement in the glass system at elevated temperature for the wallcovering product, though this is likely to cause the differences in response observed in the two tests.

Still unresolved from these experiments is the possibility that chemicals may be emitted by products at low levels (within the variability of product emissions) which adsorb to or react on stainless surfaces but not on glass. On the other hand, this may also be true of glass surfaces, relative to stainless. Ability to determine these types of effects would require currently unavailable improvements in the sensitivity of the analytical and sensory irritation test methodology used in the current study. Additionally, isolation and study of the sensory irritation receptor or receptors would be needed to determine any chemicals or combinations of chemicals emitted by products which could stimulate such a response at low levels. Finally, the relevance of any differences determined between stainless steel and all-glass systems would need to be interpreted in terms of "real-life" exposure situations.

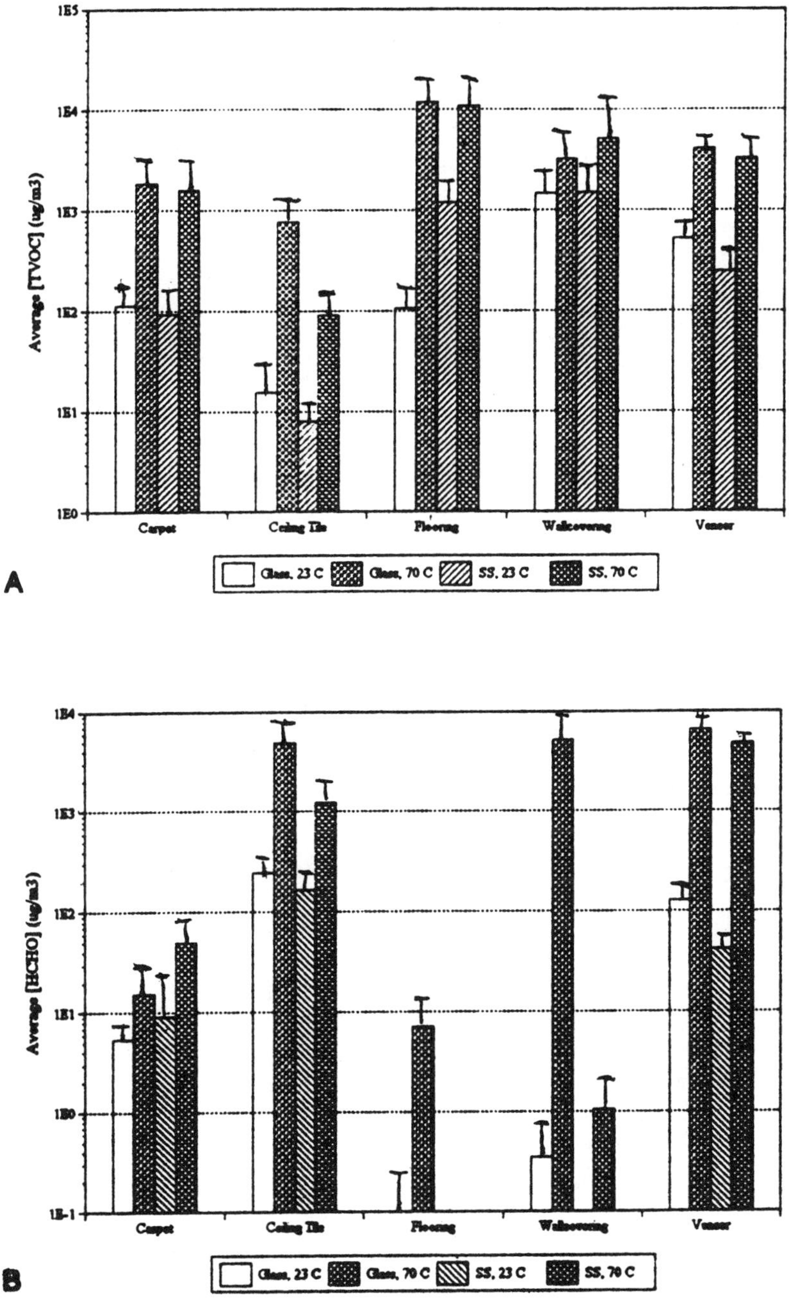

FIG. 3—*Average (and standard deviation) of the total concentration* ($\mu g/m^3$) *of VOC* (A) *and the concentration of formaldehyde* (B) *emitted from each product during the four exposure periods. For all products, increasing the temperature results in an increased emission of total VOC and formaldehyde. The emissions measured using either chamber type are generally comparable for each chamber type at similar conditions; however, some differences in formaldehyde emissions were observed from the wall-covering, and the VOC levels measured from flooring differed between the chamber types at the same temperature.*

TABLE 3—*Concentrations of total VOCs (excluding formaldehyde) and formaldehyde in the animal exposure chamber due to emissions from each product, at each exposure point, at two temperatures, for each product chamber type.*

Product Type	Exposure Period (Hours of Product Ventilation)	[TVOC], μg/m³				[HCHO], μg/m³			
		23°C		70°C		23°C		70°C	
		Glass	SS	Glass	SS	Glass	SS	Glass	SS
Carpet	E1 (1 h)	162	94	3120	4885	7	37	35	21
	E2 (4 h)	138	197	3580	738	6	*	11	86
	E3 (24 h)	43	18	100	154	6	*	10	76
	E4 (27 h)	113	60	598	557	*	*	6	15
Ceiling tile	E1 (1 h)	46	8	1072	174	333	167	9929	2045
	E2 (4 h)	12	7	1306	90	324	199	6466	887
	E3 (24 h)	*	14	373	11	166	141	1298	1207
	E4 (27 h)	4	3	357	88	148	160	1496	750
Resilient flooring	E1 (1 h)	163	1363	21623	18527	*	*	23	*
	E2 (4 h)	131	1839	13037	17692	*	*	6	*
	E3 (24 h)	30	967	9520	2858	*	*	*	*
	E4 (27 h)	106	706	3050	4723	*	*	*	*
Wallcovering	E1 (1 h)	3342	3506	2799	14854	*	*	11769	*
	E2 (4 h)	1675	1533	8519	4908	*	*	4943	*
	E3 (24 h)	484	558	677	530	*	*	1742	*
	E4 (27 h)	414	488	1064	237	*	*	1830	*
Veneer	E1 (1 h)	864	409	4653	5711	180	37	6076	2621
	E2 (4 h)	507	280	4691	2407	106	57	9529	5264
	E3 (24 h)	361	150	4131	2209	110	34	6342	5731
	E4 (27 h)	385	158	2850	2827	119	40	3891	5278

* An asterisk refers to levels below detectable limit.

TABLE 4—*Top five identified VOCs during first exposure period (one hour of product dynamic ventilation), ranked by concentration, for each product at each temperature, using glass product chamber.*

Product	23°C	Concentration, μg/m³	70°C	Concentration, μg/m³
Carpet	1. ethoxyethoxyethanol	75	1. dipropylene glycol	352
	2. acetic acid	28	2. silane, methoxytrimethyl-*	277
	3. 2,4-pentanediol, 2-methyl	20	3, tridecane	193
	4. silane, methoxytrimethyl-*	11	4. methane, diethoxy*	171
	5. formaldehyde	7	5. ethoxyethoxyethanol	124
Ceiling tile	1. formaldehyde	333	1. formaldehyde	9929
	2. 2-n-butylacrolein*	26	2. 2-n-butylacrolein*	470
	3. unidentified	9	3. methane, diethoxy*	135
	4. 1-pentanol	4	4. hexanal	97
	5. benzaldehyde	3	5. acetone	50
Resilient flooring	1. tridecane	48	1. dodecane	3447
	2. dodecane	41	2. tridecane	2820
	3. phenol	13	3. ethanol	2588
	4. benzenemethanol*	11	4. benzenemethanol*	1963
	5. cyclohexanone	10	5. 2-pentanone, 4-methyl	1558
Wallcovering	1. toluene	1589	1. formaldehyde	11769
	2. 2-pentanone, 4-methyl	701	2. 2-propanol	2534
	3. 2-butanone	514	3. tetrahydrofuran	622
	4. 1-decanol	100	4. methanol	208
	5. phenol	70	5. acetone	102
Veneer	1. hexanal	216	1. formaldehyde	6076
	2. formaldehyde	180	2. acetone	2591
	3. acetone	82	3. hexanal	977
	4. octanal*	69	4. valeraldehyde	342
	5. 1-pentanol	60	5. butanal	79

* An asterisk refers to identification based on NBS library. Other identifications made based on AQS library.

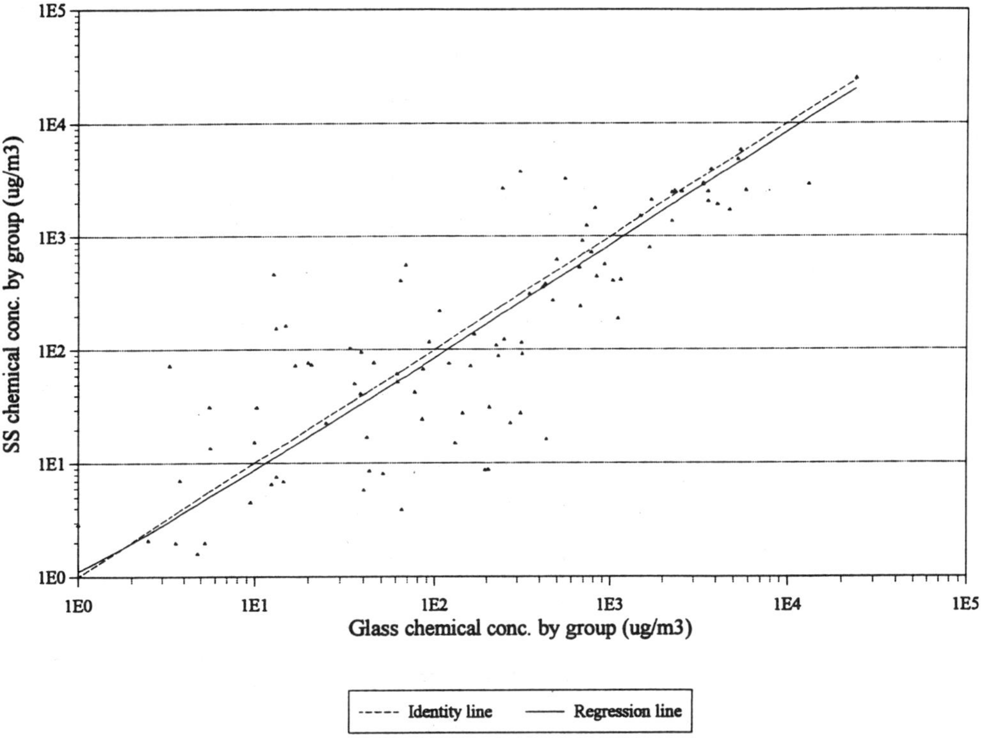

FIG. 4—*Comparison of the measured chemical concentrations from each chemical group for product testing using the stainless steel product chamber with the glass product chamber, at each exposure period. The correlation is good* (r^2 = *0.82), and the regression of the data (solid line, slope = 0.84) is very similar to the identity line (dashed line).*

Conclusions

Significant sensory irritation does not generally occur when mice are exposed to the volatile chemicals in air passed over products when products are kept at normal ambient conditions. However, sensory irritation can occur when animals are exposed to air passed over products which are heated to 70°C. This is partly attributable to the increased chemical exposure at elevated product temperature; however, increasing the temperature also changes the relative concentrations of emitted VOCs. As a result, heating of the product does not result in a test of irritation which is easily comparable to that at lower (ambient) temperature. The use of a stainless steel product chamber for sensory irritation testing of products results in respiratory responses and chemical concentrations similar to those of a glass product chamber at ambient temperature. At elevated temperature, a trend towards increased respiratory response is observed when a glass product chamber is used. However, this trend may reflect biological variability, particularly since the chemical emissions determined using either chamber type are within the limits of product variability.

References

Alarie Y. C., 1973, "Sensory Irritation by Airborne Chemicals," *CRC Critical Reviews in Toxicology,* Vol. 2, pp. 299–363.

Anderson, R. C., 1993, "Toxic Emissions from Carpets," *Proceedings,* Indoor Air '93, Helsinki, Finland, Vol. 1, pp. 651–656.

Anderson, R., 1993, "Testimony Covering Carpet Studies," paper presented at Indoor Environment '93, Baltimore, MD, 21–23 April.

Bos, P. M. J., Zwart, A., Reuzel, P. G. J., and Bragt, P. C., 1992, "Evaluation of the Sensory Irritation Test for the Assessment of Occupational Health Risk," *CRC Critical Reviews in Toxicology,* Vol. 21, No. 6, pp. 423–450.

Costa, D. L., Tepper, J. S., Moser, V. C., Mason, M. A., Roache, N., Guo, Z., and Dyer, R. S., 1993, "EPA Studies to Determine Mechanism of Complaint Carpet Effects in Mice," *Proceedings,* The Toxicology Forum 1993 Annual Summer Meeting, Aspen, CO, pp. 216–223.

Hansen, L. F., Nielsen, G. D., Tøttrup, J., Abildgaard, A., Dahl Jensen, O. F., Hansen, M. K., and Nielsen, O., 1991, "Biological Determination of Emission of Irritants from Paint and Lacquer," *Indoor Air,* Vol. 2, pp. 95–110.

Kane, L. E. and Alarie, Y., 1977, "Sensory Irritation to Formaldehyde and Acrolein During Single and Repeated Exposures in Mice," *American Industrial Hygiene Association Journal,* Vol. 38, pp. 509–522.

Kane, L. E., Barrow, C. S., and Alarie, Y., 1979, "A Short-Term Test to Predict Acceptable Levels of Exposure to Airborne Sensory Irritants," *American Industrial Hygiene Association Journal,* Vol. 40, pp. 207–229.

Kennedy, G., 1993, "Industry Studies of Complaint Carpets," paper presented at The Toxicology Forum 1993 Annual Summer Meeting, Aspen, CO, pp. 225–235.

Muller, W. J. and Black, M. S., 1995, "Sensory Irritation in Mice Exposed to Emissions From Indoor Products," *American Industrial Hygiene Association Journal,* Vol. 56, pp. 794–803.

National Institute for Occupational Safety and Health, 1990, *Manual of Analytical Methods (3rd ed.,) Method 3500,* R. Kendall, Ed. (NIOSH Pub. 84-100), Government Printing Office, Washington, DC.

Nielsen, G. D., 1991, "Mechanisms of Activation of the Sensory Irritant Receptor by Airborne Chemicals," *CRC Critical Reviews in Toxicology,* Vol. 21, No. 3, pp. 183–208.

Schaper, M., 1993, "Development of a Database for Sensory Irritants and Its Use in Establishing Occupational Exposure Limits," *American Industrial Hygiene Association Journal,* Vol. 54, No. 9, pp. 488–544.

Stott, W., 1993, "4-PCH, Odoriferant or Toxicant?" paper presented at Indoor Environment '93, Baltimore, MD, 21–23 April.

Environmental Fate

Mark E. Grismer,[1] E. Labolle,[1] T. Raihala,[2] and J. Eweis[3]

A Modified Gravimetric Method for Measuring Rates of Vapor Adsorption and Desorption on Soils. Kinetics of Toluene Adsorption/Desorption on Bentonite

REFERENCE: Grismer, M. E., Labolle, E., Raihala, T., and Eweis, J., **"A Modified Gravimetric Method for Measuring Rates of Vapor Adsorption and Desorption on Soils. Kinetics of Toluene Adsorption/Desorption on Bentonite,"** *Volatile Organic Compounds in the Environment, ASTM STP 1261,* W. Wang, J. Schnoor, and J. Doi, Eds., American Society for Testing and Materials, 1996, pp. 95–104.

ABSTRACT: In arid and semi-arid regions where net infiltration and soil moisture are relatively low, movement of volatile organic compounds (VOC) in the vadose zone may be dominated by vapor transport and adsorption mechanisms. The objective of this study is to develop the experimental methods necessary to determine the rate of isothermal adsorption and desorption, and the adsorption capacity of toluene (and other VOCs) onto different soils and minerals subject to a range of initial water contents and temperatures. The proposed method is a modification of the gravimetric method described by both Brooks and Grismer. A stainless steel chamber houses a load cell connected through a serial interface to a computer. After sealing the dry adsorbent (soil) in the chamber, the chamber air is purged using dewatered nitrogen gas before introduction of the adsorbate (VOC). The computer, through the interface, directs the load cell and records sample mass at preselected time intervals without additional supervision. The method is applied to measure rates of adsorption and desorption of toluene on bentonite. Toluene sorption rates are modeled using a second-order rate equation described by Grismer that yielded an initial rate parameter equivalent to that for water vapor adsorption despite the toluene having a much smaller adsorption capacity. The results illustrate the simplicity of the proposed method.

KEYWORDS: gravimetric method, volatile organic compounds, adsorption isotherm, desorption isotherm, bentonite, toluene, water vapor, montmorillonite, soil

Many soils have become contaminated with nonaqueous phase liquids composed of volatile organic compounds (VOCs). These compounds may enter the soil vadose zone through direct application, chemical spills, or through seepage from buried waste or storage containers. During migration of the VOCs through the soil, the compounds are subject to degradation, liquid and vapor transport, and adsorption. However, the effects of various soil properties and ambient conditions on the adsorption process, particularly vapor adsorption, have only begun to be quantified (Brookman et al. 1985). Several recent studies have considered

[1] Professor and graduate research assistant, respectively, hydrologic science, LAWR, University of California–Davis, Davis, CA 95616.
[2] Graduate research assistant, chemical engineering, University of California-Davis, Davis, CA 95616.
[3] Graduate research assistant, civil engineering, University of California–Davis, Davis, CA 95616.

equilibrium adsorption of different VOCs on soils, though little work related to adsorption kinetics can be found.

Several investigators have found that it is the high specific surface area of the mineral content of the soil that adsorbs substantial quantities of VOCs when soil moisture is absent or very low (Owens et al. 1992, Rao 1990, Smith et al. 1990). That is, the extent and rate of vapor adsorption is limited by diffusion rates and the degree to which adsorption sites are already hydrated, as water is preferentially adsorbed. It appears that for initially water-wet soils, at water contents near the vapor adsorption capacity of the soil, little, if any, additional VOC vapor adsorption occurs (Grismer 1987a, Chiou and Shoup 1985, Rhue et al. 1989, Ong and Lion 1991) and the adsorbed water may not be displaced by the liquid or gas phases of the nonaqueous-phase compound. In addition, polar molecules are more readily adsorbed as compared to nonpolar molecules. Transport of VOCs ahead of the liquid "wetting front" may also depend on the rate at which the vapor is adsorbed, as well as the wetting front speed (Grismer 1988). In this case, the kinetics of the vapor sorption-desorption reaction must be known to predict the transport of VOCs.

Clearly there is a need for experimental methods to further investigate the kinetics of sorption-desorption of VOCs at different water contents and water-vapor densities. Rao et al. (1989) states that "Additional data on equilibrium and nonequilibrium sorption of . . . contaminants need to be collected. . . . a paucity of data has limited our ability to develop conceptual models for the sorption of . . . compounds." In consideration of the need for additional data and reliable experimental methods, we developed and tested a method for measuring the rates at which VOCs are adsorbed and desorbed from soils, and the maximum adsorption capacity of the VOC onto soil under different temperature and soil-moisture conditions. The experimental apparatus is similar in concept to the gravimetric method described by Brooks (1965) and Grismer (1987a). The results presented here are preliminary as we are continuing to apply the experimental procedures using improved flow control valves and different soils and VOCs. At this point we have tested the method using toluene and bentonite.

Experimental Methods and Procedures

Experimental Apparatus

A schematic of the experimental apparatus is shown in Fig. 1. The sealable vapor-adsorption chamber is housed in a constant temperature room to maintain isothermal conditions. Industrial nitrogen with a purity of 99.99% is supplied from a cylinder in line with two drierite tubes to remove residual water vapor. A three-way valve is used to direct the dry nitrogen gas to the chamber, or to a flow stream which is subsequently split into three metered flow streams. The first and second streams are saturated with VOC and water, respectively, by sparging through gas washing bottles, and the third stream remains as pure dry nitrogen. Flowmeters enable mixing of the three streams so as to control VOC concentration and water-vapor density. The combined flow streams then enter the adsorption chamber and an equal amount exhausts through the opposite side of the adsorption chamber. Gas flow through the adsorption chamber causes a very small positive pressure to develop inside the adsorption chamber which provides additional assurance that no atmospheric gases enter the chamber.

The adsorbent is placed in containers made from Tedlar, a material that has been shown not to adsorb most VOCs. Originally, adsorbent masses were measured using a balance (Mettler PE1600) with an accuracy of 10 mg; however, due to problems with accuracy and drift, we have replaced the original balance with an isolated load cell (custom Mettler P160)

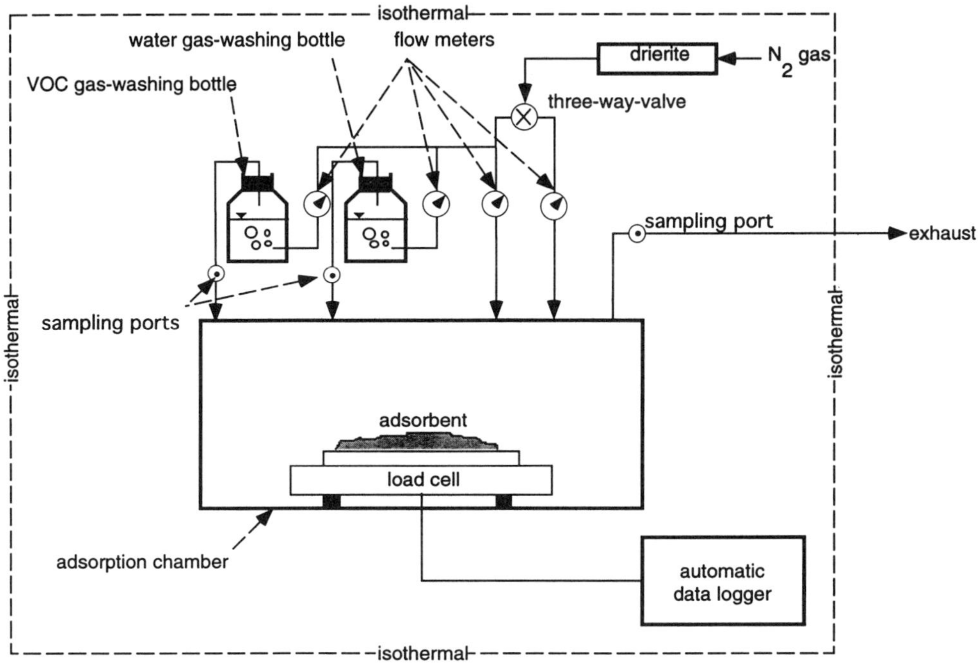

FIG. 1—*Experimental apparatus.*

having an accuracy of 1 mg. Separating the load cell from the electronic circuitry included in the original balance also precluded degradation of the internal circuit boards from the solvent vapors during measurements. Mass changes of the adsorbent are collected at preselected time intervals by an automatic data logger (personal computer) located outside of the adsorption chamber.

Procedures

Initially, the adsorbent is spread in a thin layer in a Tedlar sample container. A thin layer limits the effects of diffusion through interstitial pores on rates of adsorption. The air-dry adsorbent in its Tedlar sample container is weighed. The adsorbent is then degassed by placing it in an oven at 105°C for a minimum of 24 h. A portion of the sample is then removed, weighed, and placed back in the oven. This split sample is weighed again after several days to determine if additional water is desorbed. From this information, the oven-dried water content of the soil is determined. When beginning a sorption experiment, the dried soil is quickly transferred and sealed in the adsorption chamber and the automatic data logger started.

Approximately five litres per minute (L/min) of pure dry nitrogen is immediately passed through the adsorption chamber through the nitrogen line to purge the atmospheric water and oxygen initially in the chamber. The adsorbent typically adsorbs a small amount of atmospheric water vapor during transfer from the oven to the chamber that is removed as the water desorbs into the dry nitrogen gas stream. Nitrogen gas purging continues until the mass of the adsorbent returns to its initial value, or no change in sample mass is recorded in two hours. This procedure also allows the adsorbent adequate time to come into temperature equilibrium with the gas phase and the constant temperature of the room.

Once the adsorbent returns to its initial mass, the flow rate of nitrogen is reduced to approximately 30 mL/min and the three-way valve is changed to direct flow of VOC, or VOC-water vapor mixture through the gas-sparging bottles. This marks the start of an adsorption experiment. Flowmeters are adjusted to provide the desired mix of VOC saturated nitrogen, water-vapor saturated nitrogen, and pure nitrogen.

The adsorption experiment is allowed to run undisturbed until no additional mass gain is recorded in an eight-hour period. The desorption experiment begins when the three-way valve is switched such that the adsorbate is replaced by approximately 5 L/min of pure nitrogen that passes through the adsorption chamber. This process allows the adsorbent to desorb the adsorbate into the pure nitrogen. The desorption experiment is allowed to run undisturbed until either the adsorbent returns to its original mass or the adsorbent mass remains constant for four hours. If the adsorbent does not return to its original mass, the amount of adsorbate that did not desorb is considered the irreversible amount of adsorption.

Results

Four experiments were performed to demonstrate the reliability of the experimental methods discussed: (1) montmorillonite gassed with pure nitrogen, (2) bentonite gassed with pure nitrogen, and (3) and (4), bentonite gassed with toluene-saturated nitrogen. The first two experiments were performed to determine the efficiency of the adsorption chamber in limiting diffusion of outside air (water vapor) into the chamber. The latter two experiments were performed to demonstrate application and reproducibility of the proposed experimental technique towards measuring rates of vapor adsorption and desorption of toluene on bentonite.

The adsorbate used in this study was American Chemical Society certified toluene with a water content of 0.02%. This relatively low water content combined with the small gas volumes used in the experiments led us to believe that adsorption of this water had little effect on the experimental results. In addition, preliminary experiments indicated that the load cell adsorbs negligible amounts of toluene. The adsorbent used in this study was Whittig's bentonite that had previously been determined (Grismer 1987b) to have a specific surface area of 667 m^2/g.

Initial Desorption of Adsorbed Water

The first two experiments yielded information concerning the ability of the adsorption chamber to limit diffusion of outside air into the chamber. Approximately 0.050 kg of montmorillonite (experiment 1) and 0.050 kg of bentonite (experiment 2) were taken from the oven and placed in the adsorption chamber which was subsequently purged with dry nitrogen gas as described in the experimental procedures. Figures 2 and 3 show the loss of water, adsorbed during transfer from the oven, with time, for the montmorillonite and bentonite, respectively. Masses of adsorbed water are shown relative to oven dried mass as determined from the split samples. As shown, the montmorillonite and bentonite samples required approximately 120 and 40 min, respectively, to desorb the water adsorbed during transfer. The adsorption chamber appears to adequately limit the diffusion of outside air into the chamber. In addition, the process of purging with dry nitrogen gas removed water adsorbed during sample transfer from the oven to the chamber.

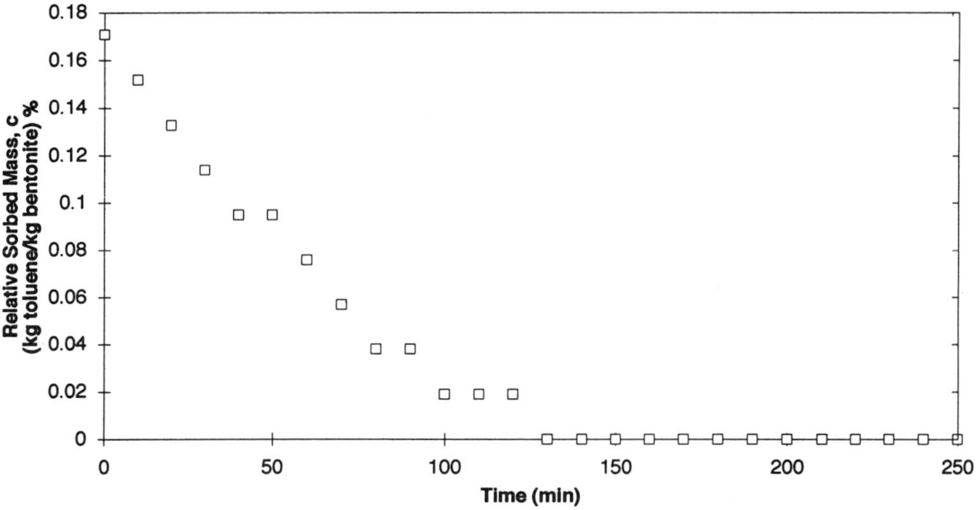

FIG. 2—*Verification of the experimental method with montmorillonite and dry nitrogen.*

Sorption and Desorption Kinetics of Toluene on Bentonite

The third and fourth experiments were performed to investigate the kinetic adsorption and desorption of toluene on bentonite. Approximately 0.050 kg of bentonite were used in each experiment. Both experiments were performed using water dry nitrogen saturated with toluene.

Results from both adsorption-desorption experiments are shown in Figs. 4 and 5. Initial water vapor adsorbed during transfer from the oven to the chamber is quickly desorbed.

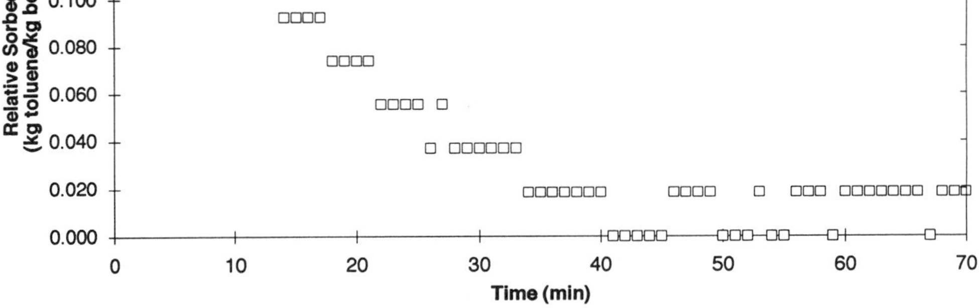

FIG. 3—*Verification of the experimental method with bentonite and dry nitrogen.*

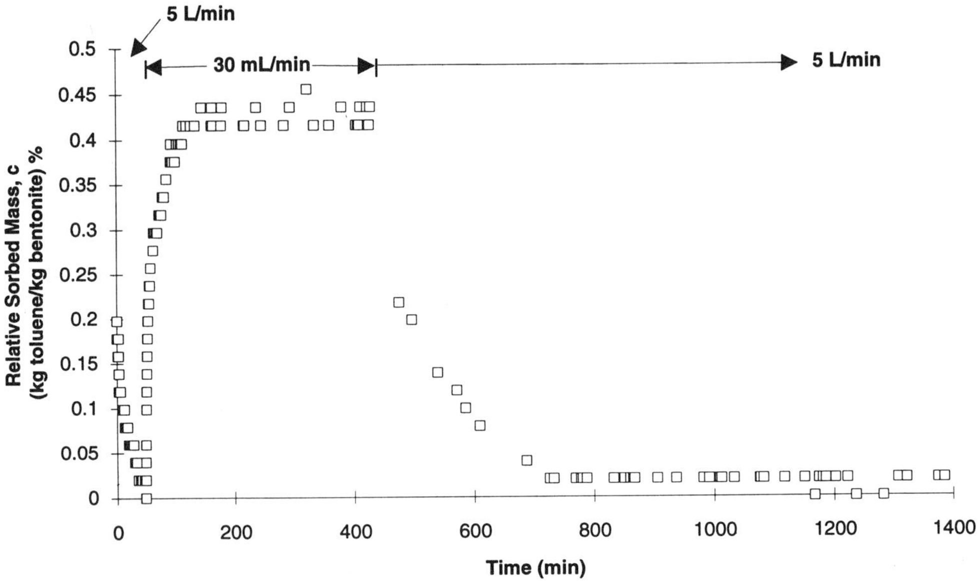

FIG. 4—*First experiment for adsorption/desorption of toluene at saturation in pure nitrogen on bentonite (25°C).*

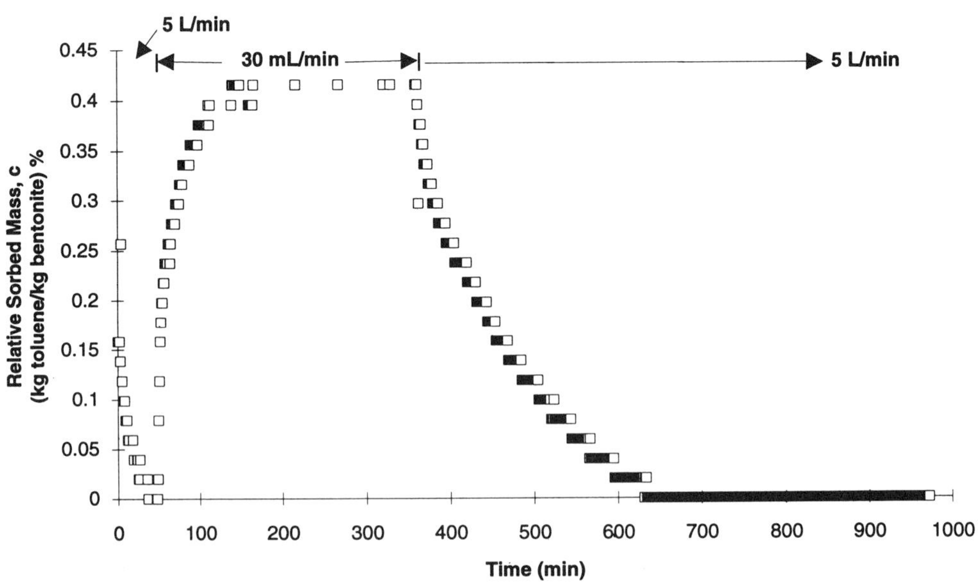

FIG. 5—*Second experiment for adsorption/desorption of toluene at saturation in pure nitrogen on bentonite (25°C).*

Adsorption experiments are marked by the start of a rapid rise in relative sorbed mass. An equilibrium adsorption concentration (capacity) is reached in approximately 70 min. Desorption experiments are marked by a rapid decrease in sorbed mass until, eventually, samples return to their approximate oven-dried mass, at which point the experiments are terminated. The discrete nature of the mass changes is due to the accuracy of the balance. Although significant changes in mass were measurable with the original balance, the more accurate load cell was an improvement. The agreement between the observed adsorption and desorption curves from the two experiments demonstrates the successful reproducibility of the experimental techniques.

Adsorption Kinetics Modeling

Through an analogy with population dynamics, it is possible to derive a quantitative tool for assessment of the observed adsorption-desorption curves. Consider the soil sample to have a finite number of adsorption sites, all sites having the same affinity for VOC vapor (toluene), the maximum number of which corresponds to the observed adsorption capacity of the VOC (toluene), c_{max} (kg sorbed/kg adsorbent). Furthermore, assume that the adsorption rate, k (min^{-1}), decreases linearly with the adsorbed mass, c (kg sorbed/kg adsorbent), as

$$k = k_{max} \left(\frac{c_{max} - c}{c_{max}} \right) \tag{1}$$

where k_{max} (min^{-1}) is the kinetic rate constant when c is zero. A differential equation for the rate of adsorption can then be written as

$$\frac{dc}{dt} = k_{max} \left(\frac{c_{max} - c}{c_{max}} \right) c \tag{2}$$

Grismer (1987a) found that for a range of clays and soils k_{max} was effectively equal to 1 min^{-1}, and that the solution to Eq 2 for $c = c_0$ at time zero using a transposed time scale (arithmetic to log) is

$$c = \frac{c_{max}}{1 + \left(\frac{c_{max} - c_0}{c_0} \right) \frac{1}{t}} \tag{3}$$

where c_0 is now a fitting parameter related to both the initial water content and the initial, or maximum adsorption rate. Dimensional consistency in Eq 3 is preserved through the factor of 1 in the expression $1/t$, where 1 has units of time.

We fit Eq 3 to the adsorption and desorption data shown in Figs. 3 and 4 using the measured values for c_{max} and computing values for c_0 based on the best nonlinear least-squares fit to the data. Figure 6 illustrates the agreement between the bentonite adsorption data and the curve given by Eq 3 for $c_0 = 4.7 \times 10^{-4}$ kg toluene/kg bentonite. The model compares well with the adsorption data. Relaxing the restriction of $k_{max} = 1$ min^{-1} would improve the overall data fit using Eq 3 of Grismer (1987a). It is interesting to note that the value of c_0 obtained for toluene adsorption on the clay is identical to that obtained by Grismer (1987a) for water-vapor adsorption. This identity suggests that the parameter c_0 in Eq 3 is independent of the adsorbate and may be taken as a constant. Assuming a constant value

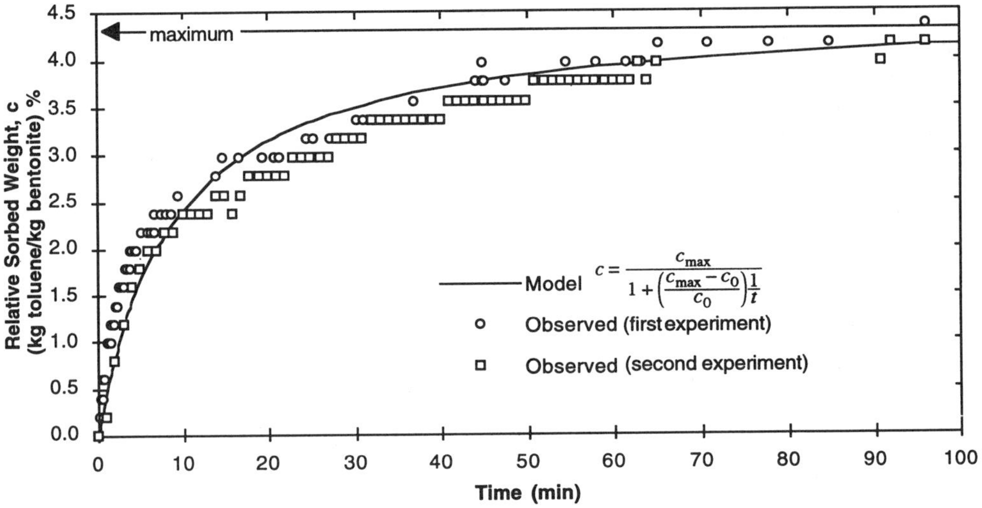

FIG. 6—*Comparison of the second-order rate model with data from the two adsorption experiments for toluene on bentonite at 30 mL/min flow rate.*

for c_0 indicates that the adsorption kinetics can be described by Eq 3 using a single equilibrium parameter, the adsorption capacity, c_{max}. Additional studies of the factors affecting c_0 are necessary to support this possibility.

During desorption, the capacity of the dry nitrogen to take on more toluene does not change with time as the nitrogen in the adsorption chamber is constantly being replenished. Therefore, the proposed conceptual model for sorption into the gas phase is inconsistent with our intuitive understanding of the desorption process. Nevertheless, for the purposes of comparing rates of adsorption with rates of desorption, we compare Eq 3 with a best-fit value $c_0 = 8.5 \times 10^{-5}$ kg toluene/kg bentonite with measured values for sorbed mass during the desorption phase of the experiments in Fig. 7. The model fits the data surprisingly well. Measured rates of desorption for toluene on bentonite are substantially slower than measured rates of adsorption.

Discussion

An experimental method, a modification of the gravimetric method described by Brooks (1965) and Grismer (1987a), to determine rates of isothermal adsorption and desorption, and the adsorption capacity of toluene (and other VOCs) onto different soils and minerals, has been developed. The experimental results demonstrate the reliability and simplicity of the proposed method. Automatic monitoring of sample masses decreases time spent supervising experiments. The three metered inlet gas streams controlling VOC concentration and water-vapor density provide a flexible method to determine rates of adsorption and desorption of VOCs at different water vapor densities. Finally, we demonstrate application of a simple second-order kinetic model to the sorption data.

The results presented herein are preliminary as we continue to apply the experimental procedures using improved equipment and different soils and VOCs. Though the significance of nonequilibrium vapor-phase adsorption of VOCs on dry soil (specifically, toluene on bentonite) towards prediction of rates of VOC transport in soils is unclear, the observed rates

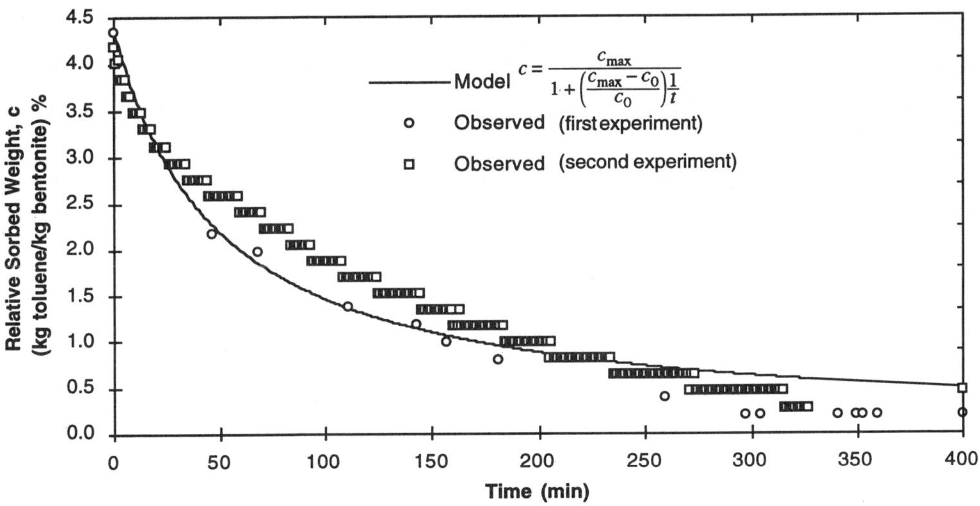

FIG. 7—*Comparison of the second-order rate model with data from the two desorption experiments for toluene from bentonite at 5 L/min flow rate.*

of adsorption and desorption for toluene on bentonite suggest that equilibrium adsorption isotherms alone may be insufficient to characterize vapor phase transport in dry soils. We expect adsorption to be largely affected by ambient soil moisture and relative humidity of the gas phase. Further sorption experiments using a modified version of the gravimetric method described here are underway and may provide data that can be used to determine sorption-desorption kinetic relationships for VOC vapor adsorption on soils at different initial water contents and water-vapor densities. In turn, these relationships should provide insight into the significance of nonequilibrium vapor-phase sorption and desorption to VOC transport and the remediation of VOC contaminated soils.

References

Brookman, G. T., Flanagan, M., and Kebe, J. O., 1985, "Literature Survey; Hydrocarbon Solubilities and Attenuation Mechanisms," *American Petroleum Institute Publication No. 4414,* Washington, DC.

Brooks, C. S., 1965, "Vapor Adsorption on Montmorillonite by Gravimetric and Chromatographic Procedures," *Soil Science,* Vol. 99, pp. 182–192.

Chiou, C. T. and Shoup, T. D., 1985, "Soil Sorption of Organic Vapors and Effects of Humidity on Sorptive Mechanism and Capacity," *Environmental Science & Technology,* Vol. 19, No. 12, pp. 1196–1200.

Grismer, M. E., 1987a, "Water Vapor Adsorption and Specific Surface," *Soil Science,* Vol. 143, No. 5, pp. 367–371.

Grismer, M. E., 1987b, "Water Vapor Adsorption and Specific Surface," *Soil Science,* Vol. 144, pp. 233–236.

Grismer, M. E., 1988, "Water Vapor Adsorption Kinetics and Isothermal Infiltration," *Soil Science,* Vol. 146, pp. 297–302.

Ong, S. K. and Lion, L. W., 1991, "Mechanisms for TCE Vapor Sorption onto Soil Materials," *Journal of Environmental Quality,* Vol. 20, No. 1, pp. 180–188.

Owens, W., Silcox, G., Lighty, J., Deng, X., Pershing, D., Cundy, V., Leger, C., and Jakway, A., 1992, "The Desorption of Toluene from a Montmorillonite Clay Adsorbent in a Rotary Kiln Environment," *Journal of the Air and Waste Management Association,* Vol. 42, pp. 681–690.

Rao, P. S. C., 1990, "Sorption of Organic Contaminants," *Water Science and Technology,* Vol. 22, No. 6, pp. 1–6.

Rao, P. S. C., Ogwada, R. A., and Rhue, R. D., 1989, "Adsorption of VOCs on Anhydrous and Hydrated Sorbents. Equilibrium Adsorption and Energetics," *Chemosphere,* Vol. 18, No. 11/12, pp. 2177–2191.

Rhue, R. D., Pennell, K. D., Rao, P. S., and Reve, W. H., 1989, "Competitive Adsorption of Alkylbenzene and Water Vapors on Predominantly Mineral Surfaces," *Chemosphere,* Vol. 18, pp. 1971–1986.

Rhue, R. D., Rao, P. S., and Smith, R. E., 1988, "Vapor-Phase Adsorption of Alkybenzenes and Water on Soils and Clays," *Chemosphere,* Vol. 17, pp. 727–741.

Smith, J. A., Chion, C. T., Kammer, J. A., and Kile, D. E., 1990, "Effect of Soil Moisture on the Sorption of TCE Vapor to Vadose Zone Soil at Picatinny Arsenal, New Jersey," *Environmental Science & Technology,* Vol. 24, No. 5, pp. 676–683.

Khaled M. Bali,[1] Mark E. Grismer,[2] and Jan W. Hopmans[2]

Outflow Methods for Evaluating the Soil Hydraulic Functional Relationships Between NAPL Pressure and Saturation in Porous Media

REFERENCE: Bali, K. M., Grismer, M. E., and Hopmans, J. W., **"Outflow Methods for Evaluating the Soil Hydraulic Functional Relationships Between NAPL Pressure and Saturation in Porous Media,"** *Volatile Organic Compounds in the Environment, ASTM STP 1261,* W. Wang, J. Schnoor, and J. Doi, Eds., American Society for Testing and Materials, 1996, pp. 105–118.

ABSTRACT: Remediation and cleanup of petroleum product contaminated ground water often require modeling of fluid transport processes when immiscible liquid phases (e.g., water and oil) are present. Modeling of such multiphase transport systems requires knowledge of the functional relationships between fluid pressures, saturations, and permeabilities. We evaluated the applicability of the multistep outflow method used in soil science to determine these functions for two porous media (loam and sand) using Soltrol 130 and water as wetting fluids.

The analytical retention and permeability functions of van Genuchten (1980) and Mualem (1976) were used, with an inverse method that has been shown to be reliable in estimating water retention and unsaturated hydraulic conductivity in soils (Eching and Hopmans 1993), to estimate soil hydraulic function parameters for Soltrol 130 and water. The water and Soltrol 130 cumulative drainage as a function of time and the equilibrium saturations were used as input to a numerical model (MLSTPM) to optimize, through an inverse solution of the Richards equation, the parameters needed for the hydraulic functions.

Optimizations were carried out for saturation paths corresponding to monotonically decreasing wetting phase saturations only. The functional relationships between oil pressures, saturations, and permeabilities in Oso-Flaco fine sand were accurately predicted from the optimized water retention curve parameters based on scaling by the ratio of interfacial tensions. However, this scaling procedure was inadequate to predict oii hydraulic function parameters from those of water in *Yolo loam.*

KEYWORDS: soil hydraulic functions, multistep method, multiphase flow, scaling

Many potentially toxic hydrocarbon compounds may result in water resource contamination when they migrate toward surface or subsurface water. The solubility of hydrocarbon compounds in water is generally very low and their vapor pressure is often relatively large. Consequently, they are often considered to move as a discrete liquid phase, or as a nonaqueous phase liquid (NAPL) and as a volatile gas phase.

The lack of information concerning the functional relationships between fluid pressures, degree of saturation, and permeabilities of coexisting phases has limited the understanding

[1] Associate Cooperative Extension Advisor, University of California Desert Research and Extension Center, Holtville, CA 92250-9615.

[2] Associate Professors, Hydrologic Science Section, Department of Land, Air, and Water Resources, University of California, Davis, CA 95616.

of NAPL movement in soil (Parker et al. 1987). Various methods have been proposed to predict the functional relationships of three-phase flow from the two-phase relationship (Parker et al. 1987; Fayers and Mathews 1984). Several researchers (Abriola et al. 1985a; Abriola et al. 1985b; Kuppusamy et al. 1987; Lenhard et al. 1988) have developed models to study the three-phase flow problems. The accuracy of simultaneous determination of NAPL and water saturations in a multiphase flow system is directly related to the accuracy of the previously evaluated soil hydraulic functions of the two-phase systems (air-water, air-NAPL, NAPL-water). What is lacking is a rapid experimental method by which two- and three-phase soil hydraulic functions can be determined.

The objectives of this work are:

(1) To evaluate the application of multistep outflow method to determine the functional relationships between liquid saturation, capillary pressure, and permeability for NAPL/air and water/air systems.

(2) To determine whether NAPL/air functional relationships can be predicted from soil-water hydraulic functions and interfacial tensions associated with NAPL, water, and air systems.

Hydraulic Properties of Two-Phase Systems

To accurately predict multiphase transport in soil, knowledge of the functional relationships between saturation, fluid pressure, and permeability is required. However, these relationships are tedious and difficult to obtain from direct measurements. Three-phase behavior is commonly estimated from two-phase measurements (Parker et al. 1987; Leverett 1941; Corey et al. 1956; Kaluarachchi and Parker 1989) and based on the capillary tube model of porous media in which interfacial tensions control fluid retention and pressure. In other words, the parameters of the functional relationships for air-water, air-NAPL, and NAPL-water are evaluated first and then scaled by interfacial tension ratios to predict the three-phase function parameters. Assuming that the soil is preferentially water wet, then NAPL followed by air—that is, the wettability follows the order water (w) > NAPL (o) > air (a) (Parker et al. 1987)—the following relationships apply for saturation-capillary pressure head relationships

$$h_a - h_w = h_{aw}(S_w^{aw})$$

$$h_a - h_o = h_{ao}(S_o^{ao}) \tag{1}$$

$$h_o - h_w = h_{ow}(S_w^{ow})$$

where h_{aw}, h_{ao}, h_{ow} refer to the air-water capillary pressure head (positive in unsaturated porous media), air-NAPL capillary pressure head, and NAPL-water capillary pressure head, respectively; the subscripts aw, ao, ow are used to designate the fluid phases present in the system; S_w^{aw} is the saturation of water in air-water system; S_o^{ao} is the saturation of NAPL in air-NAPL system; and S_w^{ow} is the saturation of water in NAPL-water system. Several empirical mathematical functions have been developed to relate capillary pressure head and relative permeability to fluid saturation in a two-phase system (Campbell 1974; van Genuchten 1980; Corey 1985). The empirical function of van Genuchten (1980) is widely used and is completely differentiable, making it useful in modeling. This function describes the saturation of water in air-water system where air is displacing water such that water saturation is gradually decreasing, and can be written

$$S(h_{aw}) = [1 + (\alpha h_{aw})^n]^{-m} \tag{2}$$

$$S = \frac{\theta - \theta_r}{\theta_s - \theta_r} \tag{3}$$

where S is the effective saturation of water in air-water system, α and n are curve shape parameters, $m = 1 - 1/n$, θ is water content at a particular air-water capillary pressure head (h_{aw}), and θ_s ($m^3\ m^{-3}$) and θ_r ($m^3\ m^{-3}$) are saturated and residual water contents, respectively. Similar equations can be written for the gradual displacement of NAPL by air (air-NAPL system) and displacement of water by NAPL (NAPL-water system). It should be noted that parameters α, n, θ_r are not fixed from one fluid pair to the next. The liquid hydraulic conductivity was also assumed to be of the form given by van Genuchten (1980) and written:

$$k_{aw}(\theta) = k_{s_{aw}} S_w^{0.5}[1 - (1 - S_w^{1/m})^m]^2 \tag{4}$$

where k_{saw} is the saturated hydraulic conductivity (cm h^{-1}) of the porous medium.

Evaluation of α, n, and θ_r for the air-water system in the porous medium can be performed using the one-step outflow method (Kool et al. 1985) as well as the multistep outflow method (Eching and Hopmans 1993; Hopmans et al. 1994). Cumulative outflow as a function of time is recorded in the one-step outflow method after applying a constant pressure at the top of a saturated core bounded at the bottom by a saturated ceramic plate. In the multistep method, several steps of instantaneous increase in pressure are used, and outflow as a function of time are obtained for each pressure step. These methods can be used to estimate both water retention and unsaturated hydraulic conductivity and we adopt the multistep outflow method here for nonaqueous systems. The objective of both outflow methods is to estimate α, n, and θ_r of the soil hydraulic functions based on experimental measurement of cumulative outflow as a function of time as the soil drains and incorporation of the experimental data into an inverse numerical solution of the appropriate transport equation. The numerical model used here optimizes for the values of α, n, and θ_r based on an inverse solution of the one-dimensional form of Richards' equation

$$c(h)\frac{\partial h}{\partial t} = \frac{\partial}{\partial z}\left[k(h)\left(\frac{\partial h}{\partial z} - 1\right)\right] \tag{5}$$

where $c = d\theta/dh$ is the water capacity, z is the vertical distance from the top of the soil core, and t is time. The numerical simulation of soil core desaturation is carried out based on the initial and boundary conditions appropriate to the system and the initial values of α, n, θ_r.

A one-dimensional version of Richards' equation is used to describe the displacement process for the time duration of the experiment. An optimum solution to the nonlinear objective function is then found which produces a set of parameters α, n, and θ_r that minimizes the difference between the observed and the simulated outflow values. Finding the optimum combination of α, n, θ_r which minimizes the objective function containing the observed and optimized outflow values is based on Marquardt's maximum neighborhood method (Marquardt 1963) commonly used in nonlinear least square optimization techniques. Detailed experimental procedures are given by Eching and Hopmans (1993). Independent measurements of the maximum moisture content, saturated hydraulic conductivity, and the hydraulic conductivity of the porous plate are required. The relative saturation and permeability functions can be determined from the optimal set of α, n, and θ_r using Eq 2–4.

For air-water systems, α generally ranges from 0.005 to 0.05 cm^{-1} and n from 1.1 to 2.8 (van Genuchten 1980). High values of α and n generally correspond to coarse-textured soils while low values correspond to fine-textured soils (Kool et al. 1985). Evaluation of α, n and θ_r for the air-NAPL system can be done similarly. However, evaluation of α, n, and θ_r for the NAPL-water system requires modification of Kool's method (Kool et al. 1985). The method involves the use of air-dry soils that have been sieved through a 2-mm screen, and packed to a uniform bulk density in brass cylinders that are 8.2 cm in diameter and 6 cm in length. The modification required involves increasing the length of the cell to leave enough room for a layer of oil above the soil sample as a nonwetting phase. Suitable ceramic plates, corresponding to the bubbling pressure associated with the oil, should be chosen. For the NAPL-water system, ceramic plates can be treated to make them hydrophobic as described by Lenhard et al. (1988). This can be achieved with the use of chlorotrimethylsilane; the ceramic plates were immersed in the chlorotrimethylsilane for two hours then rinsed with toluene followed by methanol (Lenhard et al. 1988). The chlorotrimethylsilane forms a durable Si-O-Si bond with the ceramic surface, at the same time trimethyl groups repel water molecules in preference to air. Chlorotrimethylsilane reacts violently with water (extreme caution should be exercised when dealing with this chemical). Monitoring outflow of oil with time is similar to the air-water system.

Materials and Methods

One-step and multistep outflow experiments were carried out for each of the two porous media and two fluids. Yolo loam (fine-silty, mixed, nonacidic, thermic Typic Xerorthent, 23% sand, 55% silt, 23% clay) and Oso-Flaco fine sand (100% sand) were used with distilled water and Soltrol 130[3] in these experiments. Porous media were oven dried, sieved through a 2-mm screen, and packed to a uniform bulk density (Yolo loam ρ_b = 1.39 g cm^{-3}, Oso-Flaco fine sand ρ_b = 1.52 g cm^{-3}) in 8.25 cm in diameter and 6-cm-long brass cylinders in pressure cells with 0.57-cm-thick saturated ceramic plates at the bottom (Fig. 1). The ceramic plates had an air-entry value of 1000 cm of water.

Air-water System

Three Yolo loam and five Oso-Flaco fine sand cells were saturated with distilled water from the bottom. After saturation, the saturated hydraulic conductivity of each soil core was determined using constant head method (Table 1). Pressure was applied from the top using a pressurized N_2-gas cylinder. Outflow experiments were performed using pressure increments of 30, 50, 75, 100, 200, 400, 500, 700, and 1000 cm of water. Outflow from each cell as a function of time was recorded for each pressure step until equilibrium was reached. Soil cores were then removed from the cells and weighed, then oven dried at 110°C for 48 h. The incrementally decreasing volumetric water contents were back calculated following gravimetric determination of water content at the completion of the experiment.

Air-Soltrol 130 System

The oven-dried soil or sand cores were saturated with Soltrol 130 from the bottom under suction. Pressure was applied from the top using N_2-gas and outflow measurements were

[3] A light machine oil manufactured by Phillips Petroleum Inc., Bartlesville, OK.

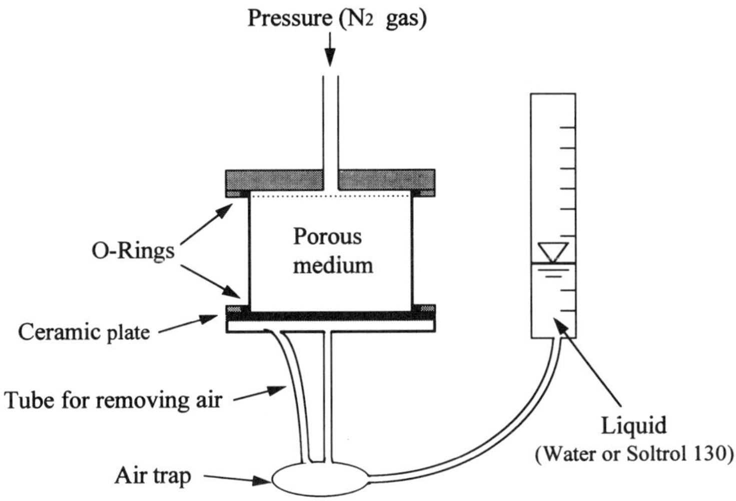

FIG. 1—*Pressure cell.*

taken following pressure increments of 3, 6, 9, 13, 15, 20, 30, 40, 70, 90, 120, 160, 200, and 400 cm of water. Outflow from each cell as a function of time was recorded for each pressure step until equilibrium was reached. Soil cores were then removed from the cells and weighed to determine the volumetric Soltrol 130 content at the end of the experiment. Volumetric oil contents during the displacement steps were calculated from the final value combined with cumulative outflow measurements.

Cumulative outflow volume and volumetric liquid content at each pressure step were used to evaluate α, n, and θ_r using MULSTPM (Eching and Hopmans 1993). Water (or Soltrol 130) retention curves were generated with various initial estimates for α, n, and θ_r for each soil core. The saturated hydraulic conductivities (K_s) and maximum liquid contents were fixed during the optimization process based on independent measurement of these parameters. The set of parameters which resulted in the highest regression correlation coefficient (r^2) and the lowest sum of squares between fitted and observed data were used. Cumulative outflow values at any given time were almost identical for the three Yolo loam samples. Cumulative outflow values were also identical for the Oso-Flaco fine sand. Similar results were obtained for the air-water system. Therefore, only data from Yolo loam core No. 3 and Oso-Flaco fine sand core No. 3 (Table 2) will be discussed in this paper.

TABLE 1—*Measured saturated hydraulic conductivity of Yolo loam and Oso-Flaco fine sand.*

Porous Medium	Liquid	Average K_s (cm h^{-1})	Θ_s (m^3 m^{-3})
Yolo loam	water	0.0895 ($\pm$0.0034)	0.468
Fine sand	water	50.4720 ($\pm$1.6322)	0.417
Yolo loam	Soltrol 130[a]	0.0480 ($\pm$0.0018)	0.468
Fine sand	Soltrol 130[a]	27.0756 ($\pm$0.8748)	0.417

[a]K_s: calculated from intrinsic permeability of the porous medium and the density (0.7536 g cm^{-3}) and viscosity (1.009 cpoise) of Soltrol at 20°C.

TABLE 2—*Optimized parameters for Yolo loam Sample 3 and Oso-Flaco fine sand Sample 3. (K_s and Θ_s were fixed in each optimization.)*

Porous Medium	Liquid	α, cm^{-1}	n	Θ_r (m^3/m^{-3})
Yolo loam	water	0.0041	1.8118	0.1798
Yolo loam	Soltrol	0.6500	1.1711	0.1428
Fine sand	water	0.0283	4.5569	0.001
Fine sand	Soltrol	0.0607	5.0000	0.070

Scaling Retention Curves

The scaling of soil water properties is a commonly used method to describe the spatial variability of soil within a heterogeneous field. Spatial variability in soil hydraulic properties can be expressed by the variability in the scaling factor (Warrick et al. 1977; Peck et al. 1977). The concept of scaling is used in this paper to describe the variability in soil hydraulic functions in a homogeneous porous media due to the variability in the wetting fluid (water versus Soltrol 130). The scaling factor is used to predict the Soltrol retention curve from the water retention curve of the same soil core. The variability in the retention curves is directly related to the variability in the wetting fluid due to each fluid's viscosity, surface tension, and density and the interactions between the soil's matrix and the wetting fluid. Neglecting interfacial contact angles, the scaling factor for the capillary pressure head h_{ao}, is given by the ratio of surface tension (σ) of Soltrol to water such that the capillary pressure head h_{ao} is related to h_{aw} by

$$h_{ao} = \frac{\sigma_{ao}}{\sigma_{aw}} h_{aw} = \alpha_h h_{aw} \tag{6}$$

where both pressure heads are expressed in cm of water head, and α_h is the scaling factor for capillary pressure head. Similarly, a scaling factor for hydraulic conductivity can be obtained from the densities (ρ) and viscosities (μ) of water and Soltrol

$$K(S_o) = \frac{\rho_o \mu_w}{\rho_w \mu_o} K(S_w) = \alpha_K K(S_w) \tag{7}$$

where α_K is the scaling factor for conductivity. Both scaling factors are temperature and fluid dependent. For clean systems, the values of α_h and α_K at 20°C for the air-Soltrol system are 0.3195 and 0.5366, respectively (distilled water σ = 72 dynes/cm, μ = 1.009 cpoise, density = 1.00 g cm^{-3}; Soltrol 130 σ = 23 dynes/cm, μ = 1.417 cpoise, density = 0.7536 g cm^{-3} at 20°C).

Results and Discussion

The relationships between liquid contents (water and Soltrol 130) and applied pressure head at equilibrium at the end of each pressure step are shown in Fig. 2*a* for Oso-Flaco fine sand (Sample 3) and in Fig. 2*b* for the Yolo loam (Sample 3). As expected, for a given volumetric liquid content, the capillary pressure (in cm of water) was lower for the Soltrol than for water. The water and Soltrol retention curves have almost the same slope for capillary pressure heads greater than 30 cm for the Oso-Flaco fine sand but not at all for the Yolo loam soil due to clay-water dispersion. When the two curves have the same slope, this

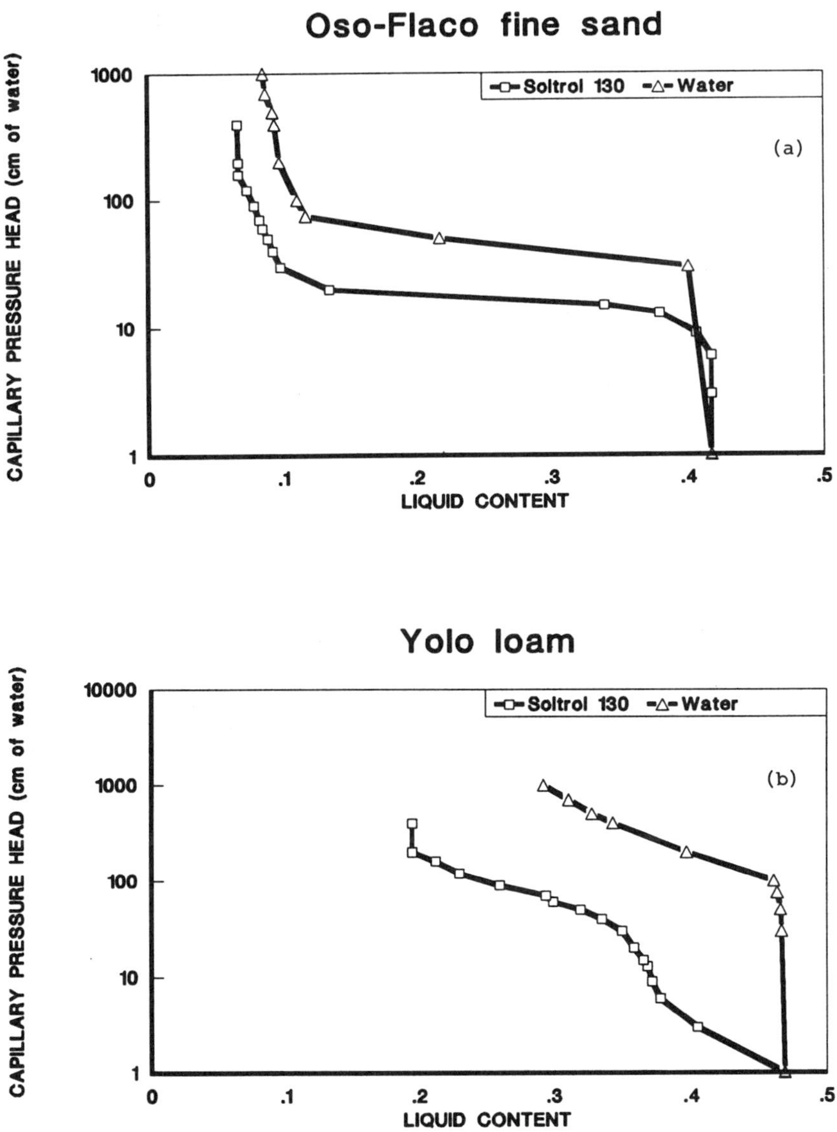

FIG. 2—*Retention curves for Oso-Flaco fine sand and Yolo loam.*

indicates that the Soltrol retention curve can be scaled or predicted from the water retention curve using Eq 6. Our assumption was that the only difference between the water and the Soltrol retention curves is due to variability in the surface tension between these two fluids. This assumption does not seem to be valid when the porous media contains clay particles that interact with the wetting fluid. This interaction results in shrinking or swelling of clay particles and subsequent changes in the pore-size distribution of the porous media. The extent of the changes in pore-size distribution and bulk density depend on the chemical composition of soil and soil-solution.

Measured and predicted retention curves from scaling by Eq 6 for the Soltrol/air system are shown in Fig. 3 for both porous media. Scaled and original Soltrol retention curves were very similar at liquid contents between 0.1 and 0.40 m^3 m^{-3} for the Oso-Flaco fine sand (Fig. 3*a*).

As predicted, the scaled Soltrol retention curve did not match the original retention curve for the Yolo loam soil (Fig. 3*b*). However, both curves had the same slope at liquid contents between 0.28 and 0.35 m^3 m^{-3}, indicating that predicting Soltrol retention curve from the

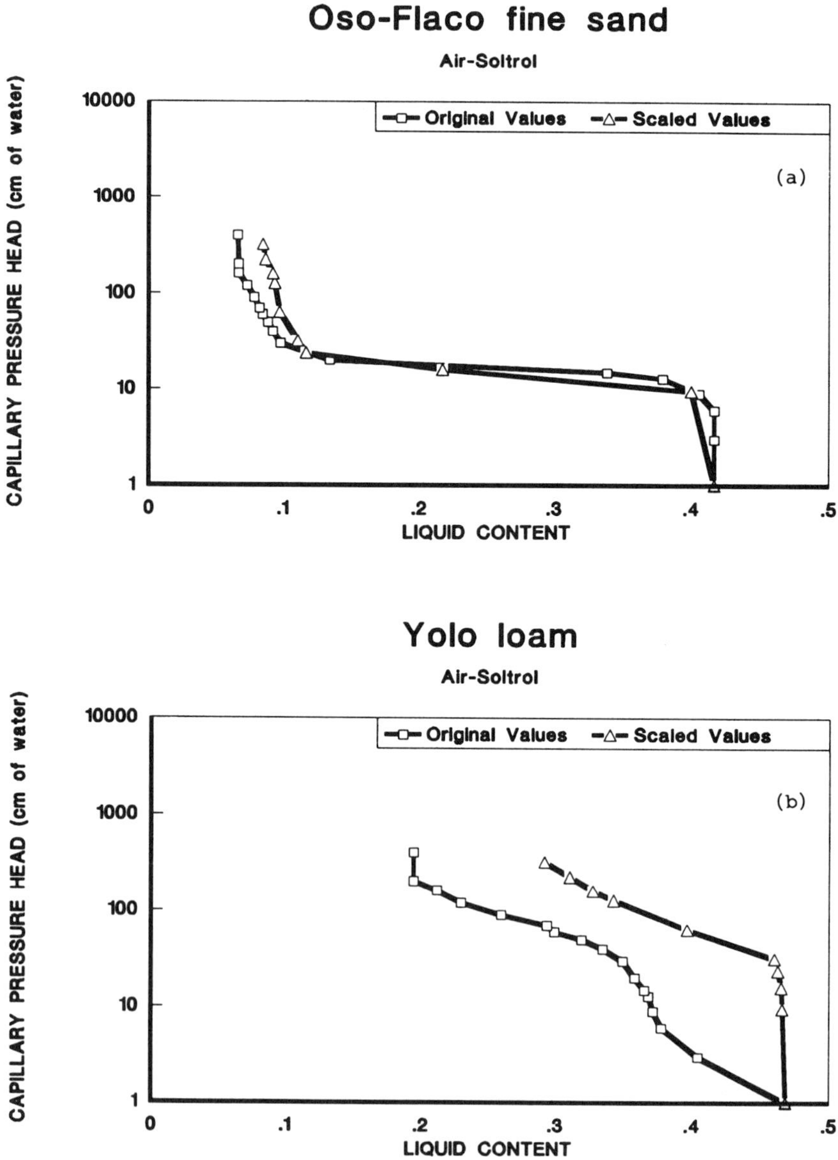

FIG. 3—*Scaled retention curves for Oso-Flaco fine sand and Yolo loam.*

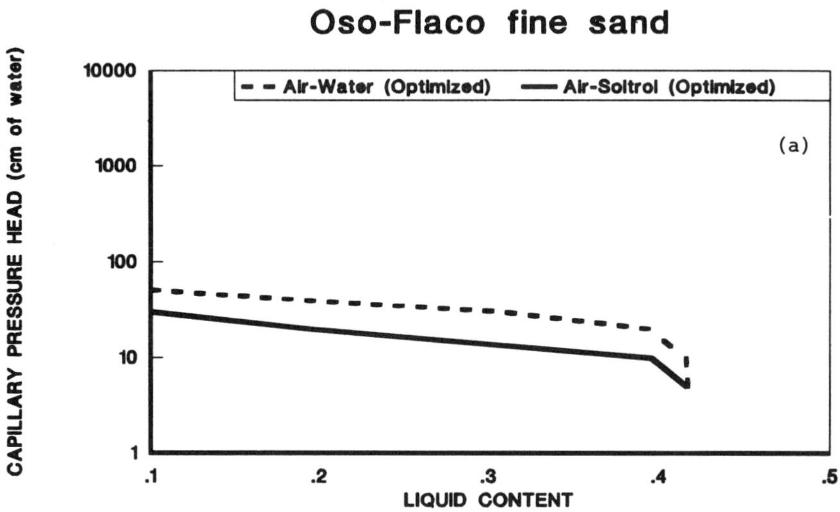

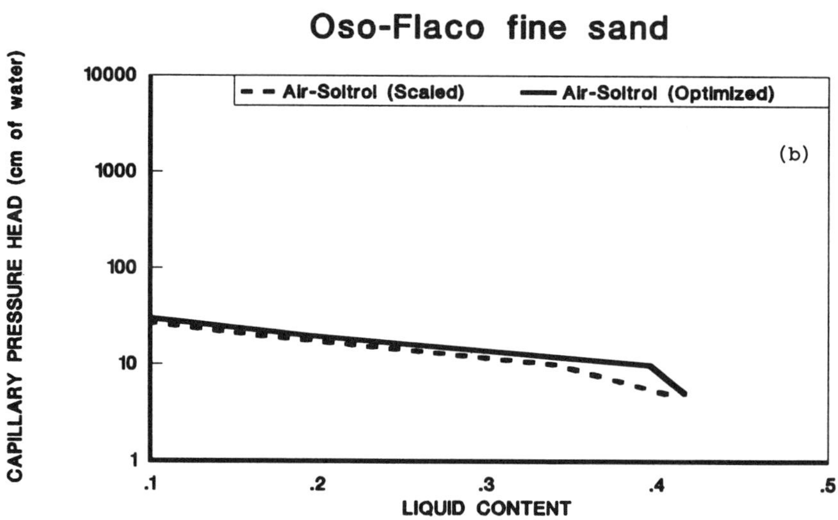

FIG. 4—*Optimized and scaled retention curves for Oso-Flaco fine sand.*

water retention curve is limited to a narrow range in liquid contents corresponding to a small range of the pore-size distribution that was drained during the desaturation process. Displacement of liquid during this limited range is not directly affected by the interactions between the clay particles and the liquid of interest. Differences between scaled and original (measured) curves (Fig. 3*b*) are due to the fact that the interaction between the clay particles and the wetting fluid are not the same for both fluids, particularly at very low or high liquid content.

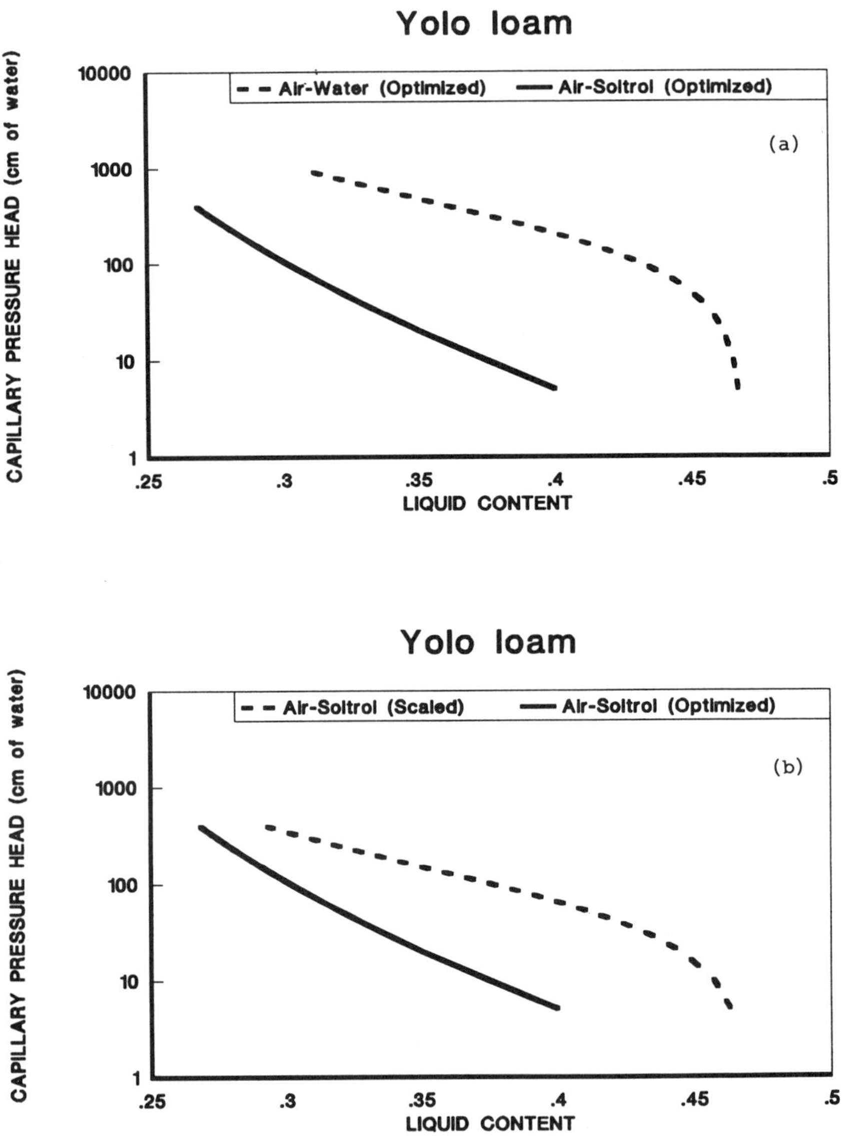

FIG. 5—*Optimized and scaled retention curves for Yolo loam.*

We assumed that the surface tension of water in the soil solution is unaffected by the porous media. While satisfactory for the Oso-Flaco sand, this assumption is poor for the Yolo loam where surface tension of the soil solution is directly affected by the chemical composition of the soil. Changes in clay-water dispersion limited our ability to predict the Soltrol retention curve from the water retention curve. Figure 4*a* shows the optimized water and Soltrol 130 curves from the multistep outflow method for the Oso-Flaco fine sand. The scaled Soltrol retention curve (from the optimized water retention curve) is shown in Fig. 4*b*. The scaled retention curve can be predicted accurately from the optimized water retention curve. Both scaled and optimized Soltrol retention functions are almost identical at liquid contents greater than 0.05 $cm^3\ cm^{-3}$.

Optimized water and Soltrol retention curves for the Yolo loam are shown in Fig. 5*a*. The scaled Soltrol retention curve did not match the optimized retention curve (Fig. 5*b*). The Soltrol retention curve cannot be predicted from the optimized water retention curve for the Yolo loam soil as was shown from the equilibrium retention data.

Figure 6*a* shows the optimized hydraulic conductivity curves for the Oso-Flaco fine sand. The Soltrol conductivity curve of Oso-Flaco fine sand can be predicted from the air-water

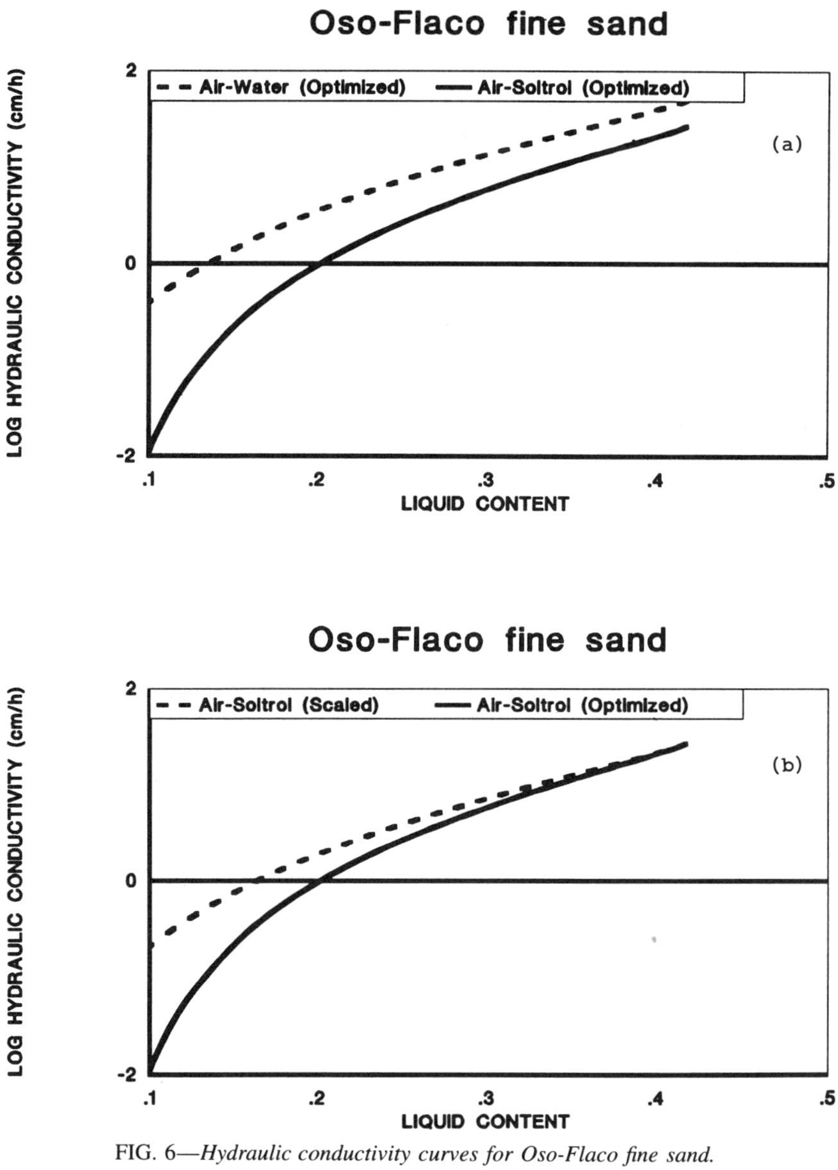

FIG. 6—*Hydraulic conductivity curves for Oso-Flaco fine sand.*

curve at liquid contents of greater than 0.2 $cm^3\ cm^{-3}$ (Fig. 6*b*). However, the scaled hydraulic conductivity curve of Yolo loam soil did not match the optimized soil-water conductivity curve (Fig. 7).

Conclusions

Our preliminary measurements of soil hydraulic parameters α, n, and θ_r using the multistep outflow method indicate that the method can be successfully applied to NAPL-air systems

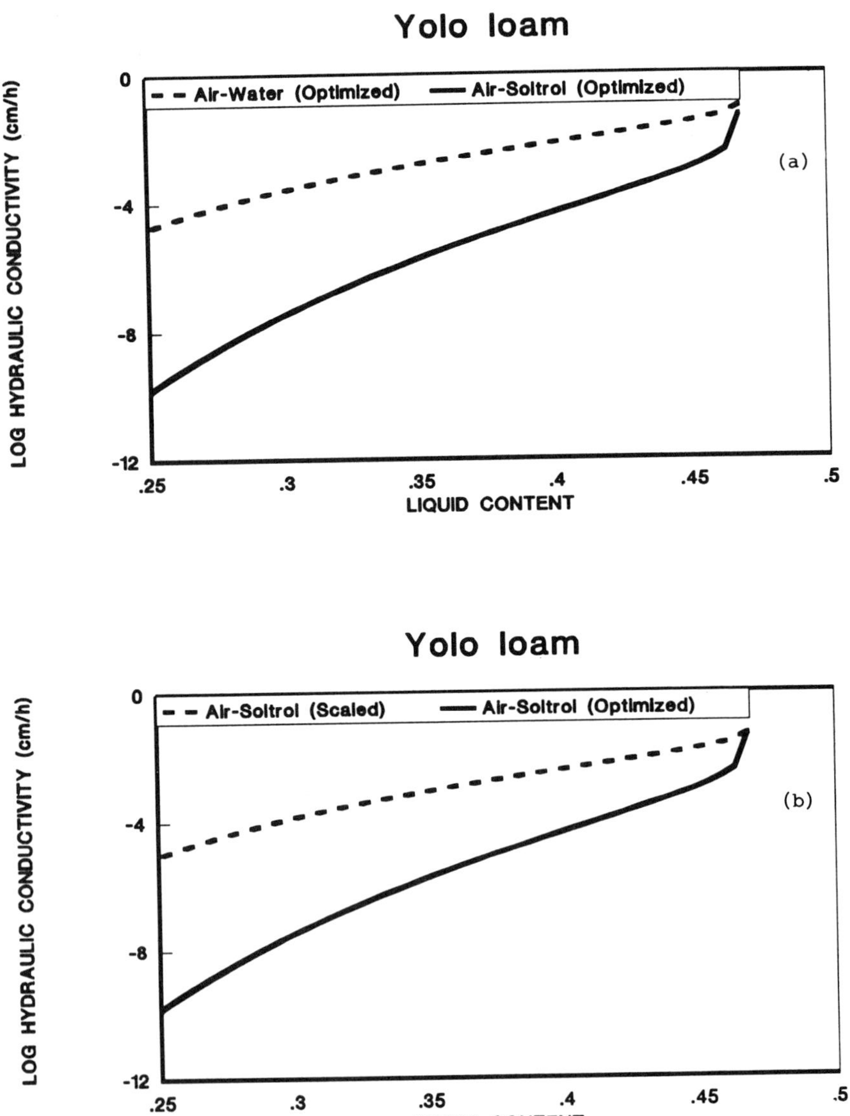

FIG. 7—*Hydraulic conductivity curves for Yolo loam.*

directly. However, estimation of the empirical parameters for NAPL saturation-pressure-conductivity relationships from scaling the air-water system parameters is successful only for the fine sands but not for the loam soils having appreciable clay mineral fractions. Scaling accounts for variability in surface tension, viscosity, and density; however, it does not account for the interactions between the wetting fluid and the soil matrix (including contact angle effects). These results suggest that scaling relationships such as Eqs 6 and 7 can be used in conjunction with results from application of the multistep outflow method to air-water systems to estimate soil hydraulic functions for NAPL-air (and potentially NAPL-water) systems. Direct estimation of the hydraulic function parameters for multiphasc flow in fine-textured soils by the multistep outflow method is recommended rather than attempting to obtain these functions from scaling relationships. The method described in this paper may not be applicable to all NAPLs. The accuracy of the method depends on the NAPL and the soil's type. Additional work in modifying the multistep method for rapid determination of soil hydraulic parameters for multiphase systems in fine-textured soils is presently underway.

References

Abriola, L. M. and Pinder, G. F., 1985a, "A Multiphase Approach to the Modeling of Porous Media Contamination by Organic Compounds: 1. Equation Development," *Water Resources Research,* Vol. 21, No. 1, pp. 11–18.

Abriola, L. M. and Pinder, G. F. 1985b, "A Multiphase Approach to the Modeling of Porous Media Contamination by Organic Compounds: 2. Numerical Simulation," *Water Resources Research,* Vol. 21, No. 1, pp. 19–26.

Campbell, G. S., 1974, "A Simple Method for Determining Unsaturated Hydraulic Conductivity from Moisture Retention Data," *Soil Science,* Vol. 117, pp. 311–314.

Corey, A. T., Rathjens, C. H., Henderson, J. H., and Wyllie, M. R. J., 1956, "Three-Phase Relative Permeability," *Transactions of the Society of Petroleum Engineering,* American Institute of Mining, Metallurgical, and Petroleum Engineers, Vol. 207, pp. 349–351.

Corey, A. T., 1985, "Mechanics of Immiscible Fluids in Porous Media," *Water Resources Publications,* Littleton, CO, 225 pp.

Eching, S. O. and Hopmans, J. W., 1993, "Optimization of Hydraulic Functions from Transient Outflow and Soil Water Pressure Data," *Soil Science Society of America Journal,* Vol. 57, pp. 1167–1175.

Fayers, F. J. and Mathews, J. D., 1984, "Evaluation of Normalized Stone's Methods for Estimating Three-Phase Relative Permeabilities," *Transactions of the Society of Petroleum Engineering Journal,* Vol. 24, pp. 230–242.

Hopmans, J. W., van Dam, J. C., Eching, S. O., and Stricker, J. N. M., 1994, "Parameter Estimation of Soil Hydraulic Functions Using Inverse Modeling of Transient Outflow Experiments," *Trends in Hydrology,* in press.

Kaluarachchi, J. J. and Parker, J. C., 1989, "An Efficient Finite Element Method for Modeling Multiphase Flow," *Water Resources Research,* Vol. 25, pp. 43–54.

Kool, J. B., Parker, J. C., and van Genuchten, M. Th., 1985, "Determining Soil Hydraulic Properties from One-Step Outflow Experiments by Parameter Estimation: I. Theory and Numerical Studies," *Soil Science Society of America Journal,* Vol. 49, pp. 1348–1354.

Kuppusamy, T., Sheng, J., Parker, J. C., and Lenhard, R. J., 1987, "Finite-Element Analysis of Multiphase Immiscible Flow Through Soils," *Water Resources Research,* Vol. 23, pp. 625–631.

Lenhard, R. J., Dane, J. H., Parker, J. C., and Kaluarachchi, J. J., 1988, "Measurement and Simulation of One-Dimensional Transient Three-Phase Flow for Monotonic Liquid Drainage," *Water Resources Research,* Vol. 24, pp. 853–863.

Leverett, M. C., 1941, "Capillary Behavior in Porous Solids," *Transactions of Society of Petroleum Engineering,* AIME, Vol. 142, pp. 152–169.

Marquardt, D. W., 1963, "An Algorithm for Least-Squares Estimation of Nonlinear Parameters," *Journal of Applied Mathematics,* SIAM, Vol. 11, pp. 431–441.

Mualem, Y., 1976, "A New Model for Predicting the Hydraulic Conductivity of Unsaturated Porous Media," *Water Resources Research,* Vol. 12, pp. 513–522.

Parker, J. C., Lenhard, R. J., and Kuppusamy, T., 1987, "A Parametric Model for Constitutive Properties Governing Multiphase Flow in Porous Media," *Water Resources Research,* Vol. 23, pp. 618–624.

Peck, A. J., Luxmoore, R. J., and Stolzy, J. C., 1977, "Effects of Spatial Variability of Soil Hydraulic Properties in Water Budget Modeling," *Water Resources Research,* Vol. 13, pp. 348–354.

Van Genuchten, M. T., 1980, "A Closed-Form Equation for Predicting the Hydraulic Conductivity of Unsaturated Soils," *Soil Science Society of America Journal,* Vol. 44, pp. 892–898.

Warrick, A. W., Mullen, G. J., and Nielsen, D. R., 1977, "Scaling Field Measured Soil Hydraulic Properties Using a Similar Media Concept," *Water Resources Research,* Vol. 13, pp. 355–362.

Susan C. Wilson[1] *and Kevin C. Jones*[1]

The Fate and Behavior of Volatile Aromatic Hydrocarbons in Sewage Sludge-Amended Soil

REFERENCE: Wilson, S. C. and Jones, K. C., **"The Fate and Behavior of Volatile Aromatic Hydrocarbons in Sewage Sludge-Amended Soil,"** *Volatile Organic Compounds in the Environment, ASTM STP 1261,* W. Wang, J. Schnoor, and J. Doi, Eds., American Society for Testing and Materials, 1996, pp. 119–123.

ABSTRACT: The fate and behavior of five commonly occurring aromatic volatile organic compounds (VOCs), namely toluene, ethyl benzene, and o-, m-, and p-xylene, were investigated in sewage sludge-amended soil. Laboratory experiments were conducted to investigate VOC concentrations in sludge-amended soil for different soils and at different sludge application rates over 31 days. Results indicated that VOC loss rates decreased with increased soil organic matter content and increased sludge application rate. VOC loss rates in spiked samples were found to be comparable to those in unspiked samples over 31 days. Results from laboratory experiments also indicated that abiotic loss processes were more important than biological degradation for the removal of VOCs from sewage sludge-amended soil. VOC loss rates were also compared in a field study where sludge was applied to grassed and plowed plots.

KEYWORDS: volatile organic compounds (VOCs), sewage sludge, soil, field study, half-lives, volatilization, biological degradation

Sewage sludge disposal to agricultural land is widely practiced in Europe and North America. Of the 1.08 million tonnes (dry weight) of sewage sludge currently produced annually in the United Kingdom, 51% is applied to agricultural land (Anderson 1992). This is expected to increase as a result of restrictions on alternative disposal routes, such as sea disposal, and also increasingly stringent waste-water treatment requirements.

Volatile organic compounds (VOCs) have been detected in sewage sludges by a number of researchers (Jacobs et al. 1987, Wilson et al. 1994). In a recent survey of twelve U.K. digested sludges collected from rural, urban, and industrial treatment works, the total VOC concentration ranged from 4.62 to 568 mg/kg dry weight (260 to 15 400 μg/L) (Wilson et al. 1994). Aromatic VOCs were present in the greatest abundance, comprising 92% of the mean total VOC concentration. Concern that some of the VOCs detected in sewage sludge may present a risk to human health and the environment when applied to agricultural land is increasing.

Few studies have investigated the fate and behavior of VOCs in sewage sludge-amended soil. Those which have been reported have used spiked sludge rather than investigating the behavior of VOCs resident in the sludge itself due to undetectable sludge VOC concentrations (Jin and O'Connor 1990, Webber and Goodin 1992). Therefore, the experiments reported

[1] Research assistant and reader, respectively, Institute of Environmental and Biological Sciences, Lancaster University, Lancaster, LA1 4YQ, U.K.

here have been designed to evaluate the behavior of VOCs in unspiked sewage-amended soils. Only aromatic VOCs (toluene, ethyl benzene, o-, m-, and p-xylene) were studied because these were generally detected in U.K. sludges at the greatest concentrations (Wilson et al. 1994). Soil samples were extracted wet using liquid-liquid extraction with n-pentane, and analyzed by gas chromatography with flame ionization detection. The method used was similar to that reported by Wilson et al. (1994) for sewage sludge samples.

Laboratory Experiments

Laboratory microcosm experiments were designed to investigate the behavior of VOCs (toluene, ethyl benzene, o-, m-, and p-xylene) in:

(*a*) different soils,
(*b*) at different sludge application rates,
(*c*) in spiked sludge-amended soil, and
(*d*) the relative importance of abiotic and biological loss processes.

The experiments used glass "pot" based microcosms which were sampled over the period of the experiment.

VOC Behavior in Different Sludge-Amended Soils

Digested sewage sludge was applied at an application rate of 100 tonnes per hectare (t/ha) to three different soils (Soil A, a ground-water gleyed brown earth situated on recent fine-textured riverine alluvium; Soil B, a peaty gleyed podsol; Soil C, a gleyed brown earth under permanent grass from a hillside) which had no previous history of sludge amendment. All soils were sieved to <2 mm and the moisture content adjusted to 20% prior to sludge addition. The soil VOC concentration was monitored over 31 days (Fig. 1). Details of the three soils are presented in Table 1. Figure 1 illustrates that the rate of VOC loss in Soil B, the soil with the highest percentage of organic matter, was significantly less than that in Soils A and C. A greater percentage of the original VOC concentration remained after 31 days in Soil B (52%) than in Soils A (23%) and C (32%).

VOC Behavior at Different Sludge Application Rates and in Spiked Samples

Digested sewage sludge was applied to soil (a gleyed brown earth with organic matter content of 9.35%) at two different rates, 50 t/ha and 100 t/ha, and the soil VOC concentration measured over time. In addition, sewage sludge-amended soil at the 50 t/ha application rate was spiked and the spike concentration monitored over time. VOC half-lives (Table 2) over 31 days were found to be similar in the 50 t/ha spiked and unspiked samples, indicating that spiking sludge-amended soil gave a representative indication of the VOC loss rates over the time scale of the experiment. However, VOC half-lives in the 100 t/ha application samples were greater than those determined for the 50 t/ha samples for all VOCs studied although the effect was less apparent for toluene. This suggests that increasing the sludge application rate decreases the rate of VOC loss.

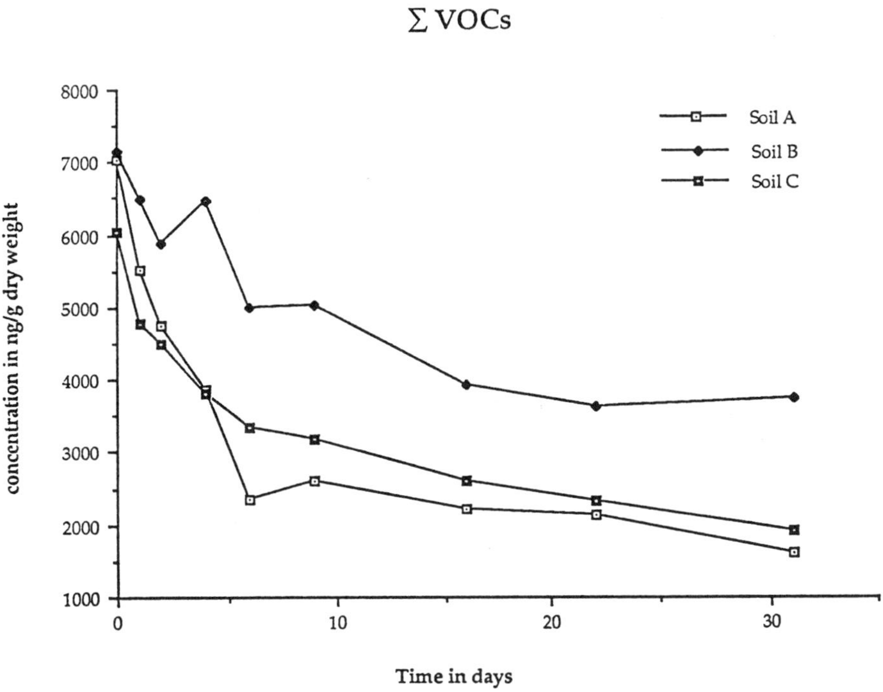

FIG. 1—*VOC concentrations in three different sludge-amended soils over 31 days.*

Relative Importance of Biological and Abiotic Loss Processes for VOC Removal from Sludge Amended Soil

Digested sewage sludge was added to unsterilized and sterilized (mercuric chloride added at 1% dry weight) soil, the sludge-amended soil placed in well-aerated glass microcosms, and the soil VOC concentrations monitored over time (Fig. 2). Figure 2 shows that little difference was detected between losses in the sterilized and unsterilized treatments over the 65 days of the experiment. Of the total VOC concentration, 92.6% was lost after 65 days from the unsterilized treatment and 88.1% from the sterilized treatment. This suggests therefore that VOC loss from sludge-amended soils is predominantly due to abiotic loss processes. Biological degradation may account for approximately 5% of the total VOC loss. Volatilization was likely to be the major loss process occurring (Jin and O'Connor 1990, Webber and Goodin 1992).

TABLE 1—*Soil characterization.*

	% Moisture	% Organic Matter	% Texture Sand	Silt	Clay	Soil	pH, (0.1 m $CaCl_2$)
Soil A	20	6.85	10	80	10	Silt/loam	5.64(4.99)
Soil B	20	54.3	peat/humus layer				4.63(4.11)
Soil C	20	9.35	40	50	10	sandy/silt/loam	5.34(4.75)

TABLE 2—*Calculated VOC half-lives (days) over 0 to 31 days.*

	50 t/ha	% Loss	100 t/ha	% Loss	Spike	% Loss
Toluene	13.7	97	14.8	98	13.1	98
Ethylbenzene	10.7	99	34.7	92	8.89	91
m- and p-xylene	8.89	100	17.3	94	9.76	92
o-xylene	10.5	99	26.7	100	9.24	99

NOTE: Half-lives have been calculated using first-order kinetic modeling of the data except for toluene, which fitted a first-order model poorly. These data were therefore fitted to a zero-order model.

Field Study

In a field study, digested sewage sludge was applied at a rate of 50 t/ha to a grassed plot and a plowed plot which had never previously had sludge amendments. The sludge was rotovated into the plowed plot. Soil VOC concentrations (toluene, o-, m-, and p-xylene) were measured over a period of 64 days for the 0-to-5-cm depth. Total VOC concentration in the plowed plot (132 ng/g dry weight at t = 0 days) declined to a concentration similar to that in the control plot within 36 days (3.86 ng/g at t = 36 days). Half-lives for individual VOCs ranged from 5 to 13 days over this time period. However, VOC concentrations (218 ng/g at t = 0 days) in the grassed plot declined more rapidly and the total VOC concentration was similar to that in the control plot eight days after sludge application (2.37 ng/g at t = 8 days). Half-lives (calculated assuming first-order kinetics) were less than one day for all VOCs in the grassed plot over eight days.

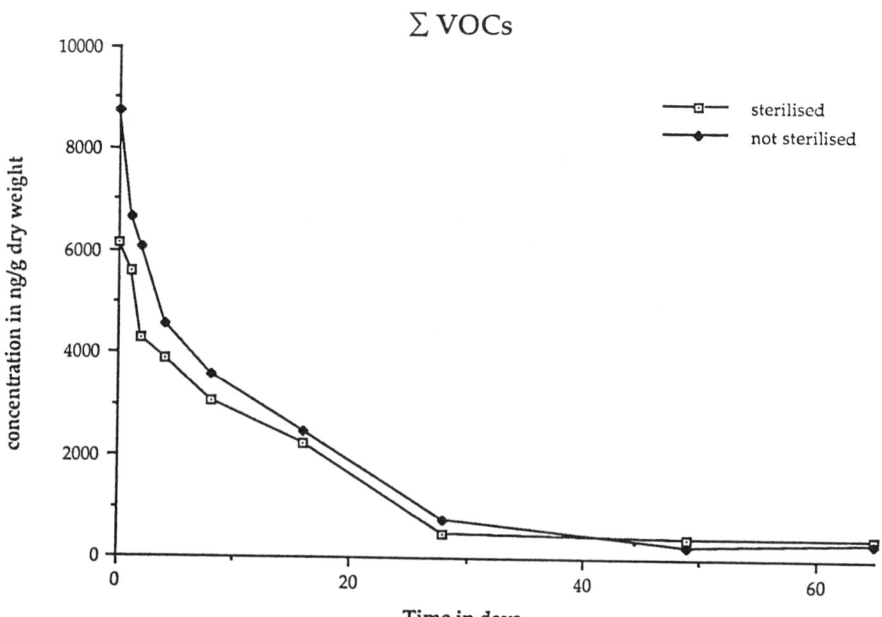

FIG. 2—*VOC concentrations in sterilized and unsterilized sludge-amended soil.*

Conclusions

Sludge amendment to soil increased soil VOC concentrations. The rate of VOC loss from soil depended on the soil properties and the sludge application rate. The rate of loss apparently decreased when the sludge application rate was increased. It appears that spiking sludge-amended soil can give a representative indication of the VOC loss rates over the period of the experiment (31 days). Abiotic loss processes were most important for VOC removal from sludge-amended soil.

A field study showed that VOC loss from sludge-amended soil was greater when sludge was applied to the grass surface, compared to the loss when the sludge was plowed into the soil. This has important implications for future sludge disposal practices to agricultural land since direct injection into the soil has been advocated as a means to reduce the nuisance effect of odor.

Note that sludge application rates in all the studies reported here were higher than rates (approximately 10 t/ha) used in present-day agricultural practice. Therefore, VOC concentrations detected in contemporary soils following sludge application might be expected to be lower than the concentrations measured here.

Acknowledgments

The authors are grateful to North West Water plc for their cooperation during sewage sludge sample collection and also to the Ministry of Agriculture, Fisheries and Food (MAFF) for funding this research on the fate and effects of organic contaminants from sewage sludge applied to agricultural land (Contract OC 8909 CSA 1972).

References

Anderson, T. A., 1992, "Impact in Scotland of UK and EC Sewage Legislation," *Journal of the Institution of Water and Environmental Management,* Vol. 6, No. 6, pp. 682–689.

Jacobs, L. W., O'Connor, G. A., Overcash, M. A., Zabik, M. J., and Rygiewicz, P., 1987, "Effects of Trace Organics in Sewage Sludges on Soil-Plant Systems and Assessing Their Risk to Humans," *Land Application of Sludge,* A. L. Page, T. G. Logan, and J. A. Ryan, Eds., Lewis Publishers, Chelsea, MI, pp. 101–143.

Jin, Y. and O'Connor, G. A., 1990, "Behaviour of Toluene Added to Sludge-Amended Soil," *Journal of Environmental Quality,* Vol. 19, pp. 573–579.

Webber, M. D. and Goodin, J. D., 1992, "Environmental Studies on the Fate of Organic Contaminants in Sludge Treated Soils," *Effects of Organic Contaminants in Sewage Sludge on Soil Fertility, Plants and Animals,* J. E. Hall, D. R. Sauerbeck, and P. L'Hermite, Eds., Commission of the European Communities, pp. 54–69.

Wilson, S. C., Burnett, V., Waterhouse, K. S., and Jones, K. C., 1994, "Volatile Organic Compounds in Digested United Kingdom Sewage Sludges," *Environmental Science and Technology,* Vol. 28, No. 2, pp. 259–266.

Environmental Measurement

Yoshio Tsuchiya[1] *and J. M. Kanabus-Kaminska*[1]

Identification and Quantification of Volatile Organic Compounds Using Systematic Single-Ion Chromatograms

REFERENCE: Tsuchiya, Y. and Kanabus-Kaminska, J. M., **"Identification and Quantification of Volatile Organic Compounds Using Systematic Single-Ion Chromatograms,"** *Volatile Organic Compounds in the Environment, ASTM STP 1261,* W. Wang, J. Schnoor, and J. Doi, Eds., American Society for Testing and Materials, 1996, pp. 127–138.

ABSTRACT: In order to determine the background level of volatile organic compounds (VOCs) in Canadian indoor air, a method of identification and quantification at a level of 0.3 $\mu g/m^3$ using systematic single-ion chromatograms (SICs) has been developed. The compounds selected for measurement included several halogenated compounds, oxygen compounds, terpenes, and C8 to C16 n-alkanes. Air samples were taken in 3-layered sorbent tubes (glass beads, Tenax TA, Ambersorb XE-340) and trapped compounds were thermally desorbed into the helium stream of a gas chromatograph/mass spectrometer (GC/MS) analytical system. Total quantities of volatile organic compounds (TVOCs) were measured using a flame ionization detector (FID). Individual compounds were analyzed by a GC/MS. For the identification of compounds in the main stream GC effluent, both the specific GC retention and mass spectra were used. For the purposes of this study, the specific retention was defined based on two primary reference compounds, benzene and p-dichlorobenzene, and several secondary reference compounds. About 50 selected SICs were routinely extracted from a total ion chromatogram (TIC) to detect and quantify compounds. For each compound, a single representative ion was selected. The specific retention was calculated from the elution time on the SIC. For quantification, ion counts under a peak in the SIC were measured. The single-ion MS response factor for some of the compounds was experimentally determined using a dynamic reference procedure. For compounds whose response factors have not been measured, the total ion counts were calculated from the single ion counts by a regression equation and the total ion counts of a peak were assumed to be proportional to the concentration of the compound in the air.

KEYWORDS: indoor air, single-ion chromatogram, gas chromatograph/mass spectrometer, background level

Nomenclature

- C Volumetric concentration, g/m^3
- F flame ionization detector (FID)
- I Ion count on MS
- R Specific retention
- Rf FID response factor, count/g
- T Elution time, min

[1] Senior research officer and technical officer, respectively, Institute for Research in Construction, National Research Council of Canada, Building M-59, Montreal Rd., Ottawa, ON, Canada K1A OR6.

Subscripts

a First reference compound
b Second reference compound
c Calibration gas
i ith component
u Unknown target compound

The background level of volatile organic compounds (VOCs) in indoor air is a key piece of information for evaluating environment and human interactions. There have been some measurements of the background level, but most were limited in the number of compounds detected and quantified or in the number of buildings studied (for example, Lewis et al. 1990, Shields and Weschler 1990). There are large-scale studies of background level measurements being planned or performed (Needham et al. 1990, Otson et al. 1992), but these works require some time to complete. The authors' laboratory has a unique opportunity to analyze VOC levels in indoor air by responding to clients' requests for VOC measurement. Such analytical work has been active since June 1986 and about 2000 analytical data files have been accumulated to date. The stored data are being reanalyzed using a method developed in the authors' laboratory.

Indoor air contains a large number of VOC components. Identification and quantification of these components are not an easy task unless the analysis is focused only on preset target compounds. There is a logical dilemma in selecting targets without knowing background levels of many components and the significance of these on environmental impact. Apart from the target analysis, when a large number of unknown compounds are analyzed, there is always uncertainty in their identification; questionable identifications are sometimes detected in publications. The uncertainty may be reduced to a certain extent by spending more time and effort; however, the art of identification could remain within that research group and contribute little in building up common scientific knowledge. This paper presents a method using SIC, for identification and quantification of many unknown compounds in indoor air. The identification method can be used by readers and the data can be accumulated to form a database for common use. In quantification using TIC, errors occur when a chromatographic peak is "impure;" use of SIC significantly reduces the possibility of misinterpreting overlapping peaks. The preliminary results were published previously (Tsuchiya and Kanabus-Kaminska 1992) using the present method, including 37 compounds in indoor air of 45 buildings. The work is continuing to expand in both number of compounds and number of buildings. This paper reports the methods used to identify and quantify the compounds.

Sampling was undertaken using a commercially available 3-layered (glass beads, Tenax TA,[2] Ambersorb XE-340) tube, 20 cm long and 4 mm inside diameter. An air sample was drawn into the tube using suction from a mechanical pump. Sampling time and sampling volume were normally 20 min and 5 L, respectively, but occasionally were increased or decreased depending on the expected VOC concentrations to avoid overloading or insufficient sampling.

Trapped VOCs were thermally desorbed into a helium stream using an Envirochem Unacon Model 810 desorber unit. The functions of the desorber unit were described by Chan

[2] Some commercial equipment is identified in this paper to adequately describe the experimental procedure. Such identification does not imply recommendation or endorsement by the National Research Council, nor does it imply that the equipment identified is the best for the purpose.

et al. (1990). The flow was split into three parts. Two 4% splits went into a FID for monitoring purposes at different preconcentration steps. One of the two 4% splits was used for quantification of the TVOC. The remaining 92% of the flow was introduced into a Hewlett Packard Model 5995 GC/MS. VOCs in the flow were cryofocused on the GC column of a phenylmethyl-silicone bonded phase fused silica capillary, 50 or 60 m long and 0.25 mm inside diameter. The oven temperature was programmed to operate from −50°C to 250°C at a rate of 10°C/min for GC separation.

The MS was operated with an electron energy of 70 eV and for a mass range of 20 to 300 M/e. The MS instrument was tuned and calibrated daily using HP mass spectrometry purity perfluorotributhylamine as standard. Multipoint calibrations of the FID and MS analyzer response factors were based on analysis of sorption tubes loaded with volumes (range 0.5 to 4000 mL) of cyclohexane standard in N_2 which were prepared and certified by Matheson. The integrity of the system and repeatability were checked periodically by spiking the tubes with aliquots of the same calibration standard. The analytical path remained consistent over long periods of time since relatively clean samples delivered by the desorption method cause little deterioration of the column or MS analyzer.

TVOCs were measured using a FID while simultaneously chromatographically fingerprinting the components using the GC/MS. The resulting analytical data were stored in the data system.

About 100 organic compounds were identified and quantified in these stored data files to determine the background level of the compounds in Canadian indoor air. The compounds included alkanes, halogenated compounds, oxygen compounds, and terpenes.

Identification of Compounds

For the identification of components in the GC effluent, both the specific GC retention of the compound and mass spectra were used. Components were initially identified by the MS using a spectra library and search program supplied by the manufacturer of the instrument. The library contained 76 000 entries, including NIH/EPA/MSDC Mass Spectra Data Base (NBS Library) and Registry of Mass Spectra Data (Wiley Library). The search program used a probability based matching algorithm.

A retention time normalization system, similar to Kovats's retention index (Kovats 1958), has been devised specifically for the purpose of the VOCs in indoor air study. In the Kovats retention index, normal alkanes are used as reference compounds. In the system introduced in the present paper, VOCs frequently found and easily identified in indoor air were used as references so that spiking with internal reference compounds was avoided. Normal alkanes, C10 and higher, were used as the references.

The primary references were benzene and p-dichlorobenzene. These compounds were frequently detected in the air samples. Among the thousands of GC/MS data files accumulated over the past seven years, approximately 400 files were analyzed in some detail. Benzene was found in 97% of these files and p-dichlorobenzene in 88%. Using the following equation, the specific retentions of other compounds were calculated

$$R_u = R_a - (R_a - R_b)(T_a - T_u)/(T_a - T_b) \quad (1)$$

where R is the specific retention, T is time of elution, and the subscripts are u for the target compound and a and b for the two reference compounds. In this paper, "elution time" is used instead of "retention time" since the time for elution included times for desorption from a sorbent tube, for transfer of compounds between the internal sorbent tubes, and the actual retention time. The specific retentions of benzene and p-dichlorobenzene are given as

30 and 60 so that the specific retentions of compounds of interest are distributed between 0 and 100.

Figure 1 shows elution times and specific retentions of the primary reference compounds and two others. As can be seen, the elution times varied with instrument conditions, but the specific retention did not significantly vary over the past seven years.

For compounds that elute outside the two reference compounds, the specific retention was calculated by extrapolation and a progressively increased standard deviation was observed. To avoid increased uncertainty in specific retention measurement, secondary references were established and their specific retentions were determined as the average values of their specific retentions calculated from the primary references. Normal alkanes, C10 and higher, were used as secondary references. Normal alkanes, from C3 to C9, were not used. The lower alkanes were frequently detected in indoor air samples; however, the identification of C3 to C9 alkanes was complicated by the presence of various branched compounds. Specific retentions of other compounds were calculated by interpolation using two references (primary or secondary). Specific retentions and the standard deviations of compounds frequently detected in the indoor air samples are listed in Table 1, in which the reference compounds are marked. The specific retention obtained in this way was linearly related to the boiling point of the compound, similar to the Kovats retention index, except for a few oxygen compounds, as shown in Fig. 2.

Peak identification and quantification were performed on SICs. An SIC can be extracted by the data system from the stored TIC data file using one of routine programs of the data system. The SIC is different from the single ion monitoring in which ion counts of only preselected single ions are recorded in a GC run. Theoretically, 281 SICs can be extracted

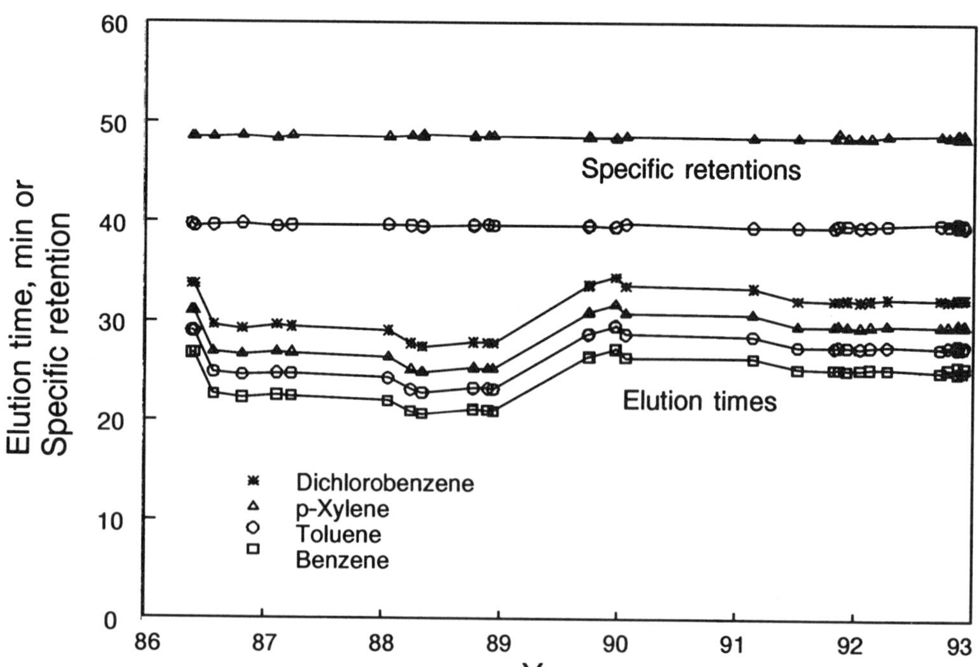

FIG. 1—*Variations of elution times and specific retentions.*

TABLE 1—*Specific retentions of VOC in indoor air.*

Compound	Specific Retention	Standard Deviation	Secondary Reference	Primary Reference	Number of Measurements
SO_2 (1)	5.16	2.243			13
CCl_3F	13.85	0.850	+		10
Ethanol	14.17	0.472			9
Acetone	15.31	0.702	+		30
CH_3CCl_2F	15.42	0.062			2
$CH_2{=}CCl_2$	17.09	0.288			7
i-Propanol	17.39	0.806			9
$C_2Cl_3F_3$	18.21	0.081			3
Methylethylketone	24.32	0.172			5
$CHCl_3$	26.11	0.159			7
Ethylacetate	26.45	0.080			2
CH_3CCl_3	28.53	0.110	+		24
Cyclohexane	29.96	0.059			3
Benzene	30.00			+	45
i-Propylacetate	30.67				1
$CHCl{=}CCl_2$	33.50	0.004			3
Methylcyclohexane	35.63	0.069			3
MiBK	37.42	0.100			2
Toluene	39.74	0.138	+		32
Hexanal	42.43	0.175			2
C8	42.52	0.155			3
$CCl_2{=}CCl_2$	43.47	0.075			13
Butylacetate	44.05				1
Ethylbenzene	47.96	0.083			5
p-Xylene	48.57	0.199	+		28
Styrene	50.45	0.061			5
o-Xylene	50.56	0.141			3
C9	50.86	0.246			4
α-Pinene	53.79	0.295			2
Camphehene	54.86	0.500			2
Trimethylbenzene	56.01	0.035			3
Trimethylbenzene	56.53	0.031			3
Trimethylbenzene	57.49	0.034			3
Trimethylbenzene	58.49	0.018			3
C10	58.62	0.275	+		14
Trimethylbenzene	60.75	0.029			2
p-Dichlorobenzene	60.00			+	38
Limonene	61.09	0.136	+		8
C11	65.67	0.059			15
Si comp. M/e = 267(2)	68.68	0.070			9
C12	72.19	0.476	+		16
Naphthalene	72.98	0.071			2
C13	78.36	0.055			7
4-Phenylcyclohexene	81.54	0.212			2
C14	84.18	0.640	+		13
C15	89.85	0.036			8
C16	94.87	0.832	+		10

NOTE: Wide peaks to ethanol and i-propanol in the respective SICs resulted in large values of standard deviation. (1) was produced from the sorbent. (2) was produced from GC column materials.

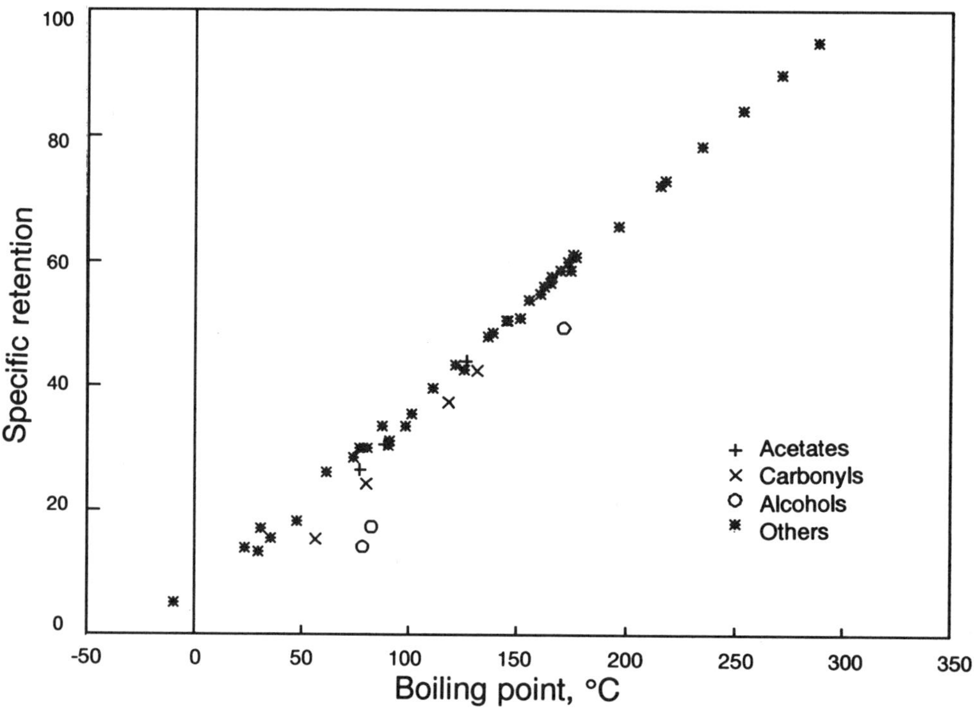

FIG. 2—*Specific retention versus boiling point of VOCs.*

from a TIC with a mass range from 20 to 300. A specific single ion was selected for each compound and, in practice, about 50 SICs were routinely extracted for identification and quantification. In SICs a target peak is usually clearly separated from other peaks. An example of a series of SICs for an air sample is shown in Fig. 3.

The specific single ion for each compound or groups of compounds frequently detected in indoor air is summarized in Table 2. In adding a new compound to this table, the relationship between the boiling point and the specific retention should agree with the trend of other compounds in Fig. 2.

When the calculated specific retention of a peak did not agree with the recorded specific retention, the identification was reexamined using one or two confirmation SICs. When they agreed, the mass spectra of the compound and the reference mass spectra retrieved from the library were compared to confirm the identification.

For halogen compounds, typical doublets (occasionally triplets or multiplets) peak appear on mass spectra due to the halogen isotopes. Two single ions composing these doublets facilitates their identification. M/e's for these doublets are listed in Table 3.

Quantification

For determining TVOC using the integrated FID signal, the response of the FID was periodically calibrated using a known volume of cylindered cyclohexane/helium mixture. The response factor, *Rf* is

$$Rf = F_c / C_c / V_c \qquad (2)$$

where F_c is the FID count in the calibration, C_c is the concentration of cyclohexane in the calibration mixture, and V_c is the volume of the calibration gas. The TVOC is

$$\text{TVOC} = F/Rf/V \tag{3}$$

where F is the FID count for the air sample and V is the volume of the air sample taken into the sorption tube.

For quantification of individual compounds, the area under a peak in the compound specific SIC was measured. Peaks in SICs were almost always well defined. To measure the total ion counts of individual compounds, the relationship between the single ion and the total ions was established by a regression analysis using values obtained when the area under a peak in the TIC was measurable with accuracy. An example of the relationship between single ion counts and total ion counts is shown in Fig. 4. In most cases, compounds were present in low concentrations and the SIC-TIC linearity was good.

The concentration of individual compounds was then assumed to be proportional to the area under a peak in the TIC and calculated by

$$C_i = \text{TVOC} \times I_i/\Sigma I \tag{4}$$

where C is the concentration in the air, I is the total ion count of a peak in the chromatogram, and the subscript i is for the ith component. ΣI is the total ion count for the entire chromatogram with the baseline value subtracted. The summation is performed on the data system.

This is an approximate method of quantification. For more accuracy, the response factors of individual compounds on the MS are being measured using reference chemicals. The approximate values will be revised using more accurately determined specific response factors.

The quantitative results showed that, in Canadian buildings, the 50th percentile TVOC concentration was approximately 1 mg/m^3. Copying machine solvent vapors were the most prominent group of compounds. Frequently detected compounds were acetone, ethanol (in homes), i-propanol, 1,1,1-trichloroethane, p-dichlorobenzene, tetrachloroethene, trichlorofluoromethane, benzene, toluene, xylenes, normal alkanes C10, C11, C12, and limonene. The lower limit of quantification of a single compound was about 0.3 μg/m^3 using the present method with the data from the authors' instrument. A newer quadrupole type GC/MS instrument had ten times higher sensitivity.

Analyzing a large number of VOCs in an air sample is a voluminous and expensive task. An analysis may not be completed for various reasons such as short time frame, cost, or insufficient technical information. When circumstances allow, further analysis of stored data files is possible. Information on specific single ions and on specific retentions is continually being updated and expanded. When quantification of a previously undetermined compound is necessary, a retroactive analysis is possible using stored data files. An example is 4-phenylcyclohexene, an irritant released from some new carpets, which was unknown five years ago. Today this compound can be detected and quantified in stored files using recent data on its characteristic SIC, specific retention, and SIC-TIC regression.

Conclusion

An analytical method for identifying and quantifying the VOCs in indoor air has been presented. The method comprises a systematic single-ion analysis of TICs. For identification of compounds, specific retentions and specific single ions were used. Quantification was

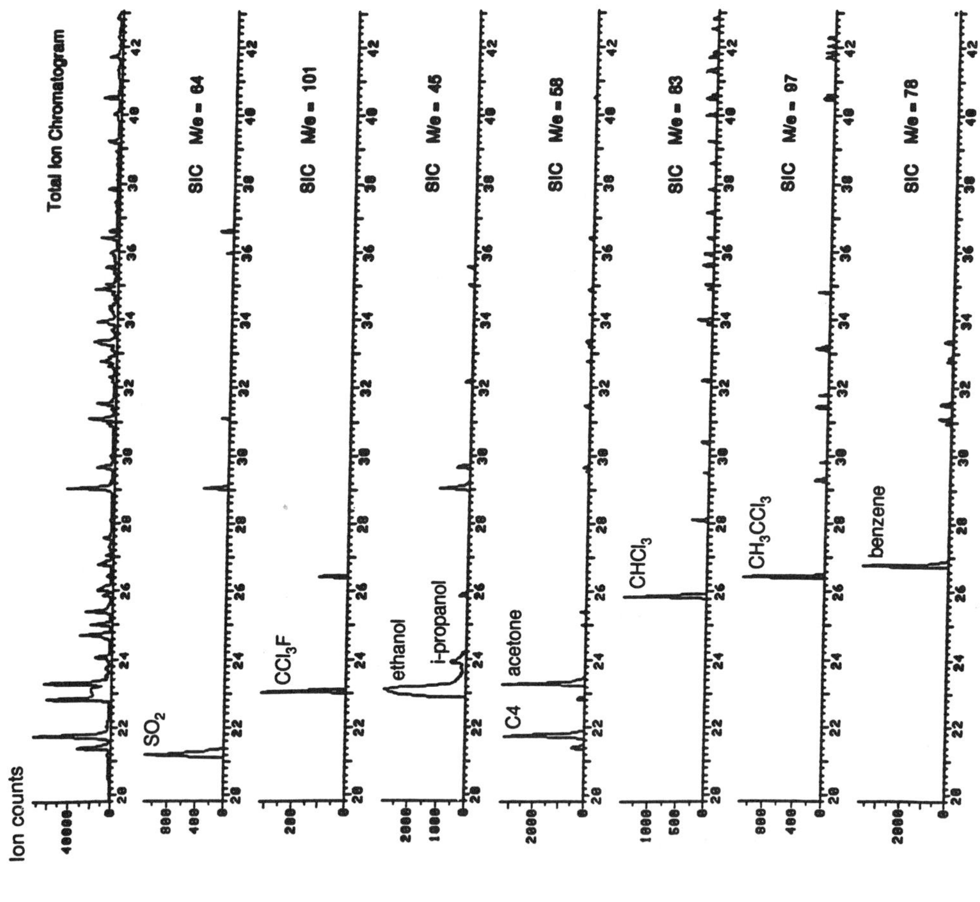
Ion counts
Total Ion Chromatogram
SIC M/e = 64
SO2
SIC M/e = 101
CCl3F
SIC M/e = 45
ethanol
i-propanol
SIC M/e = 58
C4
acetone
SIC M/e = 83
CHCl3
SIC M/e = 97
CH3CCl3
SIC M/e = 78
benzene

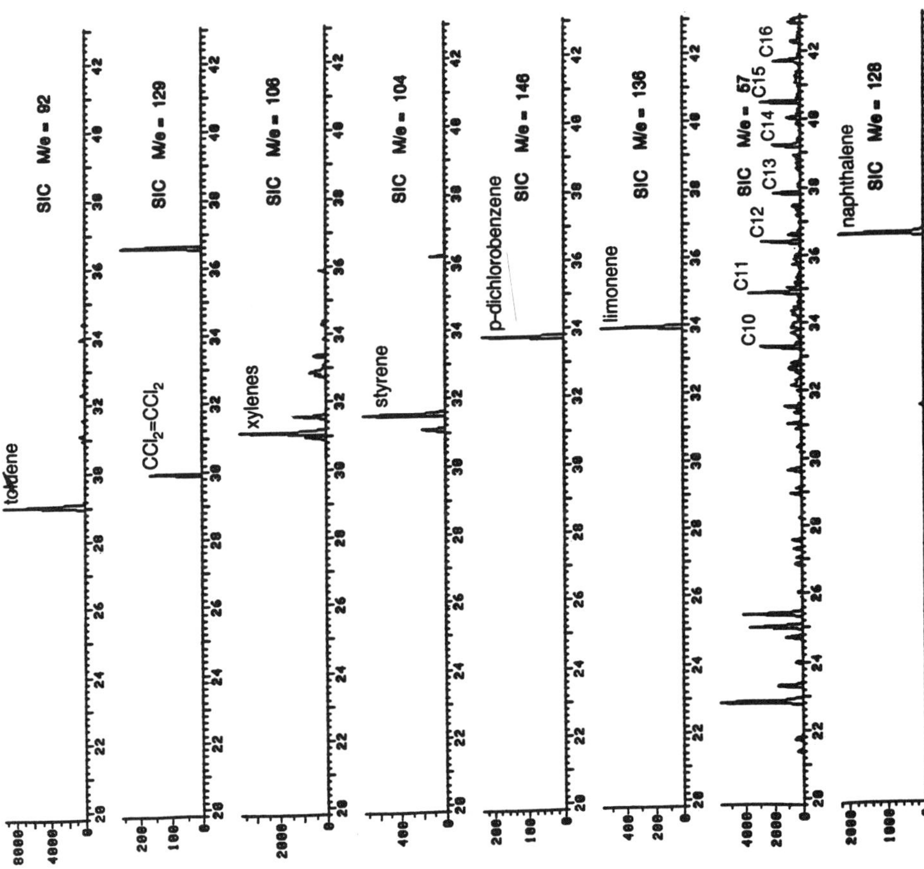

FIG. 3—*An example of a series of single-ion chromatograms.*

TABLE 2—*Significant single ions of VOC.*

M/e	Compounds
30	Formaldehyde
42	$CH_3CH_2CH_2Cl$
43	Acetates, Ketones, C3, C4, $(CH_3)_2CHCl$
44	Acetaldehyde
45	Ethanol, Acetic acid, Propyleneglycol
54	4-Vinylcyclohexene
56	Hexanal
57	n-Alkanes, i-Alkanes, Butoxyethanol
58	Acetone, i-Propanol
60	Methylformate, Acetic acid
61	$CCl_2{=}CH_2$
62	CH_2ClCH_2Cl
63	$CHCl_2CH_3$, $CH_2ClCHClCH_3$
64	C_2H_5Cl, SO_2
65	CH_2CClF_2
67	$CHCl_2F$
69	Methylcyclopentane
72	2-Butanone
73	1,3-Dioxolane
75	$CH_2ClCHClCH_2Cl$
76	$Cl(CH_3)_3Cl$, $CH_2{=}CHCH_2Cl$, CS_2
77	Chlorobenzene, Benzaldhyde
78	Benzene
79	4-Vinylcyclohexene
80	C_2H_2ClF, $CHCl{=}CHF$, $CH_2{=}CClF$
81	CH_3CCl_2F, $CH_2ClCHClF$
82	CH_3CHClF
83	$CHCl_2CHCl_2$, $CHCl_3$, Methylcyclohexane
84	CH_2Cl_2, Cyclohexane
85	CCl_2F_2
91	p-Chlorotoluene, Toluene, p-Xylene
92	Toluene
95	$CCl_2{=}CHCl$
96	$CH_2{=}CCl_2$
97	CH_3CCl_3
101	CCl_3F, CCl_2FCClF_2, CCl_2FCCl_2F
104	Styrene, 4-Phenylcyclohexene
105	Cumene, Trimethylbenzenes
106	Xylenes, Benzaldehyde
107	$BrCH_2CH_2Br$
112	Chlorobenzene
117	CCl_4, CCl_3CH_2Cl, CCl_3CHCl_2
128	Naphthalene
129	$CCl_2{=}CCl_2$
130	$CHCl{=}CCl_2$
133	$CH_2Cl{=}CCl_3$
136	α-Pinene, Camphene, Limonene
146	Dichlorobenzene
151	CCl_2FCClF_2
167	CCl_2FCCl_2F
180	Trichlorobenzenes

TABLE 3—*Doublet ions for identification of halogen compounds.*

Doublet	CFC	Compound	Specific Retention	CAS
39, **41**	1260b	$CH_2=CHCH_2Cl$	16.7	107051
39, 41	1260	$CH_2=CClCH_3$	15.9	
61, 63	1130	$CCl_2=CH_2$	17.09	75354
61, 63		CHCl=CHCl cis	25.4	156592
61, 63		CHCl=CHCl tr		156605
62, 64		CH_2ClCH_2Cl	30.0	107062
63, 65		$CHCl_2CH_3$	22.6	75343
63, 65		$CH_2ClCHClCH_3$	40.0	78875
64, 66		C_2H_5Cl	9.4	
67, 69	21	$CHCl_2F$	11.5	75434
75, 77		$CH_2ClCHClCH_2Cl$		96184
76, 78		$Cl(CH_2)_3Cl$		142289
76, 78	1260b	$CH_2=CHCH_2Cl$	16.7	107051
76, 78	1260	$CH_2=CClCH_3$	15.9	
78, 80		$(CH_3)_2CHCl$		75296
78, 80		$CH_3CH_2CH_2Cl$		540545
80, 82	1131	CHCl=CHF		460162
80, 82	1131a	$CH_2=CClF$	4.5	2317911
81, 83	141	$CH_2ClCHClF$		430579
81, 83	141b	CH_3CCl_2F	15.42	
82, 84	151a	CH_3CHClF	16.1	1615754
83, 85		$CHCl_2CHCl_2$	53.6	79345
83, 85		$CHCl_3$	26.11	
84, 86		CH_2Cl_2	17.3	75092
85, 87	142b	CH_3CClF_2	5.2	75683
95, 97		$CCl_2=CHCl$	33.50	79016
96, 98	1130	$CCl_2=CH_2$	17.09	75354
96, 98		CHCl=CHCl cis	25.4	156592
96, 98		CHCl=CHCl tr		156605
97, 99	143a	CCl_3CH_3	28.53	71556
98, 100		CH_2ClCH_2Cl	30.0	107062
98, 100		$CHCl_2CH_3$	22.6	75343
101, 103	112	CCl_2FCCl_2F	33.9	
101, 103	11	CCl_3F	12.6	75694
101, 103	113	CCl_2FCClF_2	18.21	76131
101, 103	12	CCl_2F_2	4.5	75718
101, 103	141b	CH_3CCl_2F	15.42	
102, 104	21	$CHCl_2F$	11.5	75434
107, 109		$BrCH_2CH_2Br$		106934
110, 112		$CH_2ClCHClCH_2Cl$		96184
112, 114		chlorobenzene	46.7	
112, 114		$CH_2ClCHClCH_3$	40.0	78875
116, 118	141	$CH_2ClCHClF$		430579
117, 119		CCl_3CHCl_2		76017
117, 119		CCl_4	29.2	56235
117, 119		CCl_3CH_2CL		630206
117, 119	11	CCl_3F	13.85	75694
117, 119	143a	CCl_3CH_3	28.53	71556
126, 128		p-chlorotoluene		106434
129, 131		$CCl_2=CCl_2$	43.47	127184
130, 132		$CCl_2=CHCl$	33.50	79016
131, 133		CCl_3CH_2Cl		630206
146, 148		dichlorobenzene	60.00	106467
151, 153	113	CCl_2FCClF_2	18.21	76131
164, **166**		$CCl_2=CCl_2$	43.47	127184
166, 168		$CHCl_2CHCl_2$	53.6	79345
167, 169	112	CCl_2FCCl_2F	33.9	
180, 182		trichlorobenzene	72.9	120821

NOTE: Intense ions are shown in bold. Compounds are repeatedly listed when they have more than one set of doublets.

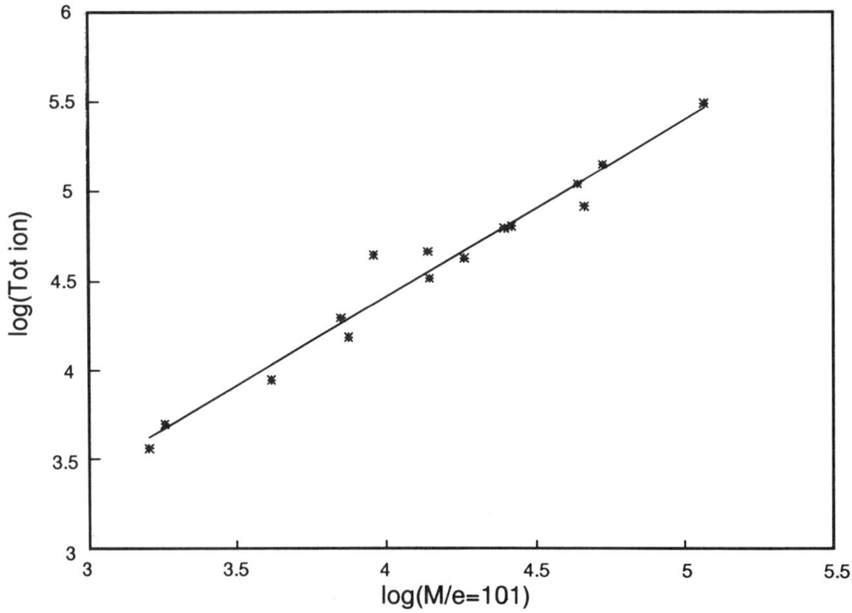

FIG. 4—*Relation between single ion count* ($M/e = 101$) *and total ion count for* CCl_3F, *measured points, and regression line.*

performed by integrating single ion counts on SICs. The advantages of this method are rapidity, reliability, low cost, and capability for retroactive analysis of stored data files.

References

Chan, C. C., Vanier, L., Martin, J. W., and Williams, D. T., 1990, "Determination of Organic Contaminants in Residential Indoor Air Using an Adsorption-Thermal Desorption Technique," *Journal of Air & Waste Managing Association,* Vol. 40, pp. 62–67.

Kovats, E., 1958, "Gas-Chromatographische Characterisierung Organischer Verbindungen," *Helvetica Chimica Acta,* Vol. 41, p. 1915.

Lewis, C. W., Zweidinger, R. B., and Stevens, R. K., 1990, "Apportionment of Residential Indoor Aerosol, VOC and Aldehydes to Indoor and Outdoor Sources," *Indoor Air '90,* Toronto, pp. 195–200.

Needham, L. L., Ashley, D. L., Hill, Jr., R. H., Turner, W. E., Patterson, Jr., D. G., Holler, J. S., and Pirkle, J. L., 1990, "A Program for Assessing Background Levels of 52 Organic Toxicants in the U.S. Population," *Indoor Air '90,* Toronto, pp. 2:453–458.

Otson, R., Fellin, P., and Whitemore, R., 1992, "A National Pilot Study on Occurrence of Airborne VOCs in Residences—Design and Progress" in *Proceedings, Measurements of Toxic and Related Air Pollutants,* 1992 EPA/AWMA International Symposium, VIP-25, Air & Waste Management Association, Pittsburgh, PA, p. 176.

Shields, H. C. and Weschler, C. J., 1990, "Three Years of Indoor/Outdoor VOC Measurements in Neenah, WI," *Indoor Air '90,* Toronto, pp. 2:771–2:776.

Tsuchiya, Y. and Kanabus-Kaminska, J. M., 1992, "Volatile Organic Compounds in Canadian Indoor Air," *Proceedings,* 5th International Jacques Cartier Conference, Indoor Air Quality, Ventilation and Energy Conservation in Buildings, Montreal, pp. 131–138.

Katie Pratt,[1] *Robert Shirey,*[2] *and Venkatachalam Mani*[2]

Solid-Phase Microextraction of VOCs in Water

REFERENCE: Pratt, K., Shirey, R., and Mani, V., "**Solid-Phase Microextraction of VOCs in Water,**" *Volatile Organic Compounds in the Environment, ASTM STP 1261,* W. Wang, J. Schnoor, and J. Doi, Eds., American Society for Testing and Materials, Philadelphia, 1996, pp. 139–146.

ABSTRACT: The measurement of very low concentrations of organic compounds in the environment has been a subject of research for many years. Recently, the sample preparation in the analysis of aqueous samples has been achieved by solid phase microextraction (SPME). This method has been shown to be fast, inexpensive, solventless, portable and automatable. SPME has several advantages over conventional liquid-liquid extraction. SPME has been shown to be a quantitative technique for volatile and semivolatile compounds from gaseous and liquid samples. The quantization by SPME is also linear over four orders of magnitude. As this method can reach a detection limit of 15 ppt (parts per trillion) for several compounds and can be easily automated, the analysis of volatile organic compounds (VOCs) in the environment, in particular, water samples, is highly advantageous by this method. This study will describe the analysis of volatile organics, BTEX, and halogenated organics in water. The lowest limit of detection, linearity, and other parameters will be discussed. The analysis of a soil sample taken near an oil storage tank for VOCs by SPME will be described. A comparison between different fibers (different types of coating materials and film thickness) using different columns will be described in this work.

KEYWORDS: volatile organic compounds (VOCs), organic compounds, volatile organics, halogenated organics, silica fiber, environment

In the analysis of organic compounds in water samples, sample preparation usually consists of purge-and-trap or headspace methods for concentrating volatiles, and liquid-liquid or solid-phase extraction for semivolatiles and nonvolatiles. These methods have various drawbacks, including cost and excessive preparation time and training. A new sample preparation technique has been developed at the University of Waterloo, Ontario, Canada, by Dr. Janusz Pawliszyn and associates. Solid-phase microextraction[3] (SPME) eliminates most of the drawbacks to preparation of water samples (Arthur et al. 1992a, 1992b, 1992c, 1993). SPME requires no solvents or complicated apparatus, can be used to concentrate volatile and semivolatile compounds in both liquid and gaseous samples, provides linear results over a wide range of analyte concentrations (down to parts per trillion for some compounds), and can be used with any gas chromatograph (GC) or gas chromatograph/mass spectrometer (GC/MS). The technique has the potential to meet detection limits specified by the U.S. Environmental Protection Agency (EPA) Methods and the Ontario Municipal/Industrial Strategy for Abatement (MISA) program.

[1] Chemist, Supelco, Canada, 1300 Aimco Boulevard, Mississauga, ON L4W5H5, Canada.
[2] Senior research chemists, Supelco, Inc., Supelco Park, Bellefonte, PA 16823.
[3] Patent pending.

In SPME, the analytes establish an equilibrium between the sample matrix and a stationary phase coated on a fused silica fiber. The fiber, containing the absorbed analytes, is thermally desorbed in the injection port of the GC, which transfers the analytes into the capillary column. Because no solvent is injected and the analytes are rapidly desorbed onto the column, narrow-bore capillary columns can be used, which helps to lower minimum detection limits.

Experimental Details

Preparation of Standards and Samples

A standard stock solution containing 31 VOCs listed in U.S. EPA method 624 was prepared at 50 ng/μL each in methanol. For analysis using direct immersion of the SPME fiber, the stock solution was spiked to a desired concentration into a capped 2 mL vial containing either 1.8 mL of saturated saltwater (28% sodium chloride) or distilled water and a mini-stirring bar. An internal standard mixture containing three compounds at 100 ng/μL in methanol was spiked into the sample to a concentration of 100 ppb. For preparation of samples for headspace sampling, the samples were similarly prepared with the exception of the volume of 1.5 mL.

Extraction and Analysis of VOCs

The analytes were extracted by inserting the 100 μm polydimethylsiloxane (PDMS) SPME fiber and holder (Supelco, Inc.) into the sample vial and exposing the fiber either to the headspace above the sample or directly into the water. The fiber was exposed to the rapidly stirred sample for 10 min. The fiber was removed from the vial and inserted into the injection port for desorption of the analytes.

The analytes were desorbed in a splitless/split injection port with the temperature set at 230°C and transferred into a 60-m × 0.25-mm × 1.5-μm VOCOL capillary column (Supelco, Inc.) installed in an INCOS 50 GC/MS/DS (Finnigan MAT Inc., San Jose, California). The column was directly interfaced with the ion source of the mass spectrometer. Helium was used as a carrier gas, at a linear velocity of 40 cm/s at 35°C. The initial oven temperature, 35°C was held for 2 min then programmed to 200°C at 5°C/min. The mass scan range was 45 to 260 at 0.6 s/scan.

Results and Discussion

Design of SPME Holder

To make SPME practical, a simple-to-use device to insert the fiber through a septa was needed. By designing a syringe-like holder that could retract and expose the fiber, the fiber can be easily exposed to samples and thermally desorbed in the injection port (Fig. 1). The holder and fiber assembly also have other design inputs to enhance precision and to make it easier to use. An adjustable depth gage controls the insertion distance of the fiber in the vial and injection port. A Z-lock is provided to keep the fiber in the exposed position for sampling and desorption. Figure 2 shows a typical extraction and desorption by SPME.

Theory and Factors Affecting SPME

In SPME, the equilibrium process is between the concentration of the analytes in solution and the concentration in the coating on the fused silica fiber. The amount of analyte absorbed

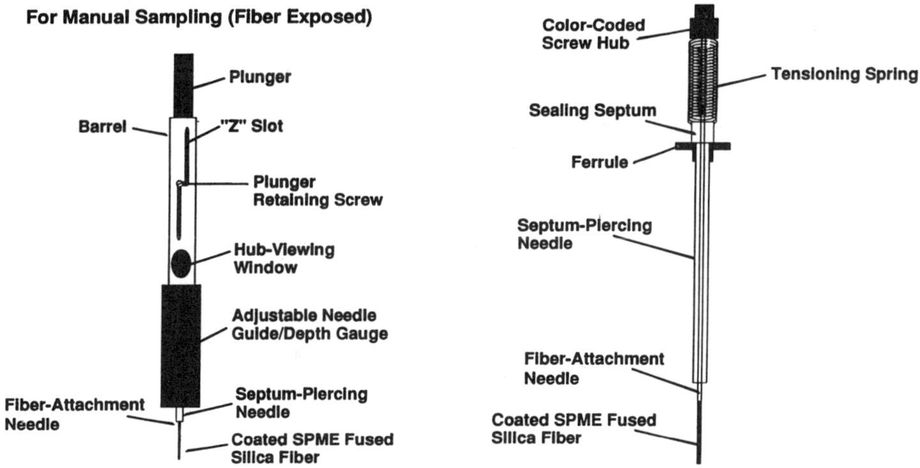

FIG. 1—*Solid-phase microextraction apparatus and fiber assembly.*

by the fiber depends on the fiber coating thickness, the distribution constant (the time to reach equilibrium) of the analyte, and the time that the fiber is exposed to the sample. In general, the distribution constants increase with increasing molecular weight and boiling point of the analytes. Analytes with low distribution constants are best retained by thick fiber coatings and shown by the work of Arthur et al. (1992a, 1992b), who based their modeling on BTEX analytes. The latter demonstrated that semivolatiles were better extracted with thin fiber coatings. The thinner coatings greatly reduced the time for equilibration (Arthur 1993). The thinner phase coatings on the fibers can also be bonded, allowing higher desorption temperatures and reducing carryover. The extraction of polar analytes can be enhanced by using a more polar fiber coating such as polyacrylate (Buchholz and Pawliszyn 1994). This work also demonstrated that modifications to the sample such as the addition of salt and pH adjustments can enhance extraction efficiency.

Extraction can be accomplished either by direct sampling (in the water) or by headspace sampling (above the sample). Equilibrium is obtained more rapidly in headspace than in

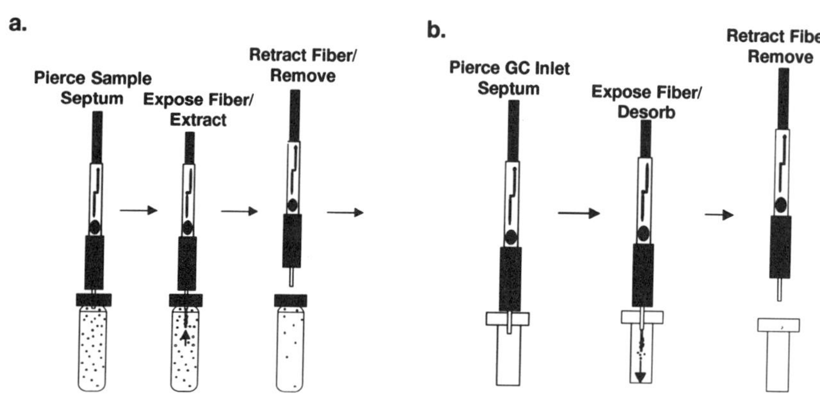

FIG. 2—*Typical extraction and desorption scheme for SPME.*

water (Zhouyao and Pawliszyn 1993). The sample matrix, the type of analytes, and the speed required for sampling can affect the decision as to the type of sampling mode.

To obtain high accuracy and precision, full equilibrium is not necessary, but consistent timing parameters are essential. Extraction time is determined by the time required to obtain good precision of the analyte with the highest distribution constant. It is also important to keep vial size, sample volume, and depth of the fiber in the sample and injection port consistent.

Optimization of Extraction

Based upon the theory and experiments of Arthur et al. (1992c), a 100 μm PDMS fiber was selected to extract the selected VOCs. Most of the analytes are at equilibrium within 10 min and the fiber has the capacity to extract analytes at concentrations between 20 and 200 ppb as recommended in U.S. EPA Method 624.

An experiment was designed to determine the better sampling mode (headspace or direct immersion) and the effect of adding salt to the sample on extraction efficiency. Table 1 lists the ion counts for each analyte obtained using the two sampling types with and without salt

TABLE 1—*Effect of salt and sampling type on extraction of 624 analytes.*

	Headspace Saturated Salt	Direct Saturated Salt	Headspace No Salt	Direct No Salt
Chloromethane	0	109	0	0
Vinyl chloride	0	1166	15	67
Bromomethane	0	106	0	0
Chloroethane	0	583	0	0
Trichlorofluoromethane	384	7242	1072	4294
1,1-dichloroethene	188	2942	538	897
Methylene chloride	621	1615	248	244
1,2-dichloroethene	369	3026	442	583
1,1-dichloroethene	1137	4839	923	949
Chloroform	1547	1945	953	958
Tetrachloroethane	2592	9020	3647	4690
Carbon tetrachloride	2079	7759	4226	6246
Benzene	5423	17531	4195	4262
1,2-dichloroethane	1206	2211	513	518
Trichloroethene	2200	5749	2643	2761
1,2-dichloropropane	1880	3820	835	865
Bromodichloromethane	2511	4613	1102	940
2-chloroethylvinyl ether	270	1091	0	0
cis-1,3-dichloropropene	4156	7084	1411	1560
Toluene	15533	26756	14477	13843
trans-1,3-dichloropropene	4156	6348	1171	1192
1,1,2-trichloroethane	1691	2198	502	455
Tetrachloroethene	4774	6313	9785	6640
Dibromochloromethane	2720	3354	770	863
Chlorobenzene	16473	17086	11383	9791
Ethyl benzene	39891	33984	51475	31510
Bromoform	3635	3576	1155	1007
1,1,2,2-tetrachloroethane	10421	7961	2274	2039
1,3-dichlorobenzene	41139	15307	39237	15707
1,4-dichlorobenzene	43745	16198	36658	18462
1,2-dichlorobenzene	41866	14060	29200	14164

(ions monitored for quantitation are listed in Fig. 3). For most of the analytes, direct immersion of the fiber and the addition of salt produced the best results.

The addition of salt enhanced the recovery for all of the analytes. This was not unexpected since the relatively nonpolar analytes become less soluble with the addition of salt.

The results between headspace sampling and direct sampling were more difficult to explain. It was expected to see better results for headspace sampling for the six less volatile compounds (the bottom six compounds in Table 1). These analytes do not reach full equilibrium by direct immersion in 10 min, while in headspace, equilibrium was more rapidly obtained. When the analytes were extracted for 15 min, the discrepancy was reduced. The lower response for the more volatile analytes using headspace was not expected.

The competition from water helps slow the rate that the analytes migrate in and out of the fiber. For the more volatile analytes, this slower rate was needed to retain the analytes. In addition, since less sample was placed in the vials for headspace sampling, there was a dilution effect that reduced analyte recovery.

The Effects of Band Broadening

There are two major sources of band broadening in gas chromatography. The first source is the injection port. The broadening is due to the rate that the sample takes to enter the column. When a solvent is injected, the broadening is the result of ineffective refocusing of the solvent on the analytical column.

In SPME there is no solvent so the cause for broadening is the time that the analyte takes to elute from the fiber and enter the column. Arthur et al. stated in their original work that cryogenics was required to analyze VOCs. Our experimentation demonstrates that cryogenic focusing is not a requirement. Figure 3 shows a typical splitless liner with a 2-mm bore liner compared with a 0.75-mm bore liner. By reducing the inlet volume, the analytes can be

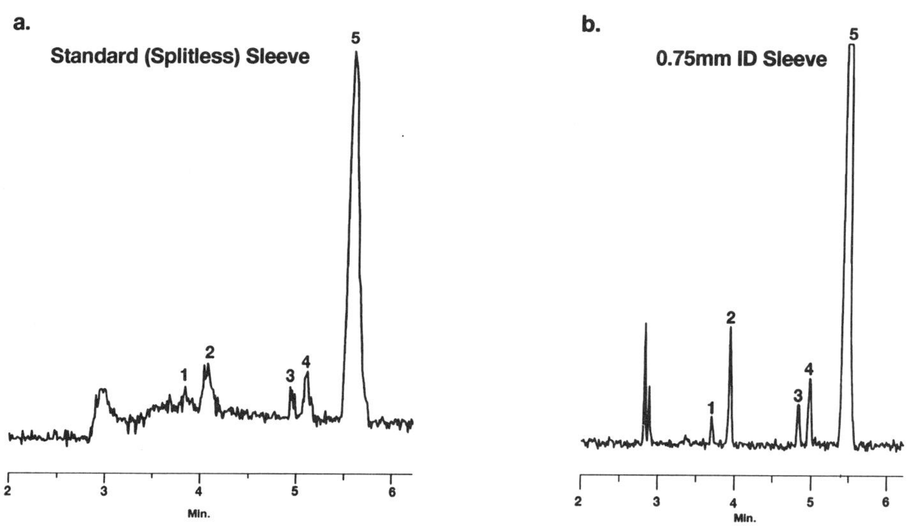

FIG. 3—*Effects of inlet sleeve inside diameter on SPME. (a) Standard 2-mm-ID splitless liner, (b) 0.75-mm-ID splitless liner. The column was a 60 m × 0.25 mm × 1.5 μm VOCOL set at a linear velocity of 40 cm/sec. The column temperature was 35°C isothermal. Each analyte at 50 ppb; (1) chloromethane, (2) vinyl chloride, (3) bromomethane, (4) chloroethane, and (5) trichlorofluoromethane.*

swept more readily onto the column. The more volatile analytes such as gaseous VOCs readily eluted off the fiber and the less volatile analytes that eluted more slowly off the fiber were focused on the column by the temperature program. This resulted in sharper, taller peaks that could be more easily quantified and resulted in lower detection and quantitation limits (see Fig. 4). In this figure all of the method 624 analytes were detected and easily quantified.

The second source of band broadening is the capillary column. A more narrow column reduces peak width. The other advantage for a narrow column is the low carrier gas flow rate. When used with mass spectrometers, jet separators are not required. This greatly improves sensitivity and eliminates band broadening from the jet separator.

Linearity and Precision

Table 2 shows the average of the response factors for each Method 624 analyte obtained from a concentration range of 10 to 200 ppb. Since response factors incorporate concentration into the formula, the %RSD is a measurement of the linearity. The %RSD for each analyte

	Compounds	m/z
1.	Chloromethane	50
2.	Vinyl chloride	62
3.	Bromomethane	94
4.	Chloroethane	64
5.	Trichlorofluoromethane	101
6.	1,1-Dichloroethene	96
7.	Methylene chloride	84
8.	1,2-Dichloroethene	96
9.	1,1-Dichloroethene	63
10.	Chloroform	83
IS	Bromochloromethane (int. std.)	130
11.	Tetrachloroethane	97
12.	Carbon tetrachloride	117
13.	Benzene	78
14.	1,2-Dichloroethane	62
15.	Trichloroethene	95
16.	1,2-Dichloropropane	63
17.	Bromodichloromethane	83
18.	2-Chloroethylvinyl ether	63
19.	cis-1,3-Dichloropropene	75
20.	Toluene	91
21.	trans-1,3-Dichloropropene	75
22.	1,1,2-Trichloroethane	83
IS	1-Bromo-2-chloropropane (int. std.)	77
23.	Tetrachloroethene	166
24.	Dibromochloromethane	129
25.	Chlorobenzene	112
26.	Ethyl benzene	91
27.	Bromoform	173
IS	Dichlorobutane (int. std.)	55
28.	1,1,2,2-Tetrachloroethane	83
29.	1,3-Dichlorobenzene	146
30.	1,4-Dichlorobenzene	146
31.	1,2-Dichlorobenzene	146

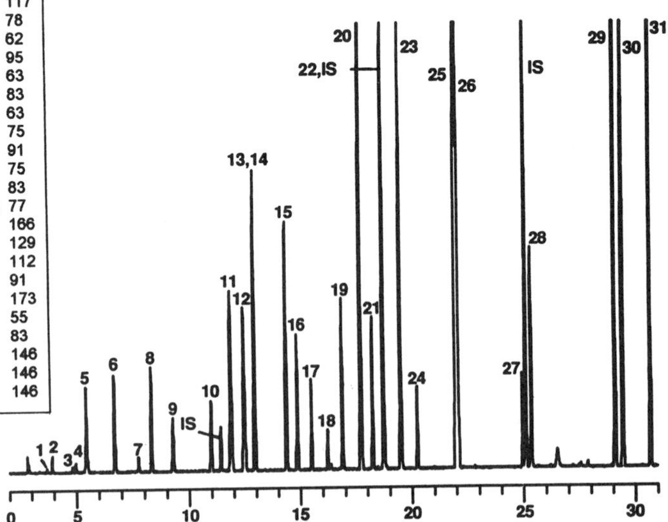

FIG. 4—*U.S. EPA Method 624 volatile analytes at 50 ppb by SPME. The column was a 60 m × 0.25 mm × 1.5 μm VOCOL set at 35°C, held for 2 min, and then programmed to 200°C at 5°C/min. The linear velocity was 40 cm/s at 35°C. The analytes and ions monitored are as shown in the box.*

TABLE 2—*Response factors for 624 analytes by SPME 100 μm PDMS fiber.*

	Response Factors, ppb									
	10	20	25	50	100	150	200	Avg.	S.D.[a]	%RSD
Chloromethane	0.45	0.27	0.47	0.45	0.38	0.37	0.44	0.40	0.07	16.4
Vinyl chloride	1.46	1.48	1.68	1.81	1.99	1.52	2.15	1.72	0.25	14.4
Bromomethane	0.48	0.36	0.46	0.43	0.50	0.52	0.51	0.47	0.05	11.1
Chloroethane	1.33	1.11	1.08	1.16	1.28	1.25	1.20	1.20	0.08	6.9
Trichlorofluoromethane	10.76	11.13	10.45	12.57	9.06	8.02	10.05	10.29	1.36	13.2
1,1-dichloroethene	5.19	5.25	4.55	5.46	4.55	4.45	4.42	4.84	0.41	8.5
Methylene chloride	0.79	0.78	0.81	0.83	1.00	0.88	0.90	0.86	0.07	8.7
1,2-dichloroethene	5.33	4.46	3.90	4.90	4.32	4.30	4.18	4.48	0.44	9.9
1,1-dichloroethene	8.83	8.96	8.31	9.07	8.48	8.16	8.18	8.57	0.35	4.1
Chloroform	9.36	8.94	7.83	9.11	8.86	8.29	7.94	8.62	0.55	6.4
Bromochloromethane (IS)										
Tetrachloroethane	17.23	19.00	17.26	19.36	14.83	17.82	14.43	17.13	1.76	10.3
Carbon tetrachloride	0.78	0.98	0.95	1.03	0.88	1.05	0.87	0.93	0.09	9.6
Benzene	1.64	1.94	1.86	1.84	1.70	2.01	1.70	1.81	0.13	7.0
1,2-dichloroethane	0.23	0.24	0.25	0.25	0.27	0.27	0.25	0.25	0.01	5.0
Trichloroethene	0.61	0.79	0.67	0.72	0.63	0.83	0.58	0.69	0.09	12.4
1,2-dichloropropane	0.44	0.51	0.47	0.44	0.45	0.47	0.44	0.46	0.03	5.5
Bromodichloromethane	0.45	0.51	0.51	0.49	0.53	0.54	0.51	0.51	0.03	5.2
2-chloroethylvinyl ether	0.17	0.17	0.19	0.17	0.20	0.19	0.19	0.18	0.01	6.8
cis-1,3-dichloropropene	0.72	0.84	0.81	0.82	0.85	0.85	0.81	0.81	0.04	5.3
Toluene	2.82	3.92	2.97	3.65	2.92	3.60	2.97	3.26	0.41	12.6
trans-1,3-dichloropropene	0.58	0.71	0.71	0.72	0.76	0.75	0.73	0.71	0.05	7.7
1,1,2-trichloroethane	0.27	0.30	0.28	0.25	0.28	0.27	0.28	0.28	0.01	5.2
1-bromo-2-chloropropane (IS)										
Tetrachloroethene	0.86	0.81	0.82	0.92	0.84	1.15	0.76	0.88	0.12	13.4
Dibromochloromethane	0.19	0.15	0.18	0.18	0.20	0.18	0.20	0.18	0.02	9.2
Chlorobenzene	0.99	1.10	0.93	1.21	0.98	1.11	0.94	1.04	0.10	9.4
Ethyl benzene	2.45	2.66	2.15	3.00	2.17	2.65	2.23	2.47	0.29	11.8
Bromoform	0.18	0.15	0.19	0.18	0.20	0.19	0.21	0.18	0.02	10.7
Dichlorobutane (IS)										
1,1,2,2-tetrachloroethane	0.49	0.40	0.46	0.45	0.48	0.44	0.48	0.46	0.03	6.3
1,3-dichlorobenzene	1.05	1.23	0.92	1.31	1.01	1.12	0.93	1.08	0.14	12.6
1,4-dichlorobenzene	1.08	1.31	0.94	1.37	1.05	1.16	0.98	1.13	0.15	13.3
1,2-dichlorobenzene	0.99	1.17	0.92	1.28	1.00	1.04	0.94	1.05	0.12	11.3

[a] S.D. = standard deviation.

was under 15%, which is within EPA requirements. Only chloromethane was slightly higher at 16% RSD. This deviation decreased to under 10% when the average was calculated from 20 ppb instead of 10 ppb.

Reproducibility or precision is also a requirement of the EPA methodology. Table 3 shows the average response factors from six extractions using three separate 100 μm PDMS fibers at 50 ppb. The specification in the EPA methodology for these analytes states that the error of precision be under 15% RSD. The results indicate that SPME meets these requirements.

Conclusions

SPME is a simple, practical, and cost-effective technique for extracting volatile compounds from water, while meeting U.S. EPA detection limit and linearity. No solvents are required in the operation. The analysis can be enhanced by using smaller volume inlet liners and by modifying the sample with salt. It was also demonstrated that direct immersion of the fiber was favorable to headspace for the analysis of U.S. EPA Method 624 analytes.

TABLE 3—*Response factors for Method 624 VOCs at 50 ppb by SPME, using a 100 μm PDMS fiber.*

	$n = 6$ Response Factor Range		$n = 6$ Response Factors		
	Min.	Max.	Avg.	S.D.	%RSD
Chloromethane	0.33	0.41	0.37	0.03	6.9
Vinyl chloride	1.68	2.01	1.87	0.11	5.8
Bromomethane	0.21	0.40	0.30	0.06	18.0
Chloroethane	1.08	1.47	1.20	0.14	11.8
Trichlorofluoromethane	12.9	18.1	15.4	2.21	14.5
1,1-dichloroethene	5.16	6.58	5.67	0.48	8.4
Methylene chloride	0.54	1.02	0.72	0.17	15.1
1,2-dichloroethene	3.80	4.45	4.18	0.23	5.6
1,1-dichloroethane	9.06	11.0	10.1	0.64	6.4
Chloroform	8.31	10.0	9.40	0.59	6.3
Tetrachloroethane	21.1	26.3	24.7	1.93	7.8
Carbon tetrachloride	1.01	1.20	1.09	0.07	6.5
Benzene	1.56	1.84	1.67	0.10	6.1
1,2-dichloroethane	0.18	0.22	0.19	0.02	8.1
Trichloroethene	0.62	0.74	0.69	0.04	5.9
1,2-dichloropropane	0.42	0.46	0.44	0.02	3.5
Bromodichloromethane	0.41	0.49	0.44	0.03	6.8
2-chloroethylvinyl ether	0.11	0.16	0.13	0.02	12.5
cis-1,3-dichloropropene	0.65	0.73	0.68	0.03	4.2
Toluene	2.71	3.60	3.14	0.30	9.5
trans-1,3-dichloropropene	0.50	0.62	0.54	0.05	8.4
1,1,2-trichloroethane	0.23	0.25	0.24	0.01	3.0
Tetrachloroethene	0.77	1.08	0.90	0.10	11.4
Dibromochloromethane	0.15	0.19	0.17	0.01	7.5
Chlorobenzene	0.89	1.08	0.99	0.07	7.5
Ethyl benzene	1.88	2.66	2.17	0.26	12.2
Bromoform	0.15	0.18	0.17	0.01	5.7
1,1,2,2-tetrachloroethane	0.42	0.50	0.47	0.03	6.5
1,3-dichlorobenzene	0.84	1.07	0.95	0.08	7.9
1,4-dichlorobenzene	0.88	1.07	0.98	0.06	6.4
1,2-dichlorobenzene	0.86	1.07	0.97	0.07	7.1

References

Arthur, C. L., Potter, D. W., Buchholz, K. D., Motlagh, S., and Pawliszyn, J., 1992a, *LC/GC,* Vol. 10, No. 9, pp. 656–661.

Arthur, C. L., Killam, L. M., Motlagh, S., Lim, M., Potter, D. W., and Pawliszyn, J., 1992b, *Environmental Science & Technology,* Vol. 26, pp. 979–983.

Arthur, C. L., Pratt, K., Motlagh, S., and Pawliszyn, J., *Journal of High Resolution Chromatography,* Vol. 15, pp. 741–744.

Arthur, C., Killam, L., Buchholz, K. Potter, D., Chai, M., Zhang, Z., and Pawliszyn, J., 1993, *Environmental Laboratory,* pp. 10–14.

Buchholz, K. and Pawliszyn, J., 1994, *Analytical Chemistry,* Vol. 66, pp. 160–167.

Zhouyao, Z. and Pawliszyn, J., *Analytical Chemistry,* Vol. 65, No. 14, July 15, pp. 1843–1852.

Christopher G. Rabideau[1] *and Kenneth R. Comey III*[1]

Estimating VOC Emissions from Waste Water Treatment Units Using a Flow-Through Flux Chamber and Chemdat7

REFERENCE: Rabideau, C. G. and Comey III, K. R., **"Estimating VOC Emissions from Waste Water Treatment Units Using a Flow-Through Flux Chamber and Chemdat7,"** *Volatile Organic Compounds in the Environment, ASTM STP 1261,* W. Wang, J. Schnoor, and J. Doi, Eds., American Society for Testing and Materials, 1996, pp. 147–155.

ABSTRACT: There is increasing concern in the petroleum industry for a practical method of accurately quantifying refinery volatile organic compound (VOC) emissions. The U.S. Environmental Protection Agency (EPA) developed the Chemdat7 emission rate model to predict VOC emissions from surface impoundments, waste-water treatment plants, and land farms for regulatory purposes. Previous studies have indicated that Chemdat7 significantly overpredicted VOC emissions. Much of the model evaluation of Chemdat7 used the EPA isolation environmental flux chamber. There are some concerns in using the EPA isolation flux chamber for measurement of fugitive emissions from surface impoundments due to its effect on ambient wind velocity, temperature, and solar radiation. This paper discusses an alternative method for quantifying VOC emissions from waste-water treatment units using a flow-through environmental flux chamber, and includes a comparison of the results from the flow-through chamber with that of the Chemdat7 model.

KEYWORDS: flux chamber, surface impoundment, Chemdat7, volatile organic compounds, emission estimation

The assessment of volatile organic compound (VOC) emissions is essential for the development of state implementation plans for the control of atmospheric ozone, and the assessment of toxic air pollutant emissions is essential for the development of toxic air emission control strategies. Additionally, this information is basic to the review of Prevention of Significant Deterioration applications, SARA Title III reports, and other Federal, state, and local agency programs involving the assessment of air pollution (EPA 1990).

Accurate, quantitative measurement of VOC emissions from surface impoundments is an important factor in meeting requirements for air toxic emissions. Estimating emissions using the U.S. Environmental Protection Agency (EPA) approved estimating techniques is more economical compared with physical measurements of the contaminants. With the enactment of the 1990 Clean Air Act Amendments, there is increasing economic concern in accurately quantifying emissions from waste-water treatment units.

Background

Historically, flux measurements from surface impoundments have been measured using isolation flux chambers. Isolation flux chambers consist of an enclosed vessel that covers a

[1] Product atmospheric scientist and chemical engineer, respectively, Texaco EPTD, 5801 South Rice, Bellaire, TX 77401.

relatively small area of a surface impoundment. Clean, dry sweep air is added to the chamber at a controlled rate and the concentration of the species of interest is measured at the outlet (EPA 1986). Though isolation flux chambers have been tested in some EPA studies, there are questions concerning their accuracy. For example, the influence of surface wind speed-induced liquid-phase turbulence on the emissions from water surfaces is unknown. The isolation flux chamber may alter the atmospheric conditions at the sampling location and influence the true emission rate (American Petroleum Institute 1990, 1993, Eklund et al. 1985, EPA 1979). Isolation flux chambers do not consider the potential changes in the local atmospheric conditions within the chamber, which may affect the transfer rates of the volatile organic compounds to the atmosphere. The sweep airflow rate may not generate the same type of flow conditions over the surface of the impoundment that would exist if the isolation flux chamber were not present. The isolation flux chamber and sweep air may also alter the air temperature and will affect the humidity of the air above the impoundment. A flow-through flux chamber was designed to investigate the effects of wind speed, temperature, and humidity on emissions from a surface impoundment.

The flow-through flux chamber was designed to obtain a suitable data set for evaluating the Chemdat7 emission rate model (EPA 1989) which is used for estimating emissions from hazardous waste treatment, storage, and disposal facilities. It is important to evaluate the capabilities of Chemdat7 to accurately predict emissions from hazardous waste sites since model-generated emission estimates are much more economical and expedient compared with direct physical measurements of the contaminants. For this paper, the flow-through flux chamber experiment was used to assess the capability of Chemdat7 to predict benzene, toluene, ethylbenzene, and xylene (BTEX) emissions from several simulated treatment units, including a waste-water disposal unit, a nonaerated flow-through waste-water unit, and a diffused air waste-water basin.

Apparatus

Emission rate measurements from the different waste-water units were made using a flow-through flux chamber system (Figs. 1 and 2). The entire system consists of a recirculation loop that passes through the flux chamber, with a continuous sample extraction system and a continuous volume system that provides humidity controls. A bypass loop and control valves determine system pressure and flow rate, while temperature is controlled by the flux chamber itself. The flux chamber consists of a stainless steel rectangular body with a polymethylmethacrylate (Plexiglas) front panel. The flux chamber is 157 cm long by 15.2 cm wide by 12.7 cm tall. The opening over the impoundment is 86.4 cm long by 15.2 cm wide. At this opening is a 5.1 cm stainless steel flange which penetrates the surface impoundment and seals any air intrusion from the surrounding surface impoundment area. The surface area of the chamber opening over the impoundment is 0.13 m^2. Ports for measuring air velocity are located on the top and the sides of the flux chamber. The velocity ports are sealed when not measuring velocities.

Ductwork is connected at each end of the chamber to a blower and a recirculation loop. This enables the wind speed to vary between 0.5 and 4.3 m/s. The chamber sweep air may also go through a humidifier to control the relative humidity inside the chamber. Chamber sweep air flow rates vary between 15 and 37 L/min, which results in residence times between 6 and 15 min. The residence times are a function of the entire flow-through flux chamber system volume. All provisions were made in the flow-through flux chamber to accurately simulate the ambient meteorological conditions inside the chamber (wind speed, temperature, and humidity), so that the local environment of the waste-water unit is not affected. For the

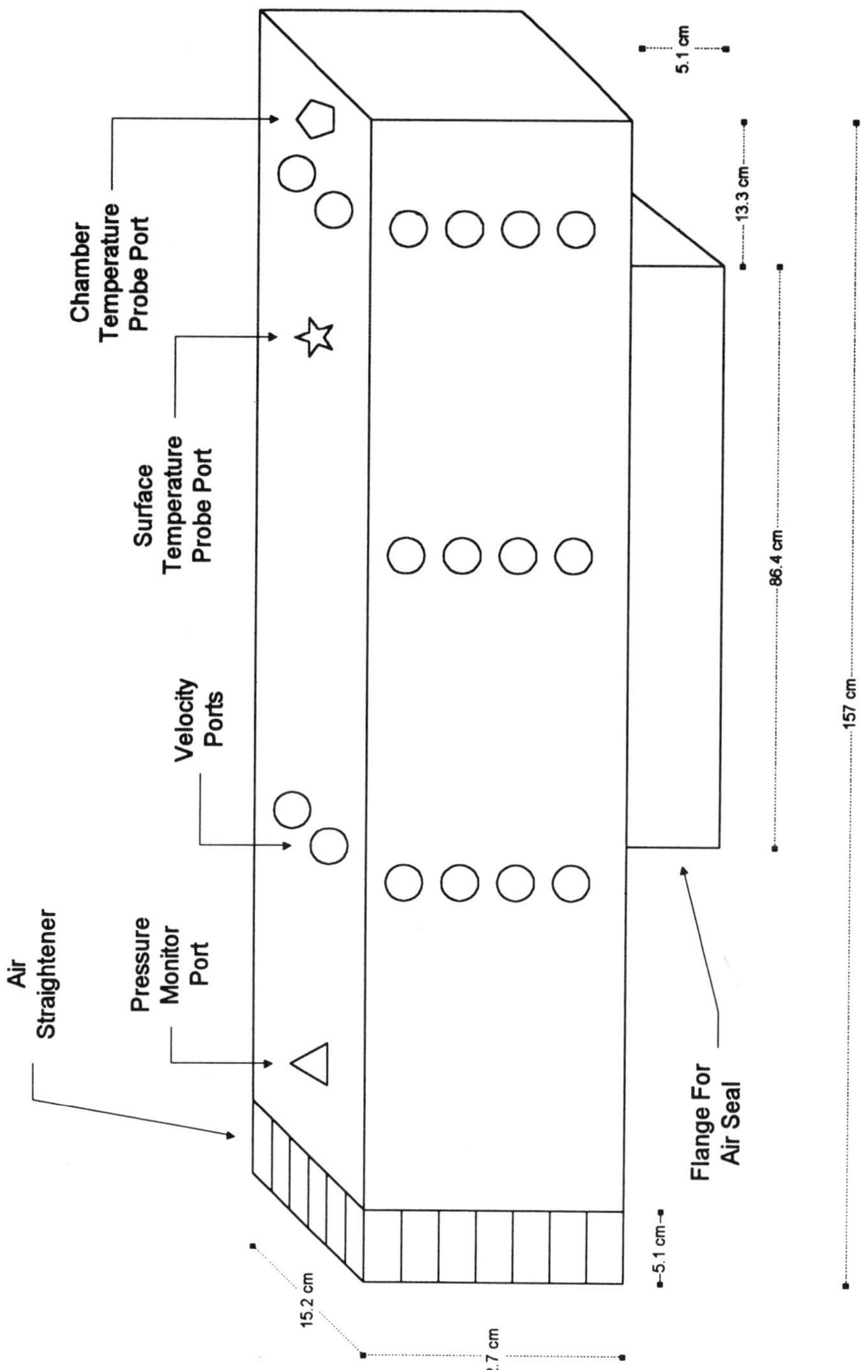

FIG. 1—*Flow-through flux chamber.*

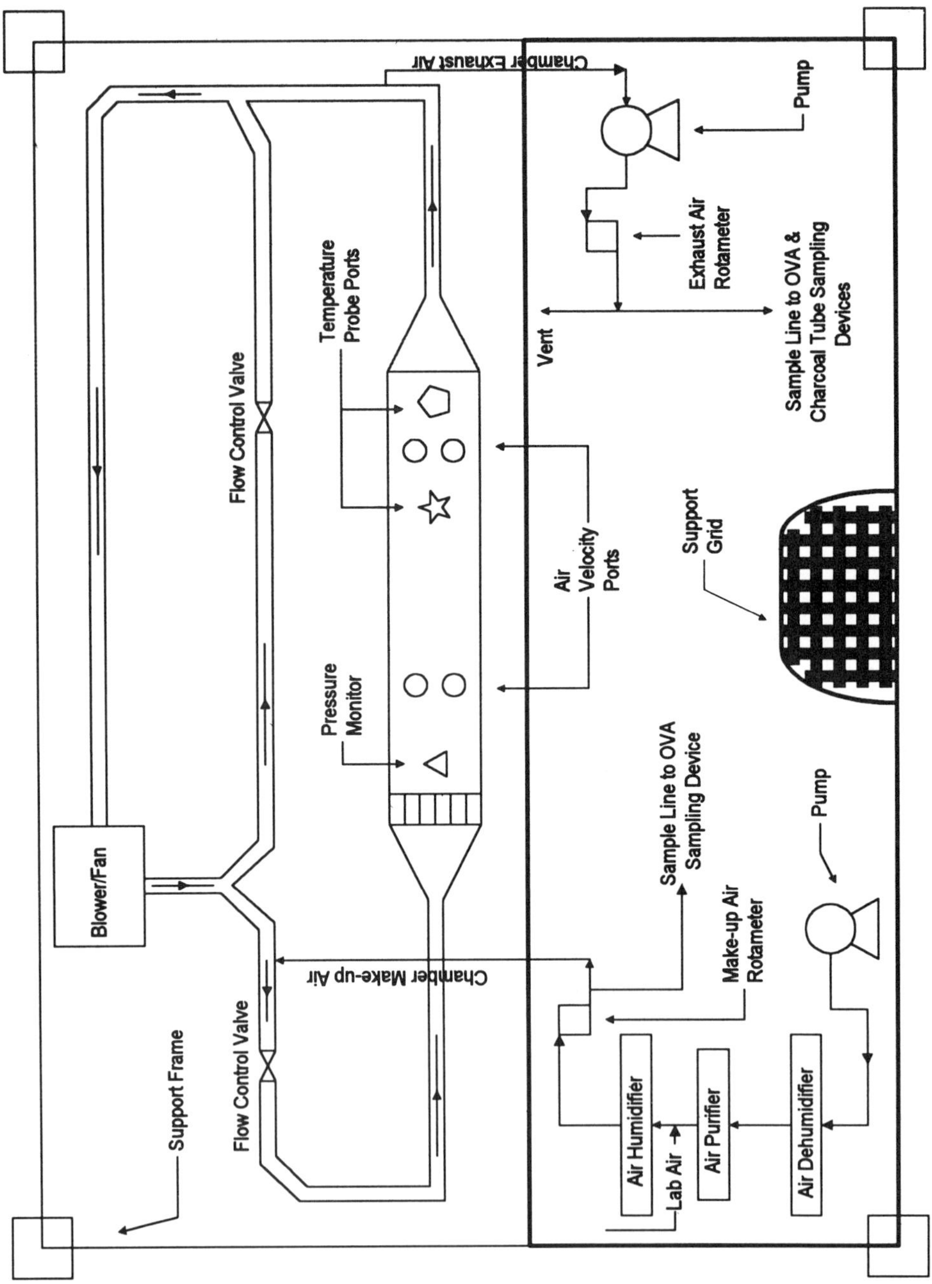

FIG. 2—*Flow-through flux chamber equipment plot plan.*

waste-water disposal unit simulations, the bench-scale basin was 0.08 m deep. For the diffused air unit simulations, the bench-scale basin was 0.3 m deep.

Operating Procedures

The flow-through flux chamber setup begins with starting the blower/fan and the organic vapor analyzer (OVA). The chamber is set onto the empty basin (pan). The chamber flange sets tightly onto the basin sides. All of the sides around the basin are taped to ensure no air leakage at the top of the pan/flange interface. The initial chamber relative humidity and temperature are recorded. The ambient relative humidity is recorded with a digital psychrometer. The waste water is prepared beforehand in the laboratory by mixing distilled water with a known mass of organic material in a sealed container. The required initial VOC concentration is used to specify the quantity of organic material needed. The average wind velocity is measured from both ends of the chamber and averaged using a hot wire anemometer, which is set using the flow control valves on the ducts. The inlet and outlet airflow streams are set to the desired flow rate. Airstreams are sampled with the OVA and charcoal tubes (further details in next section). The prepared waste water is then poured into the basin. The surface temperature of the water is recorded. Once all of the data have been collected and steady state has been reached, the inlet and outlet airflow streams are turned off.

Sampling Procedures

The two methods of sampling used in the flow-through flux chamber are an OVA analysis (one component in waste water) and charcoal tube samples (multicomponents in waste water). The OVA sampling procedure is used after one residence time. The vent line is closed, and the outlet sample line is completely opened, checking to ensure that the outlet airflow rate is constant. The inlet and exit streams are sampled using the OVA, and measured inlet and outlet flow concentrations are recorded. The procedure is repeated until the differential concentration (outlet air concentration minus the inlet air concentration) is constant for approximately three residence times. At this point we assume that constant concentrations for three residence times indicates that steady-state flow conditions have been reached.

When sampling using charcoal tubes, the flow rate on the sampling rotameter is adjusted to 5 L/min. The Tygon® tubing from the sampling flowmeter is connected to the front of the charcoal tube. Air flows through the system for two minutes. The electronic flowmeter calibrator is connected to the end of the charcoal tube. Three readings from the calibrator are averaged to get the flow rate through the charcoal tube. When sampling, charcoal tube samples are taken at least after the first retention time, the third retention time, and the last retention time (when the system reaches steady state). The OVA is used with a 20 L/min flow rate through the other sampling rotameter to determine when the system reached steady state and the last charcoal tube samples are taken. The charcoal tubes were refrigerated after sampling and analyzed by gas chromatography/flame ionization detection (GC/FID) performed by a Hewlett Packard Model 5890, 50-m PONA column.

Results

The measured emissions from the flow-through flux chamber were first compared with theoretical emissions calculated and derived from fundamental constitutive equations (that is, conservation, momentum) (Comey et al. 1993). By using the same conditions as measured in the flow-through flux chamber experiments (waste-water disposal unit simulations), a theoretical emission rate was calculated. The average measured emission rate from the flow-

TABLE 1—*Experimental test conditions.*

Wind Velocity, m/s	Ambient Temperature, °C	Relative Humidity, %	Waste-Water Flow Rates, mL/min	Residence Times, min
0.5–4.0	19.4–34.4	40–96	9–34	7–15

through flux chamber was within a factor of two of the theoretical calculations. By comparison, Chemdat7 on average overpredicted the same flux chamber emissions by 50 times. Mass balance measurements completed to check the amount of material lost in the flux chamber system indicated a recovery rate of ±4% (that is, 96% to 104%). Material could be lost due to small leaks in the system or absorption of material on the system inside walls. This demonstrates the ability of the flow-through flux chamber to accurately measure surface impoundment emission rates.

Table 1 gives the ranges of experimental conditions for the flow-through flux chamber experiments to compare with the three simulated waste-water treatment units: nonaerated waste-water disposal unit, nonaerated flow-through waste-water unit, and a diffused air disposal waste-water unit. Parameters that were thought important in the mass transfer process were wind velocity, ambient temperature, relative humidity, and residence time. Particular attention was given to meteorological conditions because the advantage of the flow-through flux chamber over the isolation flux chamber is its ability to simulate the ambient meteorological conditions.

Table 2 summarizes the comparison of predicted and measured emissions for all the waste-water treatment unit simulations. Comparing the measured flow-through flux chamber emission rates with the predicted emission rates of Chemdat7 for the nonaerated waste-water disposal unit shows that the model overpredicts the actual measured emissions by one to two orders of magnitude. This ultimately would mean emission inventories for nonaerated waste-water disposal units one to two orders of magnitude higher, resulting in higher and more costly emission controls than may be necessary. The wide range of overprediction with Chemdat7 emphasizes that the model is sensitive to some model inputs and environmental conditions and those factors need to be revised in the model.

Comparing the measured flow-through flux chamber emission rates with the predicted emission rates of Chemdat7 shows that the model tends to perform better for a nonaerated

TABLE 2—*Waste-water treatment unit results.*

		Ratio of Chemdat7 Emissions to Measured Emissions
Nonaerated Waste-water disposal Unit	maximum	279.8
	minimum	6.9
	average	51.4
Nonaerated Flow-through Waste-water unit	maximum	15.3
	minimum	0.3
	average	3.9
Diffused air Waste-water disposal Unit	maximum	19.1
	minimum	7.5
	average	12.8

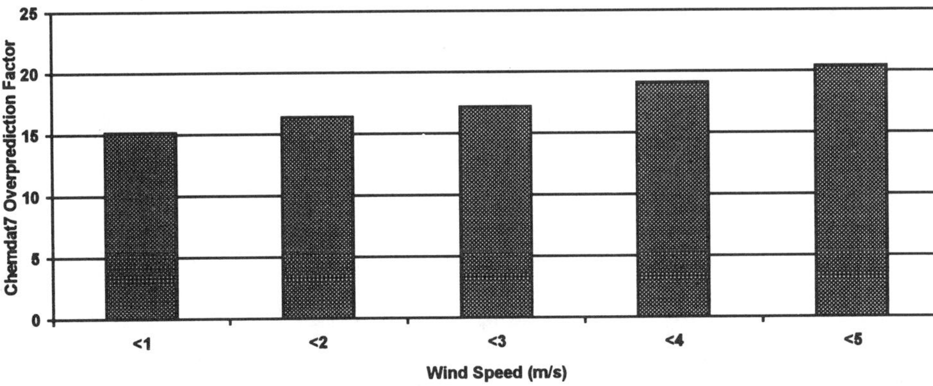

FIG. 3—*Effect of wind speed on emissions.*

flow-through waste-water unit. Still, Chemdat7 overpredicts within an order of magnitude, ultimately resulting in unnecessary higher emission inventories.

Comparing the measured flow-through flux chamber emission rates with the predicted emission rates of Chemdat7 shows that the model tends to overpredict by about a factor of 12 times the emission rates of a diffused air waste-water unit. This is not as high an overprediction as in the waste-water disposal unit simulations; however, the emission inventory would still be an order of magnitude higher than the actual emissions. This ultimately will result in higher and more costly emission controls than may be necessary.

A primary objective of this project was to quantify the effect of some atmospheric conditions at the sampling location on the emission rate. We consider, for example, the effect of wind speed, ambient temperature, and relative humidity on the extent to which Chemdat7 overpredicts measured data. These results are shown in Figs. 3, 4, and 5, respectively. Interpretation of the *x*-axis in the figures is: <1 means all numbers less than 1, <2 means all numbers between 1 and 2, <3 means all numbers between 2 and 3, etc. We found a direct

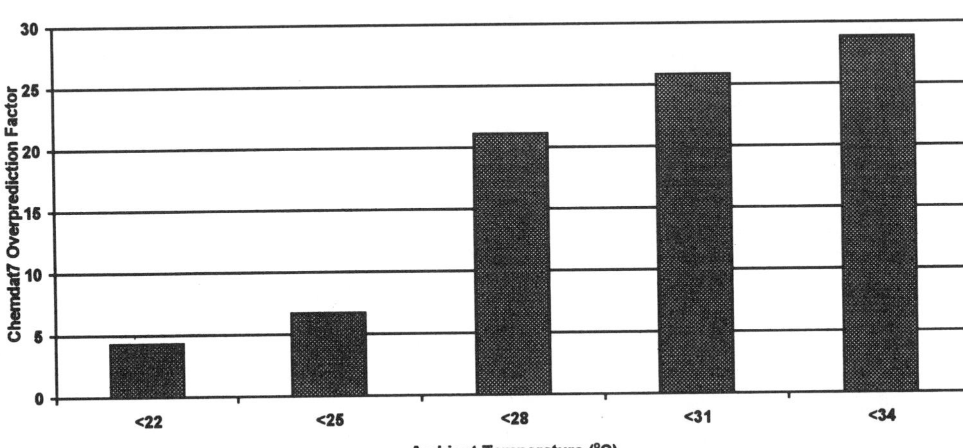

FIG. 4—*Effect of ambient temperature on emissions.*

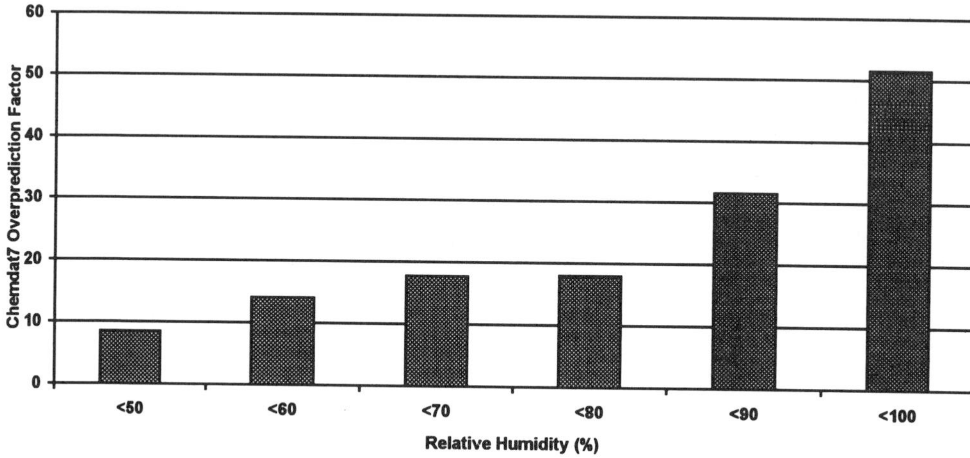

FIG. 5—*Effect of relative humidity on emissions.*

correlation between the value of each atmospheric factor and Chemdat7 model predicted emissions; that is, as each factor increased, the predicted emissions increased such that the discrepancy between measured and predicted increased. The Chemdat7 model is not as sensitive to these factors as it should be in order to be a useful tool to estimate emissions for regulatory purposes.

Future Work

Further work is planned with the bench-scale flow-through environmental flux chamber. The investigation of other Chemdat7 submodels will be completed. Additional tests are scheduled for the diffused air waste-water unit. Other submodels that will be investigated are the aerated basin and the oil film layer. Another EPA model, the Surface Impoundment Modeling System (SIMS), which has similar submodels as Chemdat7, will also be examined. (At the time of writing, EPA is updating their model to Chemdat8. Work is planned on comparing these flux chamber results with those of the new version of Chemdat.) We are also investigating the possibility of designing a portable flow-through environmental flux chamber for on-site emission measurements and tracking of emissions from an actual waste-water unit.

Conclusions

A flow-through flux chamber was designed to investigate the effects of ambient meteorological conditions (wind speed, temperature, and relative humidity) on emissions from hazardous waste treatment, storage, and disposal facilities. The flow-through flux chamber was able to replicate the ambient meteorological conditions and to measure the effects of ambient meteorological conditions on emission rates from waste-water disposal units. Relative humidity, ambient temperature and, to a lesser extent, wind velocity have been shown to affect emissions, but are not adequately accounted for in the Chemdat7 model.

Comparing calculated theoretical emission rates with the measured emission rates from the flow-through flux chamber demonstrates that the flow-through flux chamber is capable of measuring emission rates from waste-water disposal units within an order of magnitude.

When using the EPA's Chemdat7 emission rate model, it was shown that the emissions from a waste-water disposal unit can be overpredicted by one to two orders of magnitude; emissions from a diffused air disposal waste-water unit can be overpredicted by an order of magnitude; and emissions from a nonaerated flow-through waste-water unit can be simulated fairly well by Chemdat7.

Acknowledgments

The authors would like to thank Ms. Kimberly White and Ms. Sydnee Spears for completing the task of data collection; without their contribution this paper could not have been completed. We would also like to thank Dr. George Partridge for his assistance in the design of the flow-through flux chamber, Mr. Raj Karmarthi for his assistance in the data collection procedures, and Mr. Lee Gilmer for his overall guidance.

References

American Petroleum Institute, 1990, "Measurement of BTEX Emission Fluxes from Refinery Wastewater Impoundments Using Atmospheric Tracer Techniques," API Publication No. 4518.

American Petroleum Institute, 1993, "A Fundamental Evaluation of the Chemdat7 Air Emissions Model," API Publication No. 4571.

Comey, K. R., Rabideau, C. G., and Willingham, R. T., 1993, "Evaluation of a Flow-Through Environmental Flux Chamber Method for Measuring Volatile Organic Emissions from a Wastewater Surface Impoundment," *Proceedings,* Indoor Air International Volatile Organic Compounds in the Environment International Conference, London.

Eklund, B. M., Balfour, W. D., and Schmidt, C. E., 1985, "Measurement of Fugitive Volatile Organic Emission Rates," *Environmental Progress,* Vol. 4, No. 3, pp. 199–202.

U.S. Environmental Protection Agency, 1979, "A Problem with Flux Chamber Measurements of Biogenic Sulfur Emissions," EPA-600/3-79-033.

U.S. Environmental Protection Agency, 1986, "Measurement of Gaseous Emission Rates from Land Surfaces Using an Emission Isolation Flux Chamber, User's Guide," EPA Report No. EPA/600/8-86-008.

U.S. Environmental Protection Agency, 1989, "Hazardous Waste Treatment, Storage, and Disposal Facilities (TSDF)—Air Emission Models," EPA-450/3-87-026.

U.S. Environmental Protection Agency, 1990, "Background Document for the Surface Impoundment Modeling System (SIMS) Version 2.0," EPA-450/4-90-019b.

Marti M. Minnich,[1] John H. Zimmerman,[1] and Brian A. Schumacher[2]

Preparation and Analysis of Dry VOC-Fortified Soils

REFERENCE: Minnich, M. M., Zimmerman, J. H., and Schumacher, B. A., "**Preparation and Analysis of Dry VOC-Fortified Soils,**" *Volatile Organic Compounds in the Environment, ASTM STP 1261,* W. Wang, J. Schnoor, and J. Doi, Eds., American Society for Testing and Materials, 1996, pp. 156–169.

ABSTRACT: Studies were performed to assess the practicality of using dry VOC-fortified soils as performance evaluation materials (PEMs) and also to compare the effectiveness of EPA's low- and high-level purge-and-trap procedures [SW-846 Method 5030A (EPA 1992)]. Desiccator-dried soils were fortified in large batches with a combination of eight neat VOCs at 1000 ng each VOC/g soil. Three studies were performed on each of three soil types, namely, Hayesville (0.2% organic carbon, 32% clay), Kyle (0.6% organic carbon, 9% clay), and Charleston (3.8% organic carbon, 8% clay). VOC recoveries by the high-level procedure were good to excellent for all soils and compounds (98% mean recovery) and only fair by the low-level procedure (71% mean recovery). The mean relative standard deviation for all soil/compound/procedure combinations in the first study was 11%. Soil and compound properties influenced the VOC recoveries and rates of vapor losses from dry VOC-fortified soils. The poorest recoveries by the low-level procedure occurred on the soil with the highest proportion of organic matter. Rapid volatilization losses of 1,1,1-trichloroethane, benzene, trichloroethene, toluene, and tetrachloroethene were observed from 1-g samples of Kyle soil. In contrast, only trichloroethene was observed to volatilize from the Hayesville soil and no decreases in VOC concentrations were observed on the Charleston soil from 1-g samples that sat open for 20 min. No volatile losses were observed from any of the soils when dry 20-g soil samples were left open for 20 min. The technique of fortifying large batches of dry soils for PEMs as used in these studies provides good precision, but some modifications are suggested to improve the recoveries of compounds with low GC retention times on some of the soils. Also, further studies are needed to evaluate the influence of soil properties on VOC recoveries.

KEYWORDS: volatile organic compounds, soil/solid matrices, methanol extraction, quality assurance samples, performance evaluation materials

The concept of fortifying dry soils with volatile organic compounds (VOCs) to produce quality assurance samples or performance evaluation materials (PEMs) for the analysis of soil VOCs has been advanced by Hewitt et al. (1992) and Hewitt (1993). VOC-fortified dry soils will sustain slow to negligible rates of biodegradation or hydrolysis compared with moist soils. The sorption capacity of a dry soil is generally two to four orders of magnitude greater than that of the moist soil (Chiou and Shoup 1985, Poe et al. 1988, Ong and Lion 1991). Thus research is being conducted by the U.S. Environmental Protection Agency (EPA)

[1] Senior scientist and research assistant, respectively, Lockheed Environmental Systems & Technologies Company, 980 Kelly Johnson Drive, Las Vegas, NV 89119.

[2] Environmental scientist, Characterizations Research Division, U.S. Environmental Protection Agency, Las Vegas, NV 89193-3478.

to determine the stability of VOC-fortified dry soils and the rate of VOC vapor loss from dry soils.

Unlike the partitioning or absorption of (nonpolar) VOCs by organic matter that dominates sorption in moist soil, the mechanism of VOC sorption by dry soil is generally believed to be physical adsorption (Chiou and Shoup 1985, Ong and Lion 1991). Therefore, the capacity of dry soils to adsorb VOCs has been shown to be related to the soil-specific surface area rather than to the organic matter content (Poe et al. 1988, Ong and Lion 1991). Desorption of VOCs from dry soil would be expected to be very rapid for most compounds and to increase with increasing temperature. However, substantive losses from laboratory "piles" of dry soil will be slowed by diffusion through void spaces. Rates of volatile losses following desorption will depend on such factors as the total void space, continuity of void spaces, and the exposed soil surface area to volume ratio.

If homogeneous replicates of dry VOC-fortified soils can be obtained, they can be used not only as PEMs, but also to compare results of various field and laboratory soil VOC analytical methods. The experiments described here include a comparison of the extraction effectiveness of the low- and high-level purge-and-trap procedures, as described in EPA SW-846 Method 5030 (EPA 1992). In the EPA's low-level procedure, 5 mL of water is added to 1 to 5 g of soil or solid waste, and the mixture is purged to extricate VOCs. In the EPA's high-level procedure, 10 mL of methanol is added to 1 to 4 g of soil or solid waste; an aliquot of the methanol extract is added to 5 mL of water and purged as in the low-level procedure. The high-level procedure was designed to provide a method of diluting highly contaminated samples and therefore it was not planned as an equivalent procedure.

The objectives of this work were (1) to investigate the rate of vapor loss from dry, fortified soil during subsampling activities, and (2) to compare the accuracy and precision of EPA low- and high-level purge-and-trap procedures. In this set of experiments, three studies were performed on each of three soils. Dry soils were fortified by spiking with eight compounds (neat) at 1000 ng each VOC/g soil.[3] This rate is the upper end of that which can be analyzed by EPA's low-level procedure. To insure quantification by the high-level procedure at that rate, we used a soil/methanol ratio of 1 g soil/mL methanol. The first study investigated initial vapor losses and compared the low- and high-level procedures for quantifying VOCs in dry soil. The second and third studies provide supplementary data on vapor losses that might be expected while preparing VOC-fortified dry soil for low- and high-level procedures, respectively.

Materials and Experimental Procedures

Materials

Soils used in these studies are described in Table 1. The Hayesville soil (clayey, oxidic, mesic Typic Hapludult) came from Fannin County in Georgia's Blue Ridge Mountains. Soils designated "Kyle" (loamy-skeletal, carbonatic, thermic Typic Calciorthid) and "Charleston" (sandy-skeletal, carbonatic, mesic Fluventic Haplustoll) are from an altitudinal transect in the Mojave Desert, Clark County, Nevada; soil series have not been assigned in this region, but extensive soil characterization data and geomorphic descriptions of the areas have been reported (Amundson et al. 1989).

[3] The largest compound in this study, o-xylene, is approximately 1 nm^2. If only o-xylene were added to the soil at 8000 ng/g, the molecules would occupy approximately 4.5×10^{-4} m^2/g. This is approximately four orders of magnitude below the 4.3 m^2/g surface area of the Charleston soil.

TABLE 1—*Description of soils.*

Soil Designation	Horizon	Sand (%)	Clay (%)	pH Water 1:1(w:v)	pH 0.01*M* $CaCl_2$	EC (dS/m)	Surface Area,[a] (m^2/g)	Organic Carbon, (%)	Carbonate Carbon, (%)
Hayesville	Bt2	46	32	5.0	4.4	0.018	25.5	0.2	0
Kyle[b]	Bw	45	9	8.1	7.7	0.207	20.2	0.6	2.0
Charleston[b]	A	61	8	7.7	7.3	0.313	4.3	3.8	3.1

[a] Measured by three-point N_2 sorption.

[b] Extended description of these soils can be found in Amundson et al. (1989). Soils collected were from the same pedon and same streamreach as those designated "surface 3 Blackbrush" and "surface 4 fir-pine" for Kyle Canyon and Mt. Charleston soils, respectively.

Target compounds used in these studies, abbreviations for the compounds, and compound properties are listed in Table 2. Two system-monitoring compounds (SMCs), cis-1,2-dichloroethene and 1-chloro-2-fluorobenzene, were added to each sample, blank, and standard to monitor the analytical system. SMCs were selected to bracket the retention times of the target compounds without peak interference. The target compounds were purchased neat and the SMCs as 2000 μg/mL standards (Chem Services, West Chester, Pennsylvania).

Experimental Procedures

Soils were air-dried and sieved to 2 mm. One kilogram of air-dried soil was placed over indicating anhydrous $CaSO_4$ in a desiccator for 2 to 7 days just prior to fortification. Desiccator-dried soil was placed in wide-mouthed 2 L glass jars with polytetrafluoroethylene-lined lids. Soil was fortified at 1000 ng each VOC/g soil. Compounds were measured volumetrically, as calculated from liquid densities. The eight target compounds (neat) were added in order of increasing vapor pressure (o-xylene through 1,1,1-TCA) by microsyringe, injecting the compounds 4 to 5 cm below the soil surface. The jars were then sealed and tumbled end-over-end for 12 h on a rotary mixer.

Soil was subsampled by pressing the wide end of a glass weighing funnel into the fortified, bulk sample. Subsamples of approximately 1 g and 20 g were collected for the low- and high-level procedures, respectively. Samples to be analyzed by the low-level procedure were placed in preweighed 40 mL vials with modified lids, adapted to fit on the purge-and-trap autosampler (Associated Design & Manufacturing Co., Alexandria, Virginia). Samples to be analyzed by the high-level procedure were placed in 40 mL vials that were weighed after the appropriate quantity of methanol was added.

Minimum Detection Limits—Seven 1-g replicates and seven 10-g replicates of unfortified soil were prepared for each soil. The 1-g replicates were spiked with 10 ng of each compound (methanolic solution) in 5 mL water plus SMCs, added just prior to placing samples on autosampler for analysis. These samples were analyzed by the low-level procedure. The replicates analyzed by the high-level procedure were prepared to yield the same quantity (10 ng) on the detector. Methanol (10 mL) and 1000 ng of each compound (in methanolic solution) were added to the 10-g samples (100 ng/g). Aliquots (100 μL) of the extract were

TABLE 2—*Compounds and properties. GC retention times refer to those observed on analytical system described in this report.*

Compound	GC Retention Time, min	Molecular Weight, g/mol	Density, (g/mL)	Solubility, 25°C, mg/L	Vapor Pressure, 25°C, mm Hg	Log K_{ow}[a]
1,1,1-trichloroethane (1,1,1-TCA)	3.9	133.42	1.339	1495[b]	123.7[b]	2.49[b]
Benzene	4.5	78.11	0.879	1791[b]	95.19[b]	2.13[b]
Trichloroethene (TCE)	5.3	131.40	1.464	1100[b]	69.0[b]	2.42[b]
Toluene	7.3	92.13	0.867	534.8[b]	28.4[b]	2.73[b]
Tetrachloroethene (PCE)	8.2	165.82	1.623	150.3[b]	18.49[b]	3.40[b]
Ethylbenzene	10.0	106.16	0.867	161[c]	9.53[c]	3.15[c]
o-xylene	11.0	106.16	0.880	175[b]	6.6[b]	3.12[b]
1,1,2,2-tetrachloroethane (1,1,2,2-TTCA)	12.5	167.85	1.541	1100[b]	12.03[b]	3.03[b]

[a] Log octanol water partition coefficient.
[b] Howard 1990.
[c] Howard 1989.

placed in clean vials, SMCs and water were added, and the replicate extracts were then purged. The minimum detection limit by each procedure for each soil was calculated as the product of the standard deviation of the replicate analyses and the Student's "t" value for a two-sided, 99% confidence level with 6 degrees of freedom. Compound minimum detection limits as determined in this manner are given in Table 3.

Study 1. Comparison of Low- and High-Level Procedures—Fourteen samples were collected from the bulk fortified soil, seven to be analyzed by each procedure, taken alternately. As each vial was sealed, the time was noted. Unfortified soils were prepared and analyzed by each procedure as study blanks. Samples were stored in the dark at room temperature until analyzed (within 1 to 10 days).

Study 2. Volatilization from 1-g Samples—Quadruplicate 1-g samples were left in open vials in an active fume hood for 0, 5, 10, 20, and 40 min. Each set of replicates was collected in a "round" (four rounds) and the collection of samples by treatment was randomized within each round. After sitting for the specified time, samples were sealed with the modified lids, and soil weights were determined. Two vials of unfortified soil were placed in the middle of fortified samples and left open for 20 and 40 min as study blanks.

Study 3. Volatilization from 20-g Samples—Quadruplicate 20-g samples were placed in 0, 5, 10, and 20 mL methanol. A fifth set of replicates, designated "unopened control" was placed in 20 mL of methanol during sampling. Again, each set of replicates was collected in a "round" and the collection of samples by treatment were randomized within each round. Samples were sealed, weighed to determine soil weights, and returned to the hood. The lids were then removed from all but the unopened control samples, and the samples were left open for 20 min. At this point, samples were sealed and weighed to determine any methanol lost. Methanol was added to bring all samples to 20 mL. Thereafter, samples were shaken, stored, and analyzed as described for high-level samples in Study 1. The study blank consisted of 20 mL of methanol, left open for 20 min in the midst of the other open samples.

Analytical Procedures

Compounds were analyzed on a Hewlett-Packard Series II Model 5890 gas chromatograph (GC) equipped with a J&W DB 624 30 m, 0.53 mm inside diameter, fused silica column. The purge-and-trap was an O.I. Analytical Model 4460A Sample Concentrator with a MPM-16 multiple purging module autosampler. Detectors were an O.I. Analytical Model 4420 electrolytic conductivity detector and an O.I. Analytical Model 4430 photoionization detector, arranged in series. Operating conditions followed those specified in EPA SW-846 Methods 5030 and 8021 (EPA 1992). Sample concentrations were determined from equations

TABLE 3—*Minimum method detection limits (ng/g).*

	Low Level			High Level		
Compound	Hayesville	Kyle	Charleston	Hayesville	Kyle	Charleston
Benzene	4.4	3.0	2.3	42	54	45
Toluene	14.4	10.1	9.6	47	157	53
Ethylbenzene	5.8	4.5	3.5	48	101	53
o-xylene	6.9	4.8	2.7	41	129	49
1,1,1-TCA	5.1	2.7	19.1	77	19	81
TCE	4.2	2.3	3.5	32	59	33
PCE	6.1	2.6	5.1	31	62	58
1,1,2,2-TTCA	3.3	2.2	1.6	49	137	77

generated by Turbochrome software using five-point calibration curves (analyzed in triplicate). If the recovery of at least one of the SMCs was 100% ± 25%, the system was assumed to be operating in an acceptable manner.

The analysis of samples by the low-level procedure followed EPA specifications except that the modified lids allowed samples to be placed on the sparger with minimal loss of VOCs. Water (5 mL) plus SMCs (150 ng each cis-1,2-dichloroethene and 1-chloro-2-fluorobenzene in 10 μL methanol) were added by syringe through a three-way valve after the samples were secured on the autosampler.

For the analysis of samples by the high-level procedure, 100 μL of the methanol extract was placed in a clean vial and sealed with a modified lid. Thereafter, the low-level procedure was followed. However, a separate five-point calibration curve was generated, one in which all standards had 110 μL of methanol.

Statistical Procedures

The comparison of means between low- and high-level procedures for each compound on each soil were t-tested for the two-tailed hypotheses: H_o: $\bar{x}_1 = \bar{x}_2$ and H_a: $\bar{x}_1 \neq \bar{x}_2$. The ratio of the variances associated with each procedure was first compared with the appropriate F-ratio to determine the homogeneity of variances. Variances that were not significantly different at the 95% level were t-tested using a pooled variance, s_p^2 (following Zar 1984, p. 129). Variances that were significantly different were tested using an alternate t-test (following Zar 1984, p. 131).

Results

In Study 1, the chronological order of sampling was expected to demonstrate VOC losses during sampling. The time required to collect 14 samples in the first study was 5 to 10 min, depending mainly on the ease with which the soil slid through the sampling funnel. Benzene, 1,1,1-TCA, and TCE concentrations appeared to decrease during sample collection of the Kyle soil (Fig. 1). No other decreases in VOC concentrations corresponding to sample order were observed, indicating no significant losses during the 5 to 10 min sample collection time (data provided for Kyle and Charleston soils, Fig. 1).

Generally higher concentrations of the VOCs were measured by the high-level procedure as compared to the low-level procedure (Fig. 2). The differences between VOC concentrations obtained by the two methods were greater for compounds with larger GC retention times, whereas results for compounds with the lowest retention times (1,1,1-TCA and benzene) were comparable by either extraction procedure. Differences between VOC concentrations obtained by the two methods were not statistically significant (95%) for many compound/soil combinations (Table 4). Significant differences (at least 95% level) were observed for toluene on the Hayesville soil, ethylbenzene and o-xylene on the Kyle soil, and toluene, ethylbenzene, o-xylene, TCE, PCE, and 1,1,2,2-TTCA on the Charleston soil.

Recovery of the nominal fortification level of 1000 ng each VOC/g soil provided a basis for comparing the Study 1 means across compounds and soils (Fig. 2). The mean recovery of all compounds from all soils was 71% by the low-level procedure (range 25% to 149%) and 98% by the high-level procedure (range 69% to 143%). Relative standard deviations (RSDs) among replicates ranged from 2% to 24%, with an overall mean RSD of 11.0%. Data for 1,1,2,2-TTCA on the Hayesville was excluded from the above means because, for the low-level procedure, only one of the replicates was within the calibration range.

In Study 2, soil left open for up to 40 min exhibited considerable variation in VOC concentrations among the compound/soil combinations. Substantial decreases in VOC con-

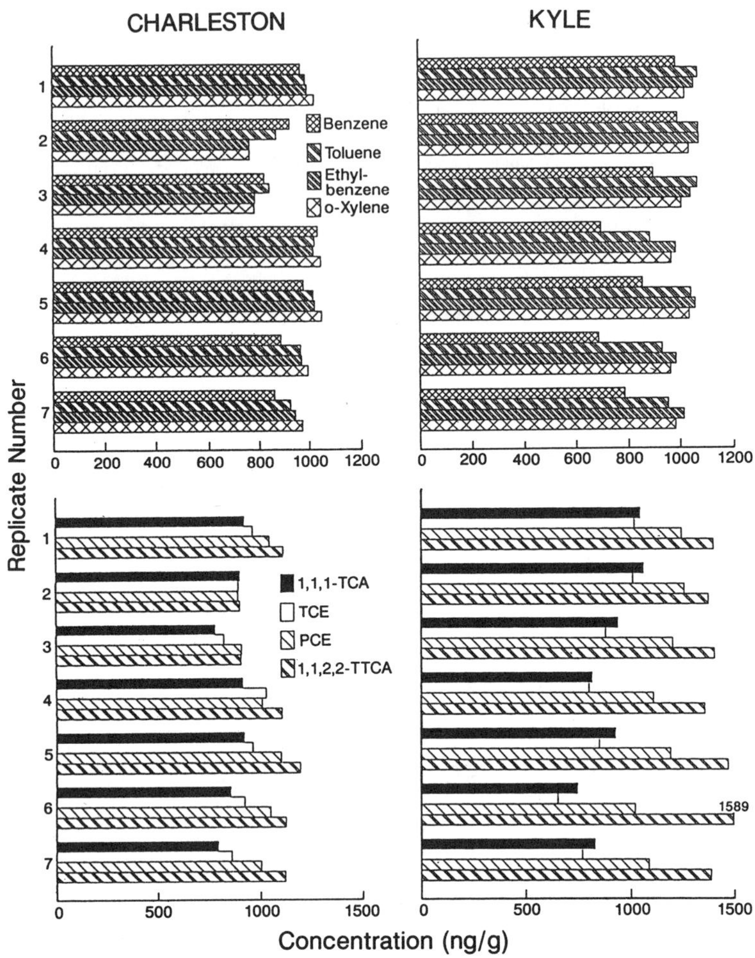

FIG. 1—*VOC concentrations (in methanol extracts) of seven replicate samples of Charleston and Kyle soils, presented in the chronological order in which samples were collected. Samples were collected in 9.5 and 5 min for the Kyle and Charleston soils, respectively.*

centrations were evident for benzene, toluene, 1,1,1-TCA, TCE, and PCE on the Kyle soil (Fig. 3). On the Hayesville soil, a considerable decrease in concentrations of benzene, 1,1,1-TCA, and TCE was noted between 20 and 40 min. In contrast, a marked increase in VOC concentrations with time for all compounds except 1,1,1-TCA was observed on the Charleston soil and for ethylbenzene, o-xylene, and 1,1,2,2-TTCA on the Hayesville soil. Subsequent studies indicated that the increase in VOC concentrations was an experimental artifact, related to the extraction time (Minnich et al., in review). While queued on the autosampler, the samples sat in water for up to 8 h, and an increasingly improved extraction by the low-level purge-and-trap resulted. Therefore, for some of the soil/compound combinations, if vapor loss occurred in Study 2, it could not be discerned because the samples were analyzed in the same order that vials were sealed, and VOC recoveries were increasing during the time samples soaked.

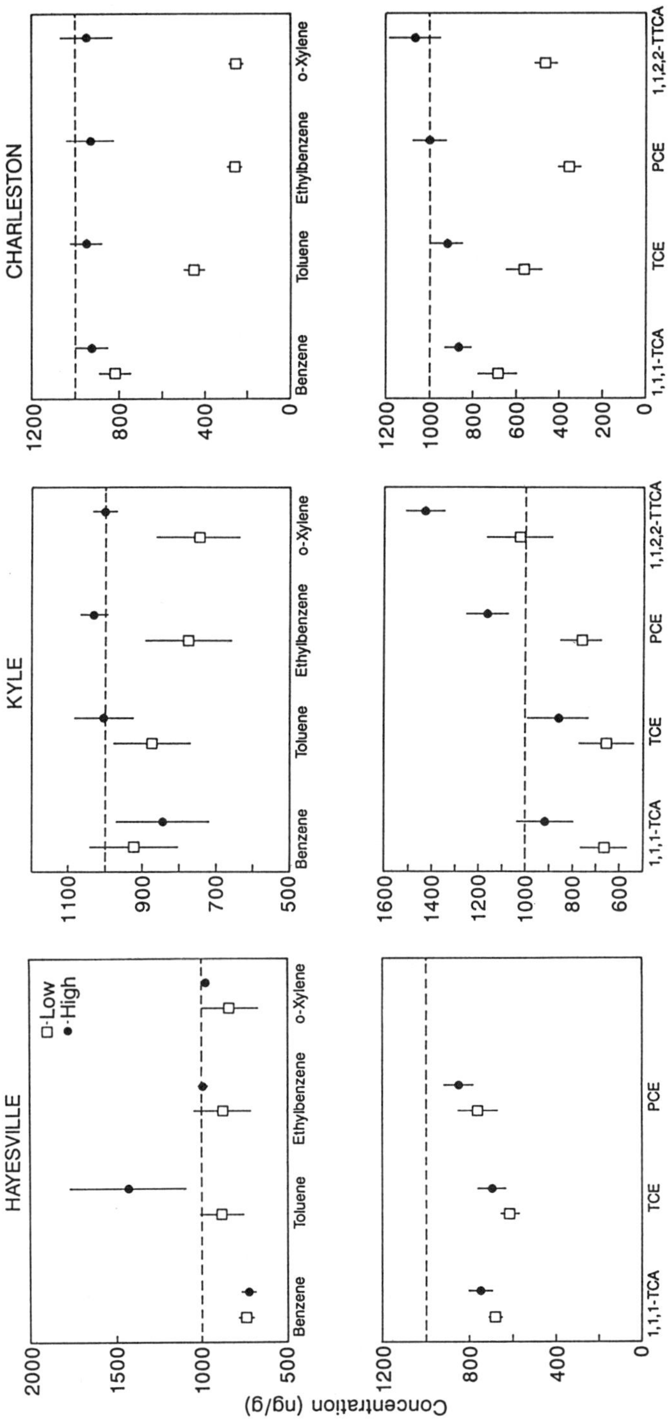

FIG. 2—Mean concentration and one standard deviation about the mean (n = 7) *of VOCs on three soils using EPA's low- and high-level purge-and-trap procedures. Note the nominal fortification level (1000 ng/g) plotted on each graph. Data for 1,1,2,2-TTCA by the low-level procedure on the Hayesville soil exceeded the calibration curve for six of the seven replicates and, therefore, the comparison for this compound was deleted.*

TABLE 4—*Variance ratios and t-test results from Study 1.*

Soil	Compound	Variance Ratio[a]	t_1[b]	t_2[c]	Degrees of Freedom	Critical $t_{a(2)=0.05}$
Hayesville	benzene	1.13	0.15	...	6	2.447
	toluene	7.70	...	3.98**	8	2.306
	ethylbenzene	60.53	...	1.84	6	2.447
	o-xylene	51.51	...	2.30	6	2.447
	1,1,1-TCA	2.53	0.81	...	6	2.447
	TCE	2.17	0.86	...	6	2.447
	PCE	1.94	0.60	...	6	2.447
	1,1,2,2-TTCA	18.97	...	0.06	7	2.365
Kyle	benzene	1.11	0.34	...	6	2.447
	toluene	1.78	0.75	...	6	2.447
	ethylbenzene	9.84	...	5.55***	7	2.365
	o-xylene	13.18	...	5.69***	7	2.365
	1,1,1-TCA	1.46	1.22	...	6	2.447
	TCE	1.26	0.88	...	6	2.447
	PCE	1.03	2.43	...	6	2.447
	1,1,2,2-TTCA	2.93	1.95	...	6	2.447
Charleston	benzene	1.09	0.81	...	6	2.447
	toluene	2.35	4.57**	...	6	2.447
	ethylbenzene	12.06	...	16.00***	7	2.365
	o-xylene	14.06	...	14.73***	7	2.365
	1,1,1-TCA	2.18	1.25	...	6	2.447
	TCE	1.40	2.48*	...	6	2.447
	PCE	2.21	5.32**	...	6	2.447
	1,1,2,2-TTCA	4.86	...	12.57***	8	2.306

[a] Ratio of variance of the seven replicates by each method. Largest variance used as numerator; values compared to critical values of $F_{0.05,6,6}$ (after Zar 1984, p. 123).

[b] Used where variance ratio less than 4.28. Significance denoted as *, **, and *** for $p = 0.05$, $p = 0.01$, and $p = 0.001$, respectively.

[c] Used where variance ratio greater than 4.28. Significance levels as noted above.

Study 3 showed no significant effect of treatment (amount of methanol) on recovery of VOCs after 20 g samples were left open 20 min. Moreover, there was no difference greater than one standard deviation around the mean between an unopened control and the zero level of methanol for any compound, even on the Kyle soil. Although Kyle soil showed volatile losses of benzene, 1,1,1-TCA and PCE in Study 2 (1-g soil samples), statistically significant volatilization was not observed in Study 3 (20-g samples). A statistically insignificant but perceptible decrease in benzene, ethylbenzene, and o-xylene concentrations at 5 mL of methanol as compared to samples with no methanol was seen on the Kyle soil (Fig. 4).

Discussion

Vapor Loss During Sampling

To achieve preservation in moist soil, loss of VOCs due to both biodegradation and volatilization must be prevented. In dry soil, the problem is largely confined to preventing vapor loss. Disturbance during sampling (that is, scooping and pouring) is likely to increase vapor losses, conceivably even more than short periods of time with soil sitting in open vials (Studies 2 and 3). For the most part, replicate samples in Study 1 demonstrated that the differences in sample concentrations were not related to sampling order (Fig. 1).

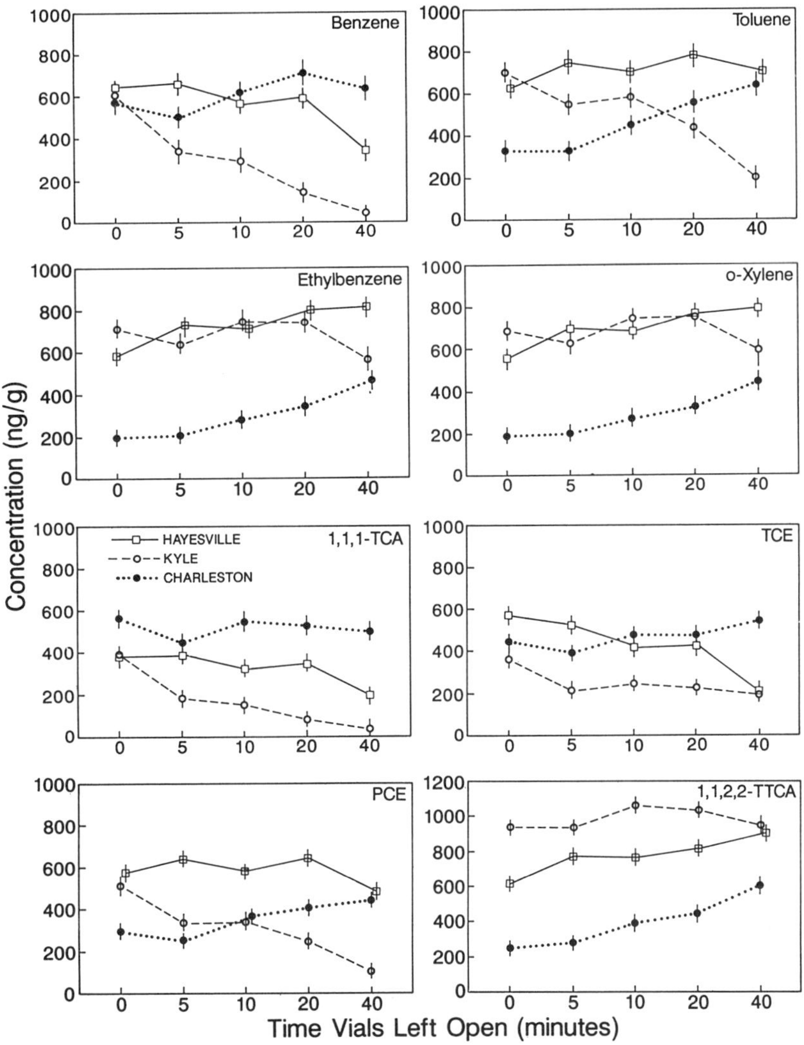

FIG. 3—*Mean VOC concentration and one standard deviation about the mean* (n = 4) *observed in 1-g fortified soil after samples sat open in hood for time indicated.*

It is assumed, however, that the headspace VOCs—the proportion of VOCs present in the vapor phase above the soil—are readily lost upon opening the jar. This initial loss of headspace VOCs is certainly followed by a continuous diffusion of VOCs from the soil to the atmosphere. Jars of fortified soil were resealed after sampling for Study 1, and then sat for up to 2 h before being opened a second and third time to obtain samples for Studies 2 and 3. Mean VOC concentrations in the control samples of Study 3 (unopened control) are compared with Study 1 concentrations in Table 5. (The Hayesville soil has been excluded because a different batch of fortified soil was used in the first study.) On the Kyle soil, the mean concentrations of all compounds except 1,1,2,2-TTCA were lower in the Study 3 control, indicating that vapor loss was occurring from the Kyle soil. On the Charleston soil,

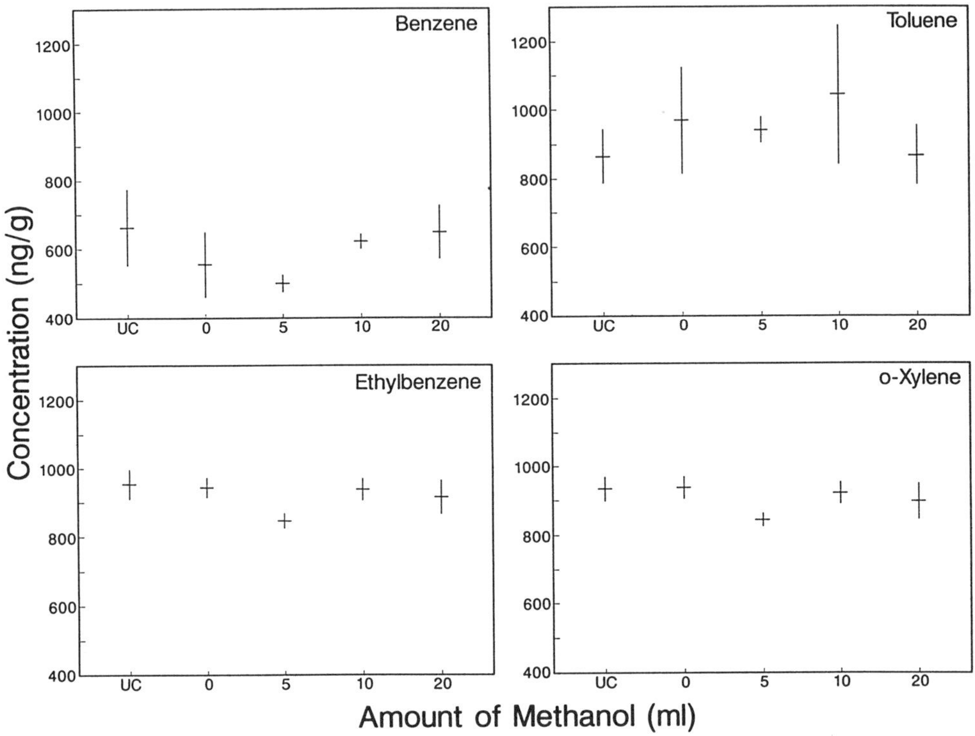

FIG. 4—*Mean VOC concentration and one standard deviation about the mean* (n = *4) after fortified Kyle soil (20-g) sat open 20 min in the amount of methanol specified. Unopened control (UC) samples were placed in 20 mL of methanol and sealed immediately.*

however, no trend of vapor losses from multiple sampling events was observed. The key difference between these soils was the organic carbon content which was relatively high in the Charleston soil and relatively low in the Kyle soil. The results imply that organic carbon was capable of sorbing the VOCs and only released the VOCs slowly.

Surface Area and Volatilization Rates

Observations on these three soils indicate that N_2 Brunauer-Emmett-Teller (BET) surface area is a poor predictor of volatilization rate, although other studies have shown that specific surface area is correlated to sorption capacity (Poe et al. 1988, Ong and Lion 1991). Note that the Hayesville and Kyle soils had similar surface areas, as measured by 3-point N_2 BET sorption (Table 1). The Hayesville soil (25.5 m^2/g) contains 32% clay (predominantly gibbsite, goethite, and kaolinite) as compared with the Kyle soil (20.2 m^2/g), which contains 9% clay (predominantly smectite, quartz, and mica). Qualitative differences between the oxides and sheet silicates could account for the similar surface areas with such dissimilar clay contents. However, the rate of vapor loss from the Kyle soil, particularly for benzene, toluene, 1,1,1-TCA, and PCE, appears much greater than the rate of vapor loss from the Hayesville soil (Fig. 3).

TABLE 5—*Comparison of mean VOC concentrations between Studies 1 and 3.*

Soil	Compound	Concentration, ng/g	
		Study 1 High	Study 3 Control
Kyle	benzene	843	662
	toluene	1001	865
	ethylbenzene	1028	954
	o-xylene	999	934
	1,1,1-TCA	914	733
	TCE	860	675
	PCE	1163	1079
	1,1,2,2-TTCA	1429	1588
Charleston	benzene	924	904
	toluene	946	943
	ethylbenzene	928	952
	o-xylene	947	983
	1,1,1-TCA	866	809
	TCE	919	942
	PCE	998	1034
	1,1,2,2-TTCA	1064	1206

In the third soil, Charleston, organic carbon was high (3.8%), but the measured surface area was only 4.3 m^2/g. The specific surface of dehydrated organic matter is generally low. For example, Jurinak and Volman (1957) reported that the surface area of a Staten peaty muck soil was 11.7 m^2/g by dry N_2 versus 264 m_2/g by glycol retention. For this study, surface area was measured by N_2 BET sorption on dry soil (dried at 80°C), selected to closely resemble the actual experimental conditions. However, two experimental observations on the Charleston soil, (1) the low VOC volatilization rate and (2) poor VOC recovery by the low-level procedure, indicated that VOCs were sorbed by, or had diffused into dehydrated organic matter. Desiccator-dried Charleston soil contained 0.7% moisture and perhaps this residual moisture content was sufficient to explain why these sorption results differ from studies that use oven dry materials. Therefore, the BET surface area measurement of soil dried at 80°C did not correspond to either the VOC desorption rate or the volatilization rate from the Charleston soil.

Vapor Loss via Minimal Amounts of Methanol

In Study 3, an interesting effect was noted for the 5 mL methanol treatment of the Kyle soil. Our intention was to investigate whether or not small amounts of methanol were sufficient as an absorbent for VOC vapors (fortified soil left open for 20 min in various amounts of methanol). The mean concentration recovered in the 5-mL methanol treatment was lower than that of all other treatments, including the treatment with no methanol present. Figure 4 presents the results for benzene, ethylbenzene, and o-xylene on the Kyle soil. We suggest that methanol vapor was displacing these compounds, similar to displacement of VOCs by water vapor observed in other studies (Chiou and Shoup 1985, Ong and Lion 1991). However, the effect was not noted on Hayesville and Charleston soils. This is further evidence that desorption/volatilization was more rapid on the Kyle soil and that sorption sites on Kyle soil were qualitatively different from the other soils.

Dry Soils as PEMs

Quality soil VOC PEMs will require: (1) homogeneous (precise) replicates, (2) slow rates of vapor loss, and (3) long-term stability. Replicate variability could arise from soil heterogeneity, differences in the amount of disturbance during sampling, and analytical noise. In these studies, RSDs were similar for both low- and high-level procedures. This indicates that 1-g soil samples were no more heterogeneous than 20-g samples for sieved, air-dried soil or, that the soil "noise" contributed less than the analytical noise. It may be possible to reduce the analytical noise by careful selection of internal standard(s) that mimic the target compounds, or perhaps, by use of one of the newer, redesigned purge-and-trap instruments that eliminate port-to-port variability and use uniform internal diameter on all fittings, valves, and tubing to avoid dead volume.

Compounds with low GC retention times are difficult to retain, particularly on soils that are low in both clay and organic matter. Chilling the soil prior to subsampling will be investigated to deter the vapor losses. The greatest losses we observed were decreases of approximately 50% for benzene and 1,1,1-TCA on the Kyle soil during the first five minutes that the 1-g samples sat unsealed (Fig. 3). However, we found no apparent difference on the Hayesville soil (for toluene, ethylbenzene, o-xylene, PCE, and 1,1,2,2-TTCA) between 1-g samples sealed immediately and replicate samples that sat open for 40 min before sealing. On the Charleston soil, it appears that any vapor losses that occurred were slight relative to the loss in extraction efficiency when samples were analyzed too quickly (within a few hours) by the low-level procedure.

The third concern, the stability of VOCs over extended periods of time, was not addressed in these studies, but is currently being investigated. Two issues should be considered: losses that may occur (degradation or slow vapor losses through leakage) and changes in extractability. Preliminary data have indicated that some of the compounds exhibit decreased recoveries by the low-level procedure with increased storage times.

Conclusions

VOC-fortified dry soils can be prepared by addition of neat compounds to achieve good reproducibility among replicates (RSDs of eight compounds on three soils ranged from 2% to 24%). VOC recoveries from the dry, fortified soils were influenced by the compound properties (for example, vapor pressure), soil properties (for example, organic carbon content), and extraction method. The most volatile compounds in this study, 1,1,1-TCA and benzene, were readily lost from the Kyle soil; that is, losses of up to 50% were observed from 1-g samples left unsealed for five minutes. In contrast, excellent recoveries of all eight VOCs in the study resulted from the methanol extraction of Charleston soil (after the EPA Method 5030 high-level procedure). Vapor losses from the Charleston soil were negligible, even in samples that sat open for 40 min. To more accurately generate VOC-fortified PEMs on all soil types by the procedure described herein, improved retention of the most volatile compounds is needed.

The EPA Method 5030 low-level purge-and-trap extraction was not as rigorous as the high-level extraction. The high-level procedure extracted a significantly greater amount of VOCs for nine soil/compound combinations and never resulted in significantly lower VOC concentrations.

Recovery by the low-level method was impeded by slow desorption of some compounds from the Charleston soil as shown by the increased concentrations of many VOCs as the soil soaked in water prior to purging. From the limited perspective of only three soils, the results implied that *dry* organic matter was capable of sorbing VOCs and only slowly released the VOCs, a conclusion that conflicts with other studies that report vapor sorption to be related to the dry soil surface area (Poe et al. 1988, Ong and Lion 1991). Unlike studies where solids are oven-dried before performing sorption tests, soil was desiccator-dried in the present study, which we believe left sufficient water to maintain a natural structure in the soil organic matter and thus the organic matter absorbed VOC vapors.

Acknowledgments

We are grateful to Dr. Murray McBride for performing the BET surface area analyses, and to the Cornell Soils Characterization Laboratory for measuring the particle size distribution, total carbon, and carbonate carbon contents of the soils. Timely and discerning quality assurance/quality control supervision was provided by Vicki Ecker. Statistical advice was provided by Bob Gerlach. Finally, we thank the many known and unknown reviewers who improved the numerous drafts of this manuscript.

The U.S. Environmental Protection Agency (EPA) through its Office of Research and Development (ORD) partially funded and collaborated in the research described here. It has been subjected to Agency review and has been approved as an EPA publication. The U.S. Government has a nonexclusive, royalty-free license in and to any copyright covering this article.

References

Amundson, R. G., Chadwick, O. A., Sowers, J. M., and Donner, H. E., 1989, "Soil Evolution Along an Altitudinal Transect in the Eastern Mojave Desert of Nevada," *Geoderma,* Vol. 43, pp. 349–371.

Chiou, C. T. and Shoup, T., 1985, "Soil Sorption of Organic Vapors and Effects of Humidity on Sorptive Mechanism and Capacity," *Environmental Science and Technology,* Vol. 19, pp. 1196–1200.

EPA, 1992, *Test Methods for Evaluating Solid Waste, SW-846,* 3rd ed., Final Update I, Office of Solid Waste and Emergency Response, U.S. Environmental Protection Agency, Washington, DC.

Hewitt, A. D., Miyares, P. H., Leggett, D. C., and Jenkins, T. F., 1992, "Comparison of Analytical Methods for Determination of Volatile Organic Compounds in Soils," *Environmental Science and Technology,* Vol. 26, pp. 1932–1938.

Hewitt, A. D., 1993, "Vapor Fortification: A Method to Prepare Quality Assurance Soil Samples for the Analysis of Volatile Organic Compounds," Abstract from the *National Symposium on Measuring and Interpreting VOCs in Soils: State of the Art and Research Needs,* Jan. 12–14, Las Vegas, NV, U.S. Environmental Protection Agency, Las Vegas, NV.

Howard, P. H., Ed., 1989, *Handbook of Environmental Fate and Exposure Data,* Vol. I, Lewis Publishers, Chelsea, MI.

Howard, P. H., Ed., 1990, *Handbook of Environmental Fate and Exposure Data,* Vol. II, Lewis Publishers, Chelsea, MI.

Jurinak, J. J. and Volman, D. H., 1957, "Application of Brunauer, Emmett and Teller Equation to Ethylene Dibromide Adsorption by Soils," *Soil Science,* Vol. 83, pp. 487–496.

Minnich, M. M., Zimmerman, J. H., and Schumacher, B. A., in review, "Extraction Methods for Recovery of VOCs from Fortified, Dry Soils."

Ong, S. K. and Lion, L. W., 1991, "Mechanisms for Trichloroethylene Vapor Sorption onto Soil Minerals," *Journal of Environmental Quality,* Vol. 20, pp. 180–188.

Poe, S. H., Valsaraj, K. T., Thibodeaux, L. J., and Springer, C., 1988, "Equilibrium Vapor Phase Adsorption of Volatile Organic Chemicals on Dry Soils," *Journal of Hazardous Materials,* Vol. 19, pp. 17–32.

Zar, J. H., 1984, *Biostatistical Analysis,* Prentice-Hall, Inc., Englewood Cliffs, NJ, 781 pp.

Alan D. Hewitt[1]

Effect of Collection and Handling Practices on Concentrations of Volatile Organic Compounds Detected in Soil Subsamples

REFERENCE: Hewitt, A. D., **"Effect of Collection and Handling Practices on Concentrations of Volatile Organic Compounds Detected in Soil Subsamples,"** *Volatile Organic Compounds in the Environment, ASTM STP 1261,* W. Wang, J. Schnoor, and J. Doi, Eds., American Society for Testing and Materials, 1996, pp. 170–180.

ABSTRACT: The losses of trichloroethylene from soil samples transferred to and from a storage bottle were evaluated by comparing values obtained using standard techniques with values obtained using a method that limits sample disruption, exposure, and volatilization losses from a subsample after a single transfer step. Results strongly suggest that volatile organic compounds (VOCs) are readily lost from contaminated soils when care is not taken to limit surface area exposure and ensure subsample isolation. For this site investigation and others using similar sample-handling protocols, VOC losses are most abundant during field collection and storage.

KEYWORDS: sample collection, sample handling, soil samples, volatile organic compounds (VOCs)

A site investigation was undertaken soon after high (ppm) levels of trichloroethylene (TCE) were found in industrial service wells located at the Cold Regions Research and Engineering Laboratory in Hanover, New Hampshire. The main objective of the investigation was to identify point sources and the extent of subsurface contamination. Under the supervision of the U.S. Army Environmental Center (U.S. AEC), numerous vadose-zone soil samples originating from split spoons were taken during Phases I and II of the field sampling exercises. Coupling research with site investigation activities authorized by government agencies (U.S. AEC, U.S. Environmental Protection Agency (EPA), and State of New Hampshire) has allowed us to study alternative sampling and analysis procedures. This report covers some of the findings for Phase II. Reports covering the research efforts associated with Phase I can be found elsewhere (Hewitt 1992, 1993, 1994, Hewitt et al. 1993).

The Phase II work plan specified that 14 boreholes would be drilled to depths of at least 15 m. These vertical holes were drilled using conventional hollow-stem augering, and subsurface soil samples were taken every 1.5 m from split spoons that had been driven ahead of the augered hole. Of the ten or more samples collected during the drilling of each borehole, only four were shipped off-site for laboratory analysis. Many of the remaining soil sample storage bottles filled by the site investigator were available for on-site analysis. Comparative studies for both the off-site laboratory and on-site analysis of the subsamples collected by the site investigators were performed with co-located subsamples (originating from the same

[1] Research physical scientist, U.S. Army Cold Regions Research and Engineering Laboratory, 72 Lyme Road, Hanover, NH 03755-1290.

split spoon) collected using a limited disruption and exposure method (described later in the Sampling Methods section).

The collection of soil samples by the site investigators followed the commonly accepted practice of filling a bottle to near capacity for shipping and storage. Protocols that require the soil to be handled twice and stored in a bottle with little headspace are suspect (Hewitt 1992, Urban et al. 1989, Spittler 1989). Often the utensils (that is, spoons, paint scrapers, trowels, spatulas and wooden tongue depressors) used during the transfer of soil to and from the storage bottle fail to maintain the sample's structure. This process allows the soil to fracture and exposes a random amount of surface area, which facilitates the release of weakly sorbed and physically trapped volatile organic compounds (VOCs). In addition, filling a bottle to near capacity inevitably soils the sealing surfaces, making leaks possible.

During Phase I of this site investigation, this transfer to and from storage bottles was shown to cause TCE concentration losses that averaged 95% compared to concentrations found in co-located subsamples obtained by a limited disruption and exposure method (Hewitt 1992, 1994). This earlier study ruled out subsampling variability (within a split spoon), holding time for samples held in vapor-tight vessels, instrumental calibration, and method of analysis as the principal reasons for this large discrepancy.

The intent of the Phase II study was to closely evaluate the losses of TCE that occur when following the site investigator's sample-handling protocols. This study specifically addressed the extent to which TCE is lost when the site investigators collect soil samples in the field, while the samples are stored, and when subsamples are transferred and weighed in preparation for off-site laboratory analysis. This evaluation highlights the TCE concentration discrepancies that exist when samples collected by this currently accepted procedure are compared with values obtained using a limited disruption and exposure method. The findings of this study apply to most hazardous waste site investigations that use the commonly accepted practice of transferring soil to and from a storage bottle prior to the determination of VOC concentrations.

Sampling Methods

Borehole Drilling

Conventional hollow-stem augering and split-spoon (60 by 4.4 cm) coring was used to obtain subsurface soils to profile VOC and total petroleum hydrocarbon (TPH) concentrations. The Phase II work plan required that these boreholes be drilled to a depth of at least 15 m (50 ft), and beyond if VOCs were detected. Field screening for VOCs was performed by partially filling a glass bottle, covering it with aluminum foil, shaking it, and then poking a hole in the cap and sampling the headspace with a hand-held instrument (PhotoVac). When photoionizable compounds were detected in the bottle's headspace, the borehole was continued to a depth where either VOCs were not detected or the hole was within 3 m (10 ft) of the water table.

Site Investigator Sample Collection

Upon the retrieval of a split spoon, the site investigators filled two identical 125 mL wide-mouth bottles equipped with polytetrafluoroethylene-lined lids, one each for the analyses of VOCs and TPH. In addition to these bottles, two wide-mouth 500 mL glass bottles were partially filled, one for geological characterization and the other for VOC screening. All bottles were labeled ahead of time, and the VOC sample bottle was filled first or simultaneously with the TPH bottle, depending on the number of site investigators on station.

The soil was transferred to these bottles from the split-spoon core using either a stainless steel paint scraper or garden trowel. Neither of these utensils fit inside the 125 mL storage bottles, so the site investigators often had to assist the transfer using plastic glove-covered fingers. During this transfer process the soil sample's native structure was fractured into randomly sized aggregates, and the storage bottle necks and brims became soiled. After filling the storage bottle, the investigators wiped the closure surfaces clean with either a gloved hand or a paper towel. Using this collection method a storage bottle was usually filled and capped in less than 30 s. Once subsampling was completed, the two soil-filled 125 mL bottles were placed in a cooler with ice until the borehole was completed and a decision was made regarding which samples would be sent for off-site laboratory analysis.

Limited Disruption and Exposure Sampling Method

Obtaining subsamples using a single, less disruptive soil subsample transfer and immediate on-site isolation from vapor losses is known as the limited disruption and exposure (LDE) method. A 10-mL tip-less plastic syringe (outside diameter 1.6 cm) was used to transfer the material with minimal structure disaggregation and exposure (Hewitt 1992, Spittler 1989, Lewis et al. 1991). This device removes an intact plug of soil from a freshly exposed surface and fits into the mouth (inside diameter 1.8 cm) of a 40-mL volatile organic compound analysis (VOA) vial. Before the syringe is pushed into the split-spoon soil core, the plunger is set, allowing the cylindrical barrel to be filled with 2 to 3 mL of material. After a subsample is extracted from the split-spoon core, the external surface of the syringe is wiped with a clean paper towel; then the plug of soil is dispensed into a VOA vial by depressing the plunger. Frequently the plugs of soil remained intact during this transfer operation. To prevent volatilization losses, upon collection the subsamples were placed into preweighed VOA vials equipped with polytetrafluoroethylene-lined silicone septum caps containing 30 mL of Type 1 water (Milli Q, Millipore Corp.). These vials were then reweighed to obtain the actual soil sample weight. Obtaining the subsample in this fashion prepared them for headspace gas chromatography (HS/GC). Subsample collection was coincident with the site investigators' subsample collection, and it took about five seconds to perform.

Experimental Design

Initially the off-site laboratory results were compared with those obtained for co-located soil samples collected by the LDE method and analyzed on site. This was followed by three on-site studies assessing the extent of TCE losses that can be attributed to each of the three main handling steps (field collection, storage, and laboratory subsample transfer) that transpire before soil subsample analysis by an off-site laboratory. All evaluations included sample storage bottles that had been filled by the site investigators. Furthermore, subsamples removed from the storage bottles remaining on site were visually classified as wet, clumped, or fractured, and the storage period was recorded. Subsamples were classified as wet if there was standing water, as clumped if there were soil aggregates (>1 cm), or as fractured if there were no soil aggregates (the sample conformed to the container's internal volume).

Of the three studies performed the first on-site compared the TCE concentrations in single subsamples removed from the site investigators' storage bottles with those of co-located subsamples collected in the field with the LDE method. The transfer of subsamples from the site investigators' storage bottles was also performed using the LDE method. This comparison assumes that the concentration of TCE in a split spoon when it was first opened was fairly homogeneous. During Phase I, when multiple subsamples were taken from several split spoons, the TCE soil concentrations varied within a factor of 3 (Hewitt 1992, 1994).

The second on-site study looked at analyte concentration stability during sample storage. Six of the site investigator bottles that had been filled with soil were randomly chosen and emptied of their contents. Three were completely washed, removing all visible soil from the bottle and cap, while the others were only partially washed, removing only the soil from the bottle's interior surface and leaving the bottle closures soiled. These bottles, each of which has a total capacity of 135 mL, were then filled with 125 mL of water, then spiked with a 1.00-mL solution containing four VOCs (trans-1,2-dichloroethylene, trichloroethylene, benzene, and toluene) and tightly capped. After five days of storage at 4°C (39°F), the bottles were opened one at a time and a 5.00-mL aliquot was transferred to a VOA vial containing 25 mL of Type 1 water for HS/GC analysis.

The last on-site study assessed the extent of VOC losses that might occur when subsamples are removed for laboratory analysis, ten subsamples were removed from two storage bottles containing clumped soils. The B15-17′ bottle was subsampled three days after collection and the G45-47′ seven days after collection. Subsamples were removed in an alternating sequence, five using a tip-less 10 mL plastic syringe using the LDE method and five using a stainless steel spatula. Those transferred with the syringe were immediately placed into a VOA vial containing 30 mL of water and sealed; subsamples transferred with a spatula were placed into an empty VOA vial and weighed, and then the 30 mL of water was added. The subsamples transferred with a spatula were exposed for three minutes to simulate the time it would take to transfer soil subsamples to a tared vessel, weigh it, and then take it to an instrument for purge-and-trap analysis.

Analysis

The sample preparation and VOC determination methods used by the off-site laboratory were methanol extraction followed by purge-and-trap gas chromatography/mass spectrometry (PT/GC/MS) analysis. Using this procedure, the off-site laboratory established a certified reporting limit of 0.23×10^{-6} g TCE/g. All on-site analyses were performed on soil subsamples weighing between 2 and 4 g, by aqueous extraction HS/GC analysis (Hewitt et al. 1991, Kaing and Grob 1986), with a detection limit of ≤ 0.003 μg TCE/g. Prior to removing headspace samples with a gastight syringe (Hamilton), the VOA vials were vigorously hand-shaken for two minutes, allowing an equilibrium condition to be reached. The HS/GC method of analysis provides results that are often in good agreement with PT/GC/MS for the analysis of VOCs in low-organic carbon soils (Hewitt et al. 1992).

Results

Table 1 gives the TCE results for the co-located soil subsamples collected by the site investigators and analyzed by an off-site laboratory and those collected by the LDE method and analyzed on site. This comparison shows that the off-site laboratory reported only one TCE concentration above their certified reporting limit that was greater than the on-site estimates. Contrarily, 25 of the 26 on-site TCE estimates that were above this certified reporting limit (0.23 μg TCE/g) were greater than the values reported by the off-site laboratory. Furthermore, 11 of these 26 on-site TCE estimates were more than an order of magnitude greater.

Table 2 contains the on-site results for samples collected with the LDE method in this field versus co-located samples obtained from the site investigators' collection bottles. The corresponding log-log plot of these co-located sample values (Fig. 1) includes the axis of theoretical agreement (solid line) and order-of-magnitude intervals about this axis (dashed lines).

TABLE 1—*Trichloroethylene concentrations for co-located subsamples collected by the site investigators and analyzed at a contract laboratory compared with samples collected by the LDE method and analyzed on-site.*

Borehole and Depth	On Site, μg/TCE/g	Off Site, μg TCE/g	$\frac{\text{Off Site}}{\text{On Site}} \times 100\%$[a]
A3-5′	<0.003	<0.23	...[b]
A33-35′	1.7	0.66	39
A53-55′	2.0	0.69	34
C4-6′	0.04	<0.23	...
C24-26′	0.22	<0.23	...
C44-46′	2.0	<0.23	<12
C69-71′	0.19	<0.23	...
D4-6′	0.89	0.98	110
D19-21′	15	2.3	15
E13-15′	0.32	<0.23	<72
E43-45′	2.4	<0.23	<9.6
F8-10′	0.93	<0.23	<25
F19-21′	7.6	2.2	29
F49-51′	2.4	<0.23	<9.6
G9-11′	1.4	<0.23	<16
G39-41′	11	<0.23	<2.1
H19-21′	6.3	<0.23	<3.7
H24-26′	82	<0.23	<0.2
H29-31′	74	2.9	<3.9
H34-36′	14	<0.23	<1.6
I19-21′	0.48	<0.23	<48
I24-26′	5.9	<0.23	<3.9
I44-46′	7.8	<0.23	<2.9
I119-121′	0.43	<0.23	<53
J19-21′	0.11	<0.23	...
J29-31′	1.6	<0.23	<14
J49-51′	0.22	<0.23	...
K23-25′	0.91	<0.23	<25
K27-29′	0.0057	<0.23	...
K45-47′	0.15	<0.23	...
L34-36′	0.18	<0.23	...
L44-46′	3.3	<0.23	<7.0
L59-61′	0.27	<0.23	<85
L64-66′	0.23	<0.23	...
M48-50′	6.7	<0.23	<3.4
M68-70′	0.34	<0.23	<68

[a] % relative agreement.
[b] On-site value was less than the contract laboratory certified reporting limit (0.23 μg TCE/g).

Figure 1 clearly shows that the samples collected in the field with the LDE method were consistently higher in TCE. In comparison, nearly half (25 out of 60, 42%) of the subsamples removed from the site investigator storage bottles were below detection (0.003 μg TCE/g), while the remainder ranged from 82% to 0.18% of the amount found for the co-located samples obtained in the field. This random pattern and range of discrepancy is consistent with the sample collection and handling comparison performed during Phase I of this site investigation (Hewitt 1992, 1994) and the results given in Table 1 for the Phase II interlaboratory comparison.

The visual classification (wet, clumped, or fractured) given to the physical state of the soils in the storage bottle appears to explain some of the randomness present when comparing

TABLE 2—*Trichloroethylene concentrations for co-located subsamples collected by the site investigators and stored between 1 and 6 days compared with concentrations established when substandards were collected in the field. All analyses performed on site.*

Borehole and Depth	LDE[a] μg TCE/g	DT and SB[b] μg TCE/g	Holding Time, days	Soil Type	$\frac{\text{DT and SB}}{\text{LDE}} \times 100\%$[c]
A29-31′	0.17	0.14	1	wet	82
A44-46′	0.015	0.0092	1	wet	61
B10-12′	5.3	0.075	1	fractured	14
B45-47′	3.0	0.019	1	fractured	0.63
C15-17′	2.5	0.24	2	clumped	9.6
C30-32′	3.8	0.20	2	clumped	5.3
C40-42′	2.4	0.87	2	clumped	36
C50-52′	2.1	0.064	2	clumped	3.0
C55-57′	1.4	0.58	2	clumped	41
D15-17′	0.95	0.75	2	clumped	79
D30-32′	8.7	0.092	2	clumped	1.1
D37-39′	10	4.9	2	clumped	49
D45-47′	2.7	0.25	2	clumped	9.3
D50-52′	1.0	0.13	2	clumped	13
E9-11′	0.41	0.017	3	clumped	4.1
E19-21′	2.0	0.057	3	fractured	2.8
E24-26′	1.7	0.633	3	wet	37
E39-41′	1.1	0.015	3	fractured	1.4
E49-51′	1.2	<0.003	3	fractured	...
E59-61′	0.15	<0.003	3	fractured	...
E64-66′	3.4	<0.003	3	fractured	...
E69-71′	0.37	<0.003	3	fractured	...
E74-76′	0.22	<0.003	3	fractured	...
E79-81′	0.63	<0.003	3	fractured	...
E104-106′	1.0	0.014	3	fractured	1.4
E114-116′	0.62	<0.003	3	fractured	...
F15-17′	1.2	0.015	3	clumped	1.2
F25-27′	8.9	0.093	3	clumped	1.0
F35-37′	4.3	0.14	3	clumped	3.3
F40-42′	2.1	0.088	3	clumped	4.2
F45-47′	1.7	0.051	3	clumped	3.0
G15-17′	2.1	0.011	4	fractured	0.52
G20-22′	38	0.21	4	fractured	0.55

these two collection and handling methods (Fig. 1). Co-located subsamples described as wet were between the 1/1 to 10/1 concentration bands, showing a mean recovery percentage of 60%. Likewise, clumped soils were between 1/1 to 100/1, with a mean recovery of 14%, and the majority (10 of 11) of fractured soils were between 10/1 to 1000/1, with a mean recovery of 4.3%. A nonparametric analysis using the Wilcoxon-Mann-Whitney test found that these three groupings were different at the 95% confidence level. This finding suggests that the TCE concentrations in the site investigator samples were inversely related to the extent that the soil structure broke apart. In soils with a texture that easily disaggregated (fractured) and conformed to fill the vessel void, an average of 96% of the TCE was liberated from the matrix. Those soil samples that were more cohesive and remained as clumps (>1 cm diameter) in the storage bottle, or were water-saturated, exposed less surface area and showed average TCE losses of 86% and 40%, respectively. The ease with which a soil breaks into small aggregates is a function of texture, grain size, moisture, and organic content. Thus,

TABLE 2—*Continued.*

Borehole and Depth	LDE[a] μg TCE/g	DT and SB[b] μg TCE/g	Holding Time, days	Soil Type	$\frac{\text{DT and SB}}{\text{LDE}} \times 100\%$[c]
G25-27′	26	0.77	4	clumped	3.0
G30-32′	75	0.85	4	clumped	1.1
G35-37′	130	4.1	4	clumped	3.2
G45-47′	31	5.9	4	clumped	19
G50-52′	17	0.031	4	fractured	0.18
G55-57′	7.3	1.8	4	fractured	25
G65-67′	0.89	0.024	4	fractured	2.7
G70-72′	4.8	0.027	4	fractured	0.56
H5-7′	1.6	<0.003	5	clumped	...
H40-42′	2.0	<0.003	5	fractured	...
H50-52′	2.5	<0.003	5	fractured	...
H55-57′	5.0	0.019	5	fractured	0.38
H60-62′	0.25	<0.003	5	fractured	...
H65-67′	0.41	<0.003	5	fractured	...
I10-12′	0.041	<0.003	6	fractured	...
I15-17′	0.097	<0.003	6	fractured	...
I30-32′	0.33	<0.003	6	fractured	...
I35-37′	0.32	<0.003	6	fractured	...
I40-42′	0.40	<0.003	6	fractured	...
I50-52′	0.41	<0.003	6	fractured	...
I55-57′	0.069	<0.003	6	fractured	...
I60-62′	0.10	<0.003	6	fractured	...
I70-72′	0.31	<0.003	6	fractured	...
I80-82′	0.29	<0.003	6	fractured	...
I90-92′	0.35	<0.003	6	fractured	...
I100-102′	0.22	<0.003	6	fractured	...
I110-112′	0.30	<0.003	6	fractured	...

[a] LDE—limited disruption and exposure method.
[b] DT&SB—double transfer and storage bottle method.
[c] % relative agreement.

it was not surprising that the contents of those storage bottles characterized as fractured often contained sandy soils, while those classified as clumped contained silts.

This data set also can be used to look for a potential storage period effect on the TCE concentrations determined for the site investigator samples. This comparison was made only for storage bottles with clumped soils, since the losses from most (24 of 36, 67%) of the fractured soils extended below the level of detection, and there were only three samples classified as wet. The sample bottles subsampled two days after collection and the sample bottles subsampled three or four days after collection also appear to create two distinct populations (Fig. 1). The majority (7 of 10) of clumped soil subsamples analyzed after being held for two days were between the 1/1 and 10/1 bands, while those subsampled after three or four days of storage showed the majority (9 of 11) to exist between the 10/1 and 100/1 bands. The means of the TCE values relative to co-located samples collected in the field for the two and three or four-day storage periods were, respectively, 27% and 4.3%. The Wilcoxon-Mann-Whitney test of these two groups also showed that they were different at the 95% confidence level. This finding shows that the amount of TCE lost from the site investigator samples may also be a function of holding time.

Table 3 shows that randomly selected storage bottles with unwashed closures failed to retain VOCs in solution as well as bottles with clean closures (95% confidence level, stu-

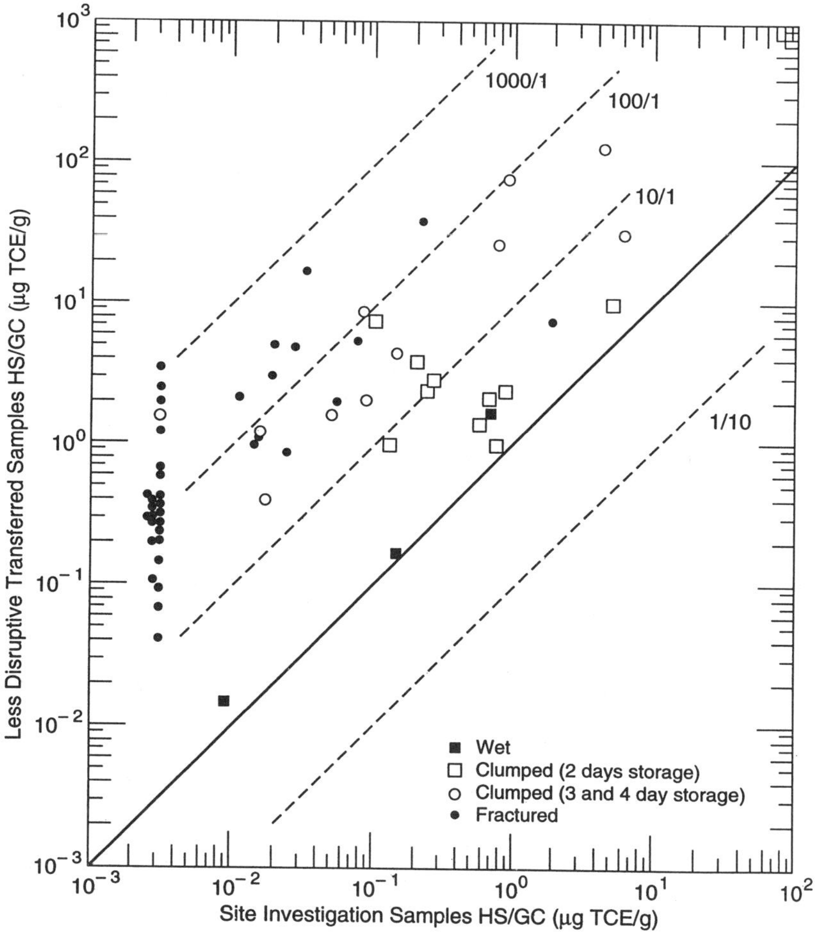

FIG. 1—*Trichloroethylene concentrations in co-located samples collected during Phase II and analyzed on site. This figure compares concentration estimates obtained from subsamples collected in the field versus subsamples obtained from investigators' collection bottles.*

TABLE 3—*Clean versus dirty closures on bottles used to store aqueous solutions containing trans-1,2-dichloroethylene (TDCE), trichloroethylene (TCE), benzene (Ben), and toluene (Tol).*

	Concentration, μg/L	
	Clean	Dirty
TDCE	480 ± 50[a]	260 ± 22
TCE	560 ± 57	320 ± 20
Ben	350 ± 32	200 ± 14
Tol	630 ± 48	370 ± 17

[a] Average and standard deviation, $n = 3$.

dent's t-test). This additional test confirms that VOC vapors escaped from the bottles filled by the site investigator.

The final test assesses the loss of TCE when the subsample is transferred by a spatula or similar device, and the sample handling and weighing does not occur in the same room as analysis. Under these circumstances a three-minute exposure period is reasonable. The results in Table 3 show significant differences (95% confidence level, students' t-test) that approach a factor of 2 when this transfer procedure was compared with the LDE method.

Discussion

Previous laboratory studies established that sample disturbance can cause VOC losses up to 80%, relative to undisturbed controls (Lewis et al. 1991, Siegrist and Jenssen 1990), and that significant (95% confidence level) differences exist when comparing subsamples transferred from one vessel to others that were either empty or contained a solvent (Jenkins and Schumacher 1987). While in agreement with the findings of these earlier efforts, the present study shows that much greater losses are likely to occur with environmental samples than seen for laboratory-prepared samples. One explanation for the greater losses is that environmental samples not only experience disruption and exposure but are also subjected to storage in vessels that are not vapor-tight. Understandably, laboratory studies would tend to overlook this potential loss mechanism since collection protocols usually specify that the storage bottle threads be cleaned before capping. The problem, as was observed during both phases of this site investigation, is twofold. First, "clean" is subjective, and second, a much greater emphasis is placed on speed than on cleanliness. Moreover, under field conditions, once a glass surface becomes soiled, it is for all practical purposes impossible to return it to a grit-free condition by quickly wiping it with either a soiled (gloved) hand or towel.

A visual inspection of all of the bottles filled by site investigators and made available for the comparison studies found grains of soil on the bottle's threads and lip (brim) and ground into the polytetrafluoroethylene lining of the cap. Although the storage bottles filled during Phase I were not visually inspected in this fashion, the sound of glass scoring could often be heard as the caps were tightened, giving a clear impression that a similar condition existed (Hewitt 1992).

Since all of the bottles had soiled closures, it is impossible to separate the losses solely attributable to the initial transfer of soil from losses caused when vapors escaped during storage. Based on the findings of this study, it appears that the extent of VOC losses during field collection and sample storage is likely to depend on soil structure and time. Figure 1 shows that losses of VOCs during the initial transfer is related to the cohesiveness of the material and that the storage time is significant. Most likely the losses that occurred during these two steps are comparable, each capable of reducing VOC concentrations by an order of magnitude or more, compared with a co-located samples collected in the field by the LDE method. When subsamples are transferred for a second time after several days of storage, the VOC losses that are anticipated for common laboratory protocols appear to be relatively small (Table 4). Clearly the potential for losses at this point have been greatly diminished since little of the easily liberated VOCs remains. On the other hand, if the soil were shipped off site inside a core liner (ASTM 1991), the LDE method would play a more critical role in the transfer of a subsample so as to maintain the sample integrity.

The initial disaggregation of soil structure and subsequent sample storage in bottles that fail to prevent vapor losses are the two greatest sources of indeterminate error that confound the determination of representative VOC concentrations in soils. Both of these sources of error are reduced when care is taken to maintain the soil structure and to transfer the subsample to a vessel that prevents VOC vapor losses during storage and analysis. Besides using

TABLE 4—*Laboratory subsampling method comparison of site investigator samples.*

	TCE Concentration, µg/g			
	B15-17′		G45-47′	
	Syringe	Spatula	Syringe	Spatula
1	0.042	0.015	0.27	0.15
2	0.020	0.021	0.24	0.15
3	0.034	0.026	0.34	0.14
4	0.025	0.023	0.51	0.17
5	0.038	0.024	0.56	0.27
	0.032[a]	0.022	0.38	0.18
	0.009[b]	0.004	0.14	0.054

[a] Average (µg/g).
[b] Standard deviation.

VOA vials prepared for HS/GC analysis, the LDE soil sample collection method described here can also be used in conjunction with bottles (VOA vials) equipped with special adapters (Associated Design & Manufacturing Co., Alexandria, Virginia) for direct connection to purging systems (Hewitt 1992) or containing methanol (Hewitt 1992, Urban et al. 1989, ASTM 1991). Both of these methods of handling subsamples are well suited for Method 8240 PT/GC/MS outlined in SW-846 (U.S. EPA 1986).

Conclusion

Investigations following the standard procedure of filling bottles for shipment and storage of soils from hazardous waste sites often underestimate the in-situ VOC concentrations by one or more orders of magnitude. The greatest losses of VOCs occur when soils become disaggregated for the first time and during storage in bottles with soiled closures. Only sample-handling protocols that limit soil structure disruption, require a single transfer step, and eliminate vapor losses during storage and analysis should be used when characterizing sites for VOC concentrations in the vadose zone.

Acknowledgments

Funding for this work was provided by the U.S. Army Environmental Center, Martin Stutz, project monitor. The author thanks Dr. Thomas M. Spittler for providing the information concerning headspace gas chromatography and the method for collecting undisturbed soil subsamples, and Philip Thorne and Thomas Ranney for critical reviews of the text.

This publication reflects the views of the authors and does not suggest or reflect policy, practices, programs, or doctrine of the U.S. Army or of the Government of the United States.

References

ASTM, 1991, "Standard Practice for Sampling Waste and Soils for Volatile Organics," ASTM Committee D-34, D 4547-91, American Society for Testing and Materials, Philadelphia.

Hewitt, A. D., 1992, "Review of Current and Potential Future Sampling Practices for Volatile Organic Compounds in Soil," *Proceedings*, 16th Annual Army Environmental R&D Symposium, Williamsburg, VA, pp. 75–82.

Hewitt, A. D., 1993, "Comparison of Site Characterization for Trichloroethylene by Headspace Gas Chromatography and Purge-and-Trap Gas Chromatography Mass Spectrometry," *Proceedings,* 1993 U.S. EPA/A&WMA International Symposium, Field Screening Methods for Hazardous Waste and Toxic Chemicals, Las Vegas, NV, pp. 195–202.

Hewitt, A. D., 1994, "Comparison of Sampling Methods for Trichloroethylene in Vadose Zone Soils," *Journal of the Association of Analytical Chemists,* Vol. 77, pp. 458–463.

Hewitt, A. D., Miyares, P. H., Leggett, D. C., and Jenkins, T. F., 1991, "Comparison of Headspace Gas Chromatography with EPA SW-846 Method 8240 for Determination of Volatile Organic Compounds in Soil," USA Cold Regions Research and Engineering Laboratory, Special Report 91-4.

Hewitt, A. D., Miyares, P. H., Leggett, D. C., and Jenkins, T. F., 1992, "Comparison of Analytical Methods for Determination of Volatile Organic Compounds in Soils," *Environmental Science and Technology,* Vol. 26, pp. 1932–1938.

Hewitt, A. D., Miyares, P. H., and Sletten, R. S., 1993, "Determination of Two Chlorinated Volatile Organic Compounds in Soils by Headspace Gas Chromatography and Purge-and-Trap Gas Chromatography Mass Spectrometry," *Hydrocarbon Contaminated Soils,* Vol. III, CRC Press, Boca Raton, FL, pp. 135–145.

Jenkins, T. F. and Schumacher, P. W., 1987, "Comparison of Methanol and Tetraglyme as Extraction Solvents for the Determination of Volatile Organics in Soil," USA Cold Regions Research and Engineering Laboratory, Special Report 87-22.

Kaing, P. H. and Grob, R. L., 1986, "A Headspace Technique for the Determination of Volatile Compounds in Soil," *Journal of Environmental Science and Health,* Vol. 21, pp. 71–100.

Lewis, T. E., Crockett, A. B., Siegrist, R. L., and Zarrabi, K., 1991, "Soil Sampling and Analysis for Volatile Organic Compounds," EPA/590/4-91/001, Technology Innovation Office, Office of Solid Waste and Emergency Response, U.S. EPA, Washington, DC.

Siegrist, R. L. and Jenssen, P. D., 1990, "Evaluation of Sampling Method Effects on Volatile Organic Compound Measurements in Contaminated Soils," *Environmental Science and Technology,* Vol. 24, pp. 1387–1392.

Spittler, T. M., 1989, Personal communication, U.S. Environmental Protection Agency, Environmental Services Division—Region 1, Lexington, MA.

Urban, M. J., Smith, J. S., Schultz, E. K., and Dickinson, R. K., 1989, "Volatile Organic Analysis for a Soil Sediment or Waste Sample," *Proceedings,* 5th Annual Waste Testing & Quality Assurance Symposium, U.S. Environmental Protection Agency, Washington, DC, pp. II-87–II-101.

U.S. Environmental Protection Agency, 1986, *Test Methods for Evaluating Solid Waste,* Vol. 1B.

Alan D. Hewitt[1]

A New Approach to Evaluating Pre-analysis Holding Times for the Determination of Volatile Organic Compounds in Soil

REFERENCE: Hewitt, A. D., **"A New Approach to Evaluating Pre-analysis Holding Times for the Determination of Volatile Organic Compounds in Soil,"** *Volatile Organic Compounds in the Environment, ASTM STP 1261,* W. Wang, J. Schnoor, and J. Doi, Eds., American Society for Testing and Materials, 1996, pp. 181–191.

ABSTRACT: In the absence of volatilization losses, the short-term concentration stabilities of benzene, toluene, trans-1,2-dichloroethylene, and trichloroethylene were assessed in soil subsamples. Previous holding time studies for this matrix failed to eliminate volatilization as a variable, making them difficult to interpret. Here, vapor-fortified soil subsamples held in sealed glass ampoules for 14 days at 22°C experienced appreciable reductions in benzene and toluene concentrations, presumably attributable to biodegradation. When the same fortified soil was held as a subsample for either headspace or purge-and-trap analyses, it showed an appreciable reduction in toluene and a complete loss of benzene over a 14-day holding period at 4°C. In contrast, fortified soils held in sealed glass ampoules at 4°C, or dispersed in methanol and held at 22°C, showed no significant analyte losses over periods of 20 and 98 days, respectively.

KEYWORDS: holding time, volatile organic compounds, soil samples, biodegradation

Soil samples collected for analyzing volatile organic compounds (VOCs) during hazardous waste site investigations are routinely shipped off-site for laboratory analysis. This makes holding time, in addition to collection and handling practices, an important variable with regard to analyte concentration. Currently, most site investigations use a soil sample collection and handling procedure that has been recommended by the U.S. Environmental Protection Agency (1986). This procedure specifies that a storage container be filled to capacity during soil sample collection; from this a subsample is removed in the laboratory for analysis. This double transfer method allows the soil structure to crumble. As a result, a considerable amount of soil surface area is exposed, promoting volatilization losses. In addition, when a container is filled to capacity with soil, the vessel closure surfaces often become covered with grit (grains of soil), thereby preventing a vapor-tight seal.

This practice, which requires multiple transfers and fails to maintain the native soil structure, has come under criticism when used for the analysis of VOCs because of the likelihood for volatilization losses (Urban et al. 1989, Spittler 1989, Siegrist and Jenssen 1990, Lewis et al. 1991, Hewitt 1992, 1994a). One approach to minimizing these losses has been to use a single transfer method that obtains the appropriate size soil subsample for instrumental analysis during the field collection exercise. Additional requirements for this procedure are rapid subsample transfer (less than ten seconds), limited structural disaggregation of the

[1] Research physical scientist, U.S. Army Cold Regions Research and Engineering Laboratory, 72 Lyme Road, Hanover, NH 03755-1290.

native substrate, and maintaining an uncompromised vessel seal. One utensil that has been used successfully for this purpose is a small coring device prepared by removing the tapered end from a 10-mL disposable syringe (Spittler 1989, Hewitt et al. 1992).

This single, less disruptive approach to soil subsample collection has shown VOC concentrations for co-located field samples that were often two orders of magnitude greater than those in subsamples from storage bottles (Urban et al. 1989, Hewitt 1992, 1994a). When using the single transfer method, researchers have placed subsamples into bottles containing methanol (MeOH), specially equipped bottles for low level (less than 1 μg VOC/g) purge-and-trap gas chromatography mass spectrometry (PT-GC-MS) analysis, or bottles prepared for headspace gas chromatography (HS-GC) analysis.

Although these studies addressed several of the problems associated with sample collection and handling, the stability of VOC concentrations when the subsamples were not immersed in MeOH was not evaluated. Studies addressing holding times have, in general, observed losses that either correlated directly with analyte vapor pressure or were independent of analyte chemistry (Jackson et al. 1991, Maskarinec et al. 1992, King 1993). Both of these trends suggest that volatilization was the dominant loss mechanism.

This study was designed to specifically see if VOC soil concentrations are stable in the absence of volatilization losses over periods of several days when they are held as subsamples prepared for either aqueous extraction PT-GC-MS or HS-GC analysis. To avoid volatilization losses during the holding time experiments, spiked soil subsamples were sealed in glass ampoules or held in vessels with closures that either had to be pierced by a syringe needle or quickly attached to a purge-and-trap system for the removal of VOCs.

Soil subsamples used in this study were spiked using a vapor fortification method (Hewitt 1993, 1994b). This method of spiking soils with VOCs is precise, does not require the injection of a carrier solvent, and is analogous to how vadose zone soils become contaminated by VOC vapors. The experiments assess the concentration stability of benzene, toluene, trans-1,2-dichloroethylene (TDCE), and trichloroethylene (TCE) in two soil matrices. These analytes are among the most frequently identified VOCs found at hazardous waste sites (Plumb and Pitchford 1985, Zarrabi et al. 1991), and are representative of compounds that biodegrade under anaerobic and aerobic conditions.

Experimental Methods

Soils

The soils used to assess analyte concentration stability were as follows: a reference matrix for the U.S. Army Environmental Center that is a composite of several soils from the Rocky Mountain Arsenal (RMA) in Denver, Colorado, and two site-specific soils collected at the Cold Regions Research and Engineering Laboratory (CRREL) (CRL) in Hanover, New Hampshire, and at Point Barrow, Alaska (PBA). The RMA soil has a sandy texture and organic carbon content of 0.053%, while the CRL and PBA soils both had a silty texture with organic carbon contents of 0.34% and 6.69%, respectively.

Soil Preparation, Treatment, and Handling

The soils were processed by air-drying, sieving through a 30-mesh screen, and thoroughly mixing. Soil subsamples were transferred into 1-mL glass ampoules using a stainless steel spatula and small plastic funnel; 2.00 g of the RMA soil was used, while 1.50 g each of the CRL and PBA soils were used.

The soil subsamples were then placed inside a large (5.6 L) desiccator with a dish of $CaSO_4$ for at least 24 h. After desiccation the $CaSO_4$ was removed, and a 60-mL glass bottle containing a spiking solution was introduced. The compositions of the solutions used for spiking the soil matrices are given in Table 1.

During this treatment method, VOC vapors impregnate the soil grain surfaces. After seven or more days of this vapor fortification treatment, the desiccator was opened, and 5-mm-diameter glass beads were placed on top of each ampoule as a temporary cap. Then the ampoules were quickly positioned in a clamp and the necks were heat-sealed using a propane torch. More complete descriptions of the vapor fortification process can be found elsewhere (Hewitt 1993, 1994b, 1994c).

Holding Time Experiments

Five holding time experiments, looking at a variety of conditions, were conducted. In the first experiment, the fortified soil was held in a sealed glass ampoule, while in all of the other experiments the fortified soils were held as subsamples prepared for either HS-GC or PT-GC-MS analysis. With the exception of the last experiment, where the soil subsamples were immersed in MeOH, water was present to promote biological degradation.

Sealed Ampoule—This experiment assessed the analyte concentration stability for the CRL soil confined in the 1-mL sealed glass ampoules. Prior to sealing the ampoules, 200 μL of ground water was added to return the moisture content (13% dry weight) to a typical level for this soil. In all, 21 subsamples were prepared, so that sets of triplicate subsamples could be sacrificed and analyzed after 0, 4, 5, 8, 11, 14, and 20 days of storage. Those held for 4, 8, and 14 days were stored at 22°C, while those held for 5, 11, and 20 days were refrigerated (4°C).

Headspace Subsamples—The second and third holding time experiments were designed to look at the stability of the VOC concentrations in subsamples that had been prepared for HS-GC analysis. On day 0, sealed ampoules containing fortified soils were placed inverted into VOA vials, which had open-faced caps and a polytetrafluoroethylene-lined silicone sep-

TABLE 1—*Fortification conditions and treatment sets.*

Stock Solution	Volume (mL) of Stock Diluted in Tetraglyme[a]	Total Volume (ml) of Fortification Solution	No. of Soil Subsample Replicates	Soils Used
Experiment 1				
A	0.1	10	7 triplicate	CRL
Experiment 2				
B	0.2	10	4 triplicate	CRL, RMA
Experiment 3				
A	0.1	10	4 triplicate	CRL
Experiment 4				
B	0.01	10	4 triplicate	CRL, RMA
Experiment 5				
C	25	50	2 triplicate	RMA, PBA

[a] Tetraethylene glycol dimethyl ether.
A—0.25 g TDCE, 0.29 g TCE, 0.18 g benzene, 0.27 g toluene in 25 mL of tetraglyme.
B—0.62 g TDCE, 0.73 g TCE, 0.44 g benzene, 0.52 g toluene in 25 mL of tetraglyme.
C—0.50 g TDCE, 0.59 g TCE, 0.35 g benzene, 0.60 g toluene in 100 mL of MeOH.

tum, and contained 30 mL of Type 1 water. After the VOA vials were closed, the holding period was initiated by breaking the ampoules and dispersing the soil via hand shaking.

For the second experiment, six replicates of each the RMA and CRL soils were prepared and repeatedly analyzed after 0, 2, 5, 9, and 14 days by removing headspace vapors with a gastight syringe. Following the analysis on day 0, the subsample sets were split so that triplicates of each soil type could be stored either refrigerated (4°C) or at room temperature (22°C).

The third experiment was performed on only the CRL soil, and subsamples were sacrificed upon analysis. Twelve subsamples were prepared so that triplicates could be analyzed after 0, 5, 8, and 14 days of storage. All samples were analyzed by HS-GC and stored at room temperature (22°C).

Low-Level PT-GC-MS Subsamples—The fourth experiment was designed to assess analyte stability for samples obtained for SW 846, Method 8240, low-level (less than 1 μg/g) PT-GC-MS analysis. On day 0, sealed ampoules of fortified soils were placed in VOA vials containing 200 μL of ground water and equipped with a purge-and-trap adapter (Associated Design and Manufacturing Company, Alexandria, Virginia, Model PT-6005-0002). The holding period was initiated after closing the VOA vials and breaking the ampoules so that the contents were dispersed. Twelve replicates were prepared of the RMA and CRL soils so that triplicate subsamples could be sacrificed and analyzed after 0, 4, 7, and 14 days of storage at 4°C.

High-Level PT-GC-MS Subsamples—The fifth experiment looked at analyte stability after fortified soils were dispersed in MeOH. On day 0, triplicate subsamples of the RMA and PBA soils in sealed ampoules were placed in VOA vials containing 20 mL of MeOH; then the ampoules were broken. The MeOH soil slurries were stored at room temperature (22°C) and repeatedly subsampled and analyzed by HS-GC after periods of 0, 30, 60, and 98 days. The preparation and handling of these QA soil subsample was consistent with the extraction procedure outlined in SW 846, Method 8240, for analysis of high levels (more than 1 μg/g) of VOCs in soil (U.S. EPA 1986).

Analysis

The subsamples from experiments 1, 2, 3, and 5 were analyzed by HS-GC. Soil subsamples held in sealed glass ampoules were prepared for analysis by opening them inside closed 40-mL VOA vials equipped with polytetrafluoroethylene-lined silicone septum caps and containing 30 mL of Type 1 water. These and all other headspace samples were shaken for two minutes prior to analysis to attain equilibrium. Samples that were refrigerated between analyses were allowed to warm to room temperature before they were agitated and analyzed. Headspace vapors were transferred from the VOA vials with gastight syringes (Hamilton) and concentrations were established by comparison to aqueous headspace standards (Hewitt et al. 1992).

The subsamples for the fourth experiment were analyzed by PT-GC-MS, following the general SW-846, Method 8240 guidelines for soils containing less than 1 μg of VOCs/g (U.S. EPA 1986). These subsamples were held in VOA vials equipped with a special adapter that allowed them to be quickly attached to a purge-and-trap system without exposing the subsample (Hewitt 1994c).

Results

Sealed Ampoules

In the first experiment, VOC-fortified CRL soil subsamples were held in sealed ampoules for various periods under refrigerated (4°C) and room temperature (22°C) storage conditions.

The results in Table 2 show that, in comparison to day 0 values, about 35% of the benzene and 20% of the toluene could not be accounted for after 14 days of storage at 22°C. In contrast, the refrigerated CRL soil subsamples showed less than a 10% relative change in concentration for these two analytes over a 20-day period. Losses of the two chlorinated analytes were less than 10% at either of the storage temperatures.

Headspace Subsamples

Tables 3*a* and 3*b* show the results for those subsamples stored as headspace samples (sealed VOA vials with 30 mL of Type 1 water). Regardless of storage temperature (22 and 4°C) or whether the subsamples were repeatedly analyzed or sacrificed upon analysis, the concentration of benzene in the CRL soil decreased in excess of two orders of magnitude over the 14-day holding period. As shown in Fig. 1, the rate of benzene loss was faster for the samples held at room temperature than for those that were refrigerated. Toluene also decreased in concentration in the CRL soil subsamples. Compared to day 0, more than 95 and 60% of the toluene concentrations were lost after 14 days under these two storage temperatures (Fig. 1).

These two aromatic compounds also appeared to decrease in the RMA soil subsamples held at 22°C. However, here the 36% and 25% decreases in benzene and toluene, respectively, over the 14-day holding period were caused by reductions in only one of the subsample triplicates.

The concentrations of both benzene and toluene in the refrigerated RMA soil subsamples and the chlorinated VOCs in all of the subsamples showed a slight decreasing trend (Tables 3*a* and *b*). The same trend was observed for a standard that was also stored in the refrigerator and analyzed along with the subsamples (data not given here). Since the aqueous standard and the soil slurry headspace subsamples behaved similarly, all showing a slow decrease in analyte concentration over the 14-day holding period, this effect was attributed to losses caused by analyte sorption by the polytetrafluoroethylene-lined VOA vial cap. This hypothesis is consistent with the losses seen for trichloroethylene and benzene from aqueous solutions in contact with polytetrafluoroethylene monitoring well casings (Parker et al. 1990, Parker and Ranney 1993).

TABLE 2—*Mean analyte concentrations (μg/g) for soil subsamples stored in sealed glass ampoules (experiment 1).*

	CRL Soil, Refrigerated (4°C)			
	Day 0	Day 5	Day 11	Day 20
TDCE	3.8 ± 0.2[a]	3.5 ± 0.2	4.1 ± 0.2	4.0 ± 0.3
Benzene	6.1 ± 0.7	5.9 ± 0.2	6.2 ± 0.2	5.7 ± 0.2
TCE	5.8 ± 0.1	5.3 ± 0.2	5.8 ± 0.2	5.8 ± 0.1
Toluene	7.4 ± 0.1	7.0 ± 0.4	7.3 ± 0.3	7.2 ± 0.4
	CRL Soil, Room Temperature (22°C)			
	Day 0	Day 4	Day 8	Day 14
TDCE	3.8 ± 0.2	3.9 ± 0.3	4.1 ± 0.1	3.8 ± 0.3
Benzene	6.1 ± 0.7	5.3 ± 0.3	4.4 ± 0.5	4.0 ± 1.4
TCE	5.8 ± 0.1	5.5 ± 0.3	5.6 ± 0.1	5.3 ± 0.2
Toluene	7.4 ± 0.1	6.5 ± 0.1	6.6 ± 0.2	6.0 ± 0.1

[a] Average and standard deviation, $n = 3$.

TABLE 3a—*Mean analyte concentrations (μg/g) for soil subsamples stored in VOA with 30 mL of water and repeatedly analyzed (experiment 2).*

	Day 0	Day 2	Day 5	Day 9	Day 14
			REFRIGERATED (4°C)		
RMA Soil					
TDCE	4.6 ± 0.2[a]	4.4 ± 0.2	4.2 ± 0.1	4.2 ± 0.1	4.1 ± 0.1
Benzene	22 ± 1.1	20 ± 1.0	20 ± 0.3	20 ± 0.3	20 ± 0.6
TCE	8.7 ± 0.5	8.1 ± 0.4	7.5 ± 0.2	7.5 ± 0.1	7.3 ± 0.2
Toluene	24 ± 1.4	23 ± 1.3	21 ± 0.4	22 ± 0.6	20 ± 0.6
CRL Soil					
TDCE	3.7 ± 0.2	3.6 ± 0.1	3.5 ± 0.1	3.3 ± 0.1	3.2 ± 0.1
Benzene	7.1 ± 0.2	6.2 ± 0.2	6.1 ± 0.2	4.8 ± 0.4	<0.1
TCE	6.0 ± 0.2	5.6 ± 0.2	5.4 ± 0.2	4.9 ± 0.1	4.8 ± 0.3
Toluene	14 ± 0.4	12 ± 0.4	12 ± 0.5	10 ± 0.7	5.2 ± 1.1
			ROOM TEMPERATURE (22°C)		
RMA Soil					
TDCE	4.6 ± 0.2	4.6 ± 0.1	4.4 ± 0.2	4.3 ± 0.1	4.2 ± 0.1
Benzene	22 ± 1.1	21 ± 0.7	20 ± 0.7	14 ± 11	14 ± 12
TCE	8.7 ± 0.5	8.4 ± 0.2	8.1 ± 0.4	7.6 ± 0.1	7.6 ± 0.1
Toluene	24 ± 1.4	23 ± 0.9	23 ± 1.5	20 ± 4.4	18 ± 4.6
CRL Soil					
TDCE	3.7 ± 0.2	3.7 ± 0.1	3.6 ± 0.1	3.5 ± 0.1	3.3 ± 0.1
Benzene	7.1 ± 0.2	6.3 ± 0.1	5.7 ± 0.1	<0.1	<0.1
TCE	6.0 ± 0.2	5.6 ± 0.1	5.4 ± 0.1	4.8 ± 0.2	4.4 ± 0.3
Toluene	14 ± 0.4	12 ± 0.2	12 ± 0.5	1.8 ± 0.9	<0.3

[a] Average and standard deviation, $n = 3$.

Low-Level PT-GC-MS Subsamples

The results in Table 4 again show that benzene and toluene in the CRL soil experienced the greatest losses. Relative to day 0, more than 95% and 70%, respectively, of these two aromatic VOCs were lost after 14 days of storage at 4°C (Fig. 2). Concentration reductions were less than 22% for the two chlorinated VOCs in either soil and for the two aromatic VOCs in the RMA soil.

High Level PT-GC-MS Subsamples

Table 5 contains the results for the repeated analyses of two sets of triplicate soil subsamples dispersed in MeOH. The analyte concentrations in these methanolic slurries showed no change in concentration over a 98-day holding period when stored at 22°C.

TABLE 3b—*Mean analyte concentrations (μg/g) for CRL soil subsamples stored in VOA with 30 mL of water at 22°C and sacrificed upon analysis (experiment 3).*

	Day 0	Day 5	Day 8	Day 14
TDCE	6.4 ± 0.3[a]	6.5 ± 0.1	6.1 ± 0.1	6.2 ± 0.2
Benzene	11 ± 0.6	7.1 ± 0.7	2.0 ± 2.7	<0.1
TCE	10 ± 0.6	9.6 ± 0.2	8.0 ± 1.1	7.2 ± 0.4
Toluene	12 ± 0.6	9.4 ± 0.8	2.7 ± 2.7	<0.1

[a] Average and standard deviation, $n = 3$.

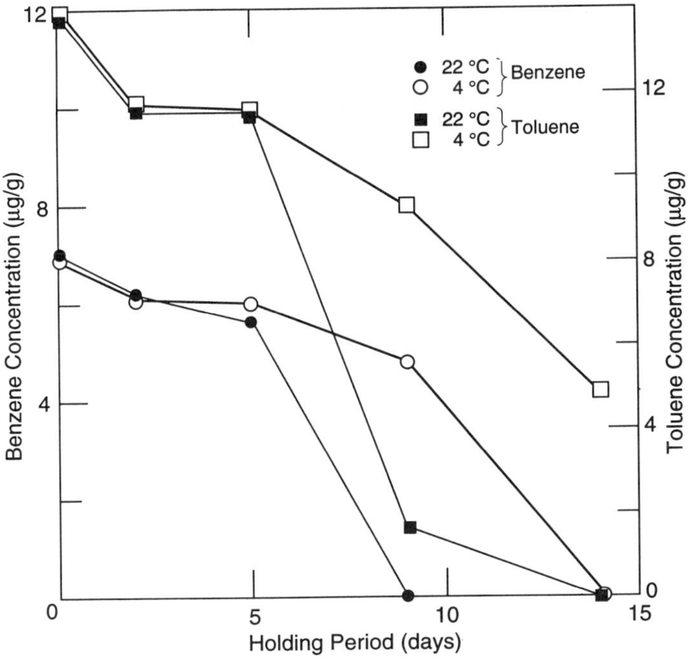

FIG. 1—*Stability of benzene and toluene concentrations* (µg/g) *in the CRL soil subsamples prepared for HS-GC analysis.*

Discussion

These four compounds degrade microbially at different rates under different environmental conditions. Labile hydrocarbons such as benzene and toluene degrade rapidly in the presence of aerobic heterotropic microorganisms (Suflita 1989). Under aerobic conditions, chlorinated

TABLE 4—*Mean analyte concentrations (ng/g) for soil subsamples stored in VOA with purge-and-trap adapter cap (experiment 4).*

	Refrigerated (4°C)			
	Day 0	Day 4	Day 7	Day 14
RMA Soil				
TDCE	19.9 ± 1.4[a]	17.2 ± 1.6	19.2 ± 0.5	19.0 ± 0.3
Benzene	67.0 ± 3.9	58.8 ± 6.0	61.9 ± 0.8	57.1 ± 2.6
TCE	31.1 ± 2.9	24.4 ± 3.4	27.1 ± 1.3	25.5 ± 1.4
Toluene	60.9 ± 5.1	50.3 ± 7.9	56.5 ± 2.3	48.3 ± 4.3
CRL Soil				
TDCE	9.52 ± 0.46	8.96 ± 0.23	9.59 ± 0.32	9.2 ± 1.2
Benzene	12.7 ± 0.35	12.7 ± 0.20	7.41 ± 1.8	ND[b]
TCE	13.8 ± 0.70	12.6 ± 0.15	12.4 ± 0.67	11.9 ± 1.5
Toluene	33.1 ± 0.47	31.9 ± 0.97	26.5 ± 3.8	10.0 ± 2.5

[a] Average and standard deviation, $n = 3$.
[b] ND = not detected.

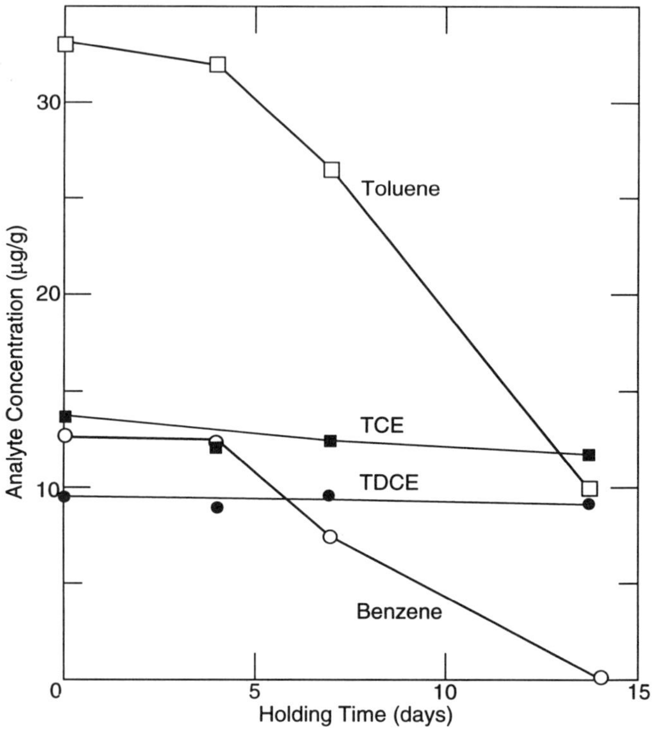

FIG. 2—*Stability of benzene, toluene, TDCE, and TCE concentrations in CRL soil subsamples prepared for PT-GC-MS analysis.*

TABLE 5—*Analyte concentrations (µg/g) for soil subsamples dispersed in 20 mL of MeOH (experiment 5).*

	Room Temperature (22°C)			
	Day 0	Day 30	Day 60	Day 98
RMA Soil				
TDCE	13 ± 0.8[a]	13 ± 1.2	12 ± 1.1	12 ± 1.3
Benzene	15 ± 0.4	15 ± 0.3	15 ± 0.4	14 ± 0.6
TCE	17 ± 0.5	18 ± 1.8	17 ± 0.5	16 ± 0.6
Toluene	22 ± 0.7	20 ± 0.5	22 ± 1.5	20 ± 1.8
PBA Soil				
TDCE	39 ± 1.4	39 ± 0.6	35 ± 3.7	38 ± 1.7
Benzene	42 ± 0.9	41 ± 0.8	38 ± 1.9	38 ± 1.3
TCE	59 ± 1.7	60 ± 1.4	56 ± 1.4	58 ± 1.9
Toluene	68 ± 2.2	71 ± 3.0	64 ± 4.8	65 ± 6.0

[a] Average and standard deviation, $n = 3$.

aliphatic organic compounds resist degradation because of their oxidized state (Russell et al. 1992). The analyte stability characteristics observed during this study were consistent with holding time studies for natural waters and with studies of biological degradation under aerobic conditions (Fig. 2). Maskarinec et al. (1990) observed that, in general, chlorinated aliphatic compounds were more stable than non-chlorinated aromatic hydrocarbons. Roe et al. (1989) found benzene to be more rapidly degraded than toluene.

Furthermore, the rates of degradation of these two hydrocarbons increased when subsamples were prepared for analysis by either HS-GC or PT-GC-MS, most likely because of the increased amount of oxygen that was available once the ampoules had been broken inside of the VOA vials.

Both the activity of microorganisms and the amount of total organic carbon (TOC) in soils vary widely; however, often they are correlated. This, combined with the sensitivity of microbial activity to temperature, should result in VOC concentration stabilities that depend on the soil type and temperature. Thus, it is not surprising that the RMA soil with 0.053% TOC showed greater fortified analyte stability than the CRL soil, which was taken from the top horizon and contained 0.34% TOC (Tables 4 and 5). Moreover, analyte concentration stability improved at the lower storage temperature (Fig. 1).

These experiments successfully prevented volatilization losses, as shown by the stability of the two chlorinated compounds and by benzene and toluene in the RMA soil matrix. In particular, trans-1,2-dichloroethylene, the compound with the highest vapor pressure, was remarkably stable during these different tests.

Overall, these experiments show that VOC losses, presumably caused by biological degradation, are likely to depend on analyte, soil type, holding time and storage temperature. In terms of providing protocols for all soils and VOCs of concern, the findings of this study show that analyte concentrations were constant over a 14-day period only after they were dispersed in MeOH or sealed in a glass ampoule and held at 4°C. The latter method, while impractical for field applications, shows that these analytes were not likely lost in the subsequent experiments because of sorption onto the soil. Keeping in mind that soil samples should only be transferred once for the most representative VOC concentrations, site investigators must place subsamples into vessels that are suitable for analysis or that contain a solvent in which VOCs are very soluble (Hewitt 1992, 1994c). Clearly, if samples are prepared for either HS-GC or low-level PT-GC-MS (not immersed in MeOH), they should be analyzed within a couple of days.

For practical reasons a holding time of less than 14 days will seldom be achieved when samples are shipped off-site for laboratory analysis. Sample preservation is the most reasonable way to maintain representative VOC concentrations in soils when holding times of several days cannot be avoided. If MeOH cannot be used, then another chemical preservative will be necessary. When subsamples are taken in the field for either HS or low-level purge-and-trap using only a single transfer step, it would be necessary to introduce a preservative prior to collection to prevent biodegradation. Another potential advantage that might be gained by using chemical preservatives is that refrigeration may no longer be necessary.

Conclusion

Holding time studies for VOCs in a soil matrix without confounding volatilization losses found that concentration stability depended on subsample protocol configuration, analyte, soil type, temperature, and holding time. Soil subsamples prepared for either HS-GC or PT-GC-MS analysis and held without chemical preservatives to inhibit biological degradation will often fail to maintain representative concentrations of aromatic hydrocarbons when held for several days at 4°C. When soil samples are dispersed in MeOH, holding time and storage temperature (22 or 4°C) appear to be of little concern.

Acknowledgments

Funding for this work was provided by the U.S. Army Environmental Center (formerly the U.S. Army Toxic and Hazardous Materials Agency), Martin H. Stutz, project monitor. The author thanks Marrianne Walsh and Louise Parker for critical reviews of the text, and Philip Thorne for assistance in the analysis of samples.

This publication reflects the views of the author and does not suggest or reflect policy, practices, programs, or doctrine of the U.S. Army or of the Government of the United States. The contents of this report are not to be used for advertising or promotional purposes. Citation of brand names does not constitute an official endorsement or approval of the use of such commercial products.

References

Hewitt, A. D., 1992, "Review of Current and Potential Future Sampling Practices for Volatile Organic Compounds in Soil," *Proceedings,* 16th Annual Army Environmental R&D Symposium, June, Williamsburg, VA, pp. 75–82.

Hewitt, A.D., 1993, "Feasibility Study of Vapor Fortification Preparation of Volatile Organic Compound Soil Samples," U.S. Cold Regions Research and Engineering Laboratory Special Report 93-5, Hanover, NH.

Hewitt, A. D., 1994a, "Comparison of Sampling Methods for Trichloroethylene in Vadose Zone Soils," *Journal of the Association of Official Analytical Chemists,* Vol. 77, No. 2, pp. 458–463.

Hewitt, A. D., 1994b, "Preparation of Spiked Soils for Volatile Organic Compound Analysis by Vapor Fortification," *Journal of the Association of Official Analytical Chemists,* Vol. 77, No. 3, pp. 735–737.

Hewitt, A. D., 1994c, "Concentration Stability of Four Volatile Organic Compounds in Isolated Soil Subsamples," U.S. Cold Regions Research and Engineering Laboratory Special Report 94-6, Hanover, NH.

Hewitt, A. D., Miyares, P. H., Leggett, D. C., and Jenkins, T. F., 1992, "Comparison of Analytical Methods for Determination of Volatile Organic Compounds in Soils," *Environmental Science and Technology,* Vol. 26, No. 10, pp. 1932–1938.

Jackson, J., Thomey, N., and Dietlein, L. F., 1991, "Degradation of Hydrocarbons in Soil Samples Analyzed within Accepted Analytical Holding Times," *Proceedings,* 5th Outdoor Action Conference on Aquifier Restoration, Ground Water Monitoring, and Geophysical Methods, May 13–16, Las Vegas, NV, pp. 567–576.

King, P., 1993, "Evaluation of Sample Holding Times and Preservation Methods for Gasoline in Fine-grained Sand," *Proceedings,* National Symposium on Measuring and Interpreting VOCs in Soils: State of the Art and Research Needs, Jan. 12–14, Las Vegas, NV, Tab 19.

Lewis, T. E., Crockett, A. B., Siegrist, R. L., and Zarrabi, K., 1991, "Soil Sampling and Analysis for Volatile Organic Compounds," EPA/590/4-91/001, U.S. Environmental Protection Agency, pp. 1–24.

Maskarinec, M. P., Johnson, L. H., Holladay, S. K., Moody, R. L., Bayne, C. K., and Jenkins, R. A., 1990, "Stability of Volatile Organic Compounds in Environmental Water Samples During Transportation and Storage," *Environmental Science and Technology,* Vol. 24, No. 11, pp. 1665–1670.

Maskarinec, M. P., Bayne, C. K., Jenkins, R. A., Johnson, L. H., and Holladay, S. K., 1992, "Stability of Volatile Organics in Environmental Soil Samples," Office of Scientific and Technical Information, ORNL/TM-12128, Oak Ridge, TN, pp. 1–81.

Parker, L. V. and Ranney, T. A., 1993, "Effect of Concentration on Sorption of Dissolved Organics by Well Casings," U.S. Cold Regions Research and Engineering Laboratory Special Report 93-8, Hanover, NH.

Parker, L. V., Hewitt, A. D., and Jenkins, T. F., 1990, "Influence of Casing Materials on Trace-level Chemicals in Well Water," *Ground Water Monitoring Review,* Vol. 10, No. 2, pp. 146–156.

Plumb, R. H., Jr. and Pitchford, A. M., 1985, "Volatile Organic Scans: Implications for Ground Water Monitoring," *Proceedings,* National Water Well Association/American Petroleum Institute Conference on Petroleum Hydrocarbons and Organic Chemicals in Ground Water, Nov. 13–15, Houston, TX, pp. 13–15.

Roe, V. D., Lacy, M. J., Stuart, J. D., and Robbins, G. A., 1989, "Manual Headspace Method to Analyze for the Volatile Aromatics of Gasoline in Groundwater and Soil Samples," *Analytical Chemistry,* Vol. 61, No. 6, pp. 2584–2585.
Russell, H. H., Matthews, J. E., and Sewell, G. W., 1992, "TCE Removal from Contaminated Soil and Groundwater," EPA/540/S-92/002, U.S. Environmental Protection Agency, pp. 1–16.
Siegrist, R. L. and Jenssen, P. D., 1990, "Evaluation of Sampling Method Effects on Volatile Organic Compound Measurements in Contaminated Soils," *Environmental Science and Technology,* Vol. 24, No. 9, pp. 1387–1392.
Spittler, T. M., 1989, Personal communication, U.S. Environmental Protection Agency, Environmental Services Division–Region 1, Lexington, MA.
Suflita, J. M., 1989, "Transport and Fate of Contaminants in the Subsurface," U.S. Environmental Protection Agency, Cincinnati, OH, and Ada, OK, Technical Report No. EPA/625/4-89/019, pp. 1–67.
Urban, M. J., Smith, J. S., Schultz, E. K., and Dickinson, R. K., 1989, "Volatile Organic Analysis for a Soil Sediment or Waste Sample," *Proceedings,* 5th Annual Waste Testing and Quality Assurance Symposium, U.S. Environmental Protection Agency, Washington, DC, pp. II-87–II-101.
U.S. Environmental Protection Agency, 1986, "Test Methods for Evaluating Solid Waste," Vol. 1B, Washington, DC.
Zarrabi, K., Cross-Smiecinski, A. J., and Starks, T., 1991, "A Review of Existing Soil Quality Assurance Materials," *Proceedings,* Second International Symposium, Field Screening Methods for Hazardous Waste and Toxic Chemical, Feb. 12–14, Las Vegas, NV, pp. 235–252.

Oldrich Mazac,[1] *Ivan Landa,*[1] *John R. Rohde,*[2] *William E. Kelly,*[2] *and Jan H. Blaha*[3]

Experiences with a New Soil Gas Technique for Detecting Petroleum Pollution

REFERENCE: Mazac, O., Landa, I., Rohde, J. R., Kelly, W. E., and Blaha, J. H., **"Experiences with a New Soil Gas Technique for Detecting Petroleum Pollution,"** *Volatile Organic Compounds in the Environment, ASTM STP 1261,* W. Wang, J. Schnoor, and J. Doi, Eds., American Society for Testing and Materials, 1996, pp. 192–200.

ABSTRACT: This paper presents field experiences obtained with a new technology for detecting petroleum pollution in soil and ground water based on "in situ" determination of hydrocarbon concentrations in soil air. Ecoprobe is a new soil gas device from RS-Dynamics in the Czech Republic. The rugged waterproof device is equipped with a built-in computer-controlled semiconductor sensor. Three case histories are presented that demonstrate the use of the equipment under typical conditions. Two case histories present the use of the device under typical field conditions; the third case history compares results from the Ecoprobe and a commercial photoionization detector (PID) device.

KEYWORDS: hydrocarbon, soil gas, ground water, pollution, detection

Almost everywhere there are service stations, pipelines, and storage tanks which are sources or potential sources of petroleum leakage and, hence, sources of water and groundwater pollution by hydrocarbons. Pollution sources can be classified as either point sources in the cases of petroleum or gas leaks from storage tanks, or line sources as in the case of leakages from petroleum pipelines.

Petroleum pollution in the Czech Republic is significant. For example, leakage in the area of the International Airport at Prague has caused large-scale ground water pollution. During remediation between 1975 and 1990, more than 326 000 L of petroleum product was pumped from the local ground water. In nearby Bratislava, the capital of the Slovak Republic, more than 89 million L of petroleum product were pumped from the ground water in the Slovnaft–Bratislava area from 1973 to 1989.

Increasingly, hydrocarbon pollution is being found around older service stations as a result of leaking fuel storage tanks. Pollution is typically detected by various indirect methods such as fluorescence, botanic, geophysical, atmochemical, and methods based on measurement of changes in redox-potential and dissolved oxygen content. In this paper attention is concentrated on a specific variant of the atmochemical method, that is, soil-gas surveying utilizing the newly developed Ecoprobe device from RS-Dynamics.

For detection of contamination in soil air, devices can be divided into three groups according to the type of detector:

[1] Principals, ECOLAND, Box 512, 111 21 Prague 1, Czech Republic.

[2] Associate professor and professor, respectively, Department of Civil Engineering, University of Nebraska–Lincoln, Lincoln, NE 68588-0531.

[3] Principal, RS-Dynamics, Box 512, 111 21 Prague 1, Czech Republic.

(*a*) photoionization detector (PID).
(*b*) flame ionization detector (FID), and
(*c*) colorimetric tube (CT).

The Ecoprobe device cannot be classified as any one of these since it uses a new type of semiconductor sensor based on the quantum effect which was designed especially for the Ecoprobe device. The sensor does not require any gas for operation and, because it operates at low temperature, it suffers no side effects from combustion for any part of the measured spectrum.

The most recent version of the device is the Ecoprobe 30, which provides for simultaneous three-parameter measurement with graphic output of:

(*a*) soil gas concentration,
(*b*) soil gas temperature, and
(*c*) soil structure (permeability).

Optimal combination of these three parameters provides the user with the advantage of immediate integral information about the contaminant, the contaminant origin, and the soil structure. A computer-controlled semiconductor sensor makes the device sensitive (better than 0.5 ppm by volume) and provides excellent zero stability (better than 0.01 ppm) plus a large dynamic range. Computerized data collection is supported by the system's graphic software, which can produce correlation graphs of concentration level, the temperature of the measured gas, and the response time, which varies from 5 to 15 s.

The design concept was to provide an integral system that utilizes a tested measuring method in a field device with real-time graphical output, and graphic interpretation software that provides for transmission of data via modem if desired. The unit's rugged waterproof construction, using military grade components, permits reproducible operation under the harshest of field conditions.

Features of the Device

A major advantage of the device, shown in Fig. 1, is the capability to do in situ surveys (Fig. 2) without the necessity for laboratory measurements. Three parameters are measured simultaneously with direct graphic output on the device screen. Zero stability is better than 0.01 ppm, with a detection limit of 0.1 ppm, and a dynamic range from 0.5 to 40 000 ppm (100 000 ppm optional). Data can be transferred directly or via modem and the graphic software in the instrument is capable of printing correlation graphs of the three measured parameters to the screen or to an optional bubble-jet or dot matrix printer. Automatic temperature correction provides results which are independent of the soil gas temperature within the range of −20 to 60°C. Individual multipoint calibrations for three hydrocarbons are stored in the device memory with computer-controlled self-calibration for the calibration check and recalculation. The built-in vacuum pump and rechargeable battery provide approximately 120 measurements per charge. The device is compact (140 × 100 × 160 cm), lightweight (2.9 kg), and includes a battery charger and computer accessories.

Modes of Operation

Three modes of operation are possible.

FIG. 1—*The Ecoprobe device (140 × 100 × 160 cm), probe, and carrying case.*

FIG. 2—*Making a field measurement with the Ecoprobe.*

Soil Contamination Measurement

This is the primary application mode for which the system was designed. In this mode it is possible to identify pollution sources, their extent and intensity, and the dominant paths of pollutant movement. Field surveys can be performed easily by sampling a network of drill holes. As with other methods for soil gas measurement, the results can be represented as relative concentration levels if actual contaminants are not identified. With a specific contaminant a multipoint calibration standard may be used to develop absolute measurements.

In contrast to other soil gas measurement methods, because all of the extracted soil gas is tested with temperature being measured in real time, the device is not susceptible to variations due to soil structure, gas temperature, ambient temperature, condensation of the measured gas on the sampling tube, time effects of the measured gas response, and many other disturbances which could otherwise cause significant fluctuations in measured results. The multisampling measuring method using the computer-controlled zeroing procedure, multiparameter measurement, and graphic data interpretation ensures that short peaks will not be missed when sampling in wet clays or other impervious soils because all of the soil gas is being tested during extraction. Unlike other methods, results will not be disturbed when sampling a line of boreholes with varying soil permeability along the line, again because of continuous testing and the ability to distinguish natural methane associated with higher temperatures, and because soil gas temperature is measured during the sampling.

Integral Measurement of Air Contamination

In this mode the device can be used to take air samples. The resulting measurement gives the hydrocarbon contamination in the air in parts-per-million volume.

Leakage Detection

In this mode the device is used to make continuous measurements of leakage intensity along a pipe fitting or around rotating seals or shafts. Results are given in numeric or in graphic form.

Operation of the Device

The built-in vacuum pump transports gas or vapor through the reaction chamber housing the semiconductor sensor; the sensor does not require any operating gas. Calibration is provided for three hydrocarbons using three individual cubic spline recalculations that ensure a best-fit approximation over a large dynamic range.

The measuring process consists of three presetable stages: reset, pre-integration, and integration. During the reset stage the built-in software continuously monitors the exponential decay of the sensor time-response and halts the reset procedure as soon as the process is satisfactorily stabilized. This ensures perfect cleaning of the reaction chamber that might otherwise eventually become contaminated due to a high previously measured concentration. In this way the sensor is prepared to measure on the most sensitive range with no systematic error. Reset time is, however, dependent on previous concentration values and may vary within the interval of about 15 to 50 s.

In the pre-integration stage, several important operations to ensure perfect zero stability are automatically performed and the sampled air is transported to the reaction chamber. This stage accounts for soil gas flow.

In the integration stage the input sensor signal is processed at the rate of one sample per second. The device stores all measured values and determines the numerical average and peak value as an average of the three highest values. The results are stored and identified by the line and point numbers of the measurement network. At this time, the device also stores the measured gas temperature and calculates the relative rising and falling slopes of the measured sample's envelope to characterize the soil permeability. In the continuous survey mode these processes are repeated until the operator halts the survey.

Data Storage and Handling System

The device's internal data management system provides automatic data handling that organizes the measured data along lines and grids. The number of lines and the number of stations on each line are flexible.

It is possible to insert additional lines and points within the standard network. The software allows the operator to jump from one line to another, observe the data, add a point, skip a point, and so on. Storage is provided for more than 30 000 measured values. The operator can choose to save only concentrations or all values including temperatures, calculated slopes, station values describing the value time response, or any combination of these values. All measured data can be transferred easily for further processing either directly or via modem. The system software Ecoprobe provides for data acquisition in graphic form and for conversion to other formats for easy import to many common graphic processors.

Case Histories

Prague, Czech Republic

A soil gas contour map shown in Fig. 3 was developed for a case of hydrocarbon contamination at an AGIP service station site located in a heavily populated area of Prague in the Czech Republic. The measurements were made along four nearly parallel (L1 to L4) and one perpendicular (L5) line with soil air sampled at a constant depth of about 1 m below the surface. The survey was conducted during a check for possible fuel leakage from the station's fuel tanks. A background concentration was calculated from the observed values along each line. The contours are given as means of the ratio of observed to background concentration values. The absolute concentration values in ppm are shown in boxes. The background concentration value for line L4 is 25 ppm, and for lines L1, L2, L3, and L5 it is 10 ppm. Two anomalous zones were found. The western zone extends in a nearly north-northeast to south-southwest direction corresponding to the direction of ground-water flow. At the center of the anomaly the ratio exceeds 100 and the absolute concentration is about 1500 ppm. The eastern zone extends in a northwest to southeast direction corresponding to the strike of an underlying fault system. The maximum anomaly is about 1000 times the background value and the absolute concentration exceeds 17 000 ppm. The fuel tanks appear to be the source of both observed anomalous pollution zones.

Lincoln, Nebraska

Evidence of petroleum contamination was encountered during installation of a sewer line in April 1988. No petroleum storage tanks were at the site but tanks had been removed approximately five years earlier. The tanks had been removed because of problems associated

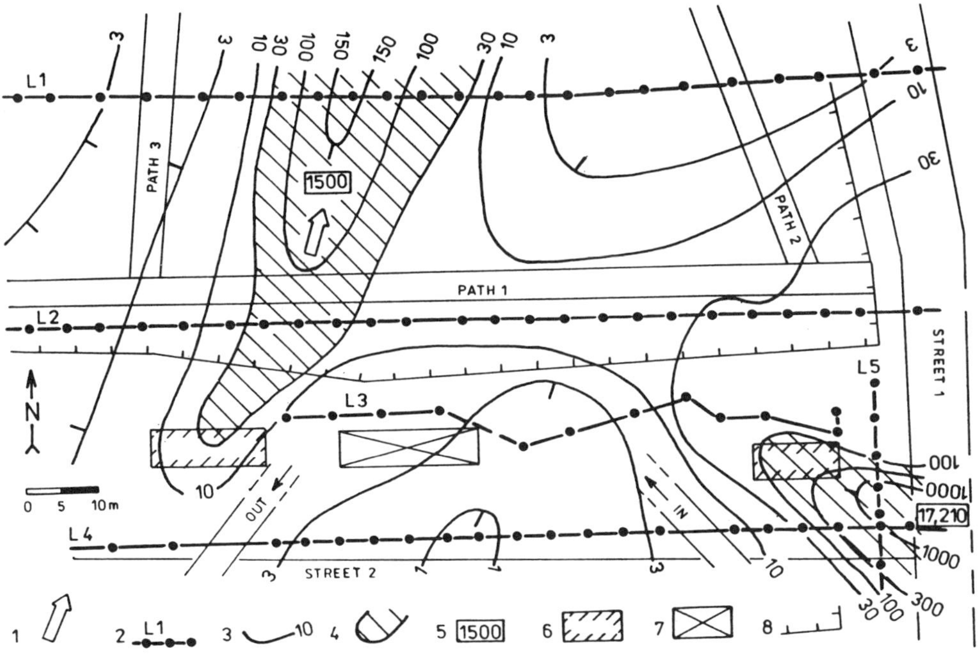

FIG. 3—*Concentration contours for total hydrocarbons in soil gas at a service station site (Prague, Czech Republic). Definitions for legend: (1) direction of ground-water flow, (2) location and number of lines, including observed points, (3) relative concentration contours, (4) anomalous zones, (5) absolute concentration value (ppm) in the anomaly centers, (6) fuel tanks, (7) pump island, (8) fence.*

with the presence of water in the gasoline that suggested some leakage may have occurred at that time.

Hydrogeologic conditions were investigated in 1988 by HWS Consulting of Lincoln, Nebraska. In that investigation, nine boreholes were drilled and soil vapor was analyzed using a HNU photoanalyzer. Water samples were also collected and analyzed for petroleum constituents.

In June of 1993 the site was resurveyed using the Ecoprobe and a Photovac device. The Photovac device is a widely used PID instrument. The main purpose of this survey was to compare the two devices. The results of the gas sampling along a profile where contamination was encountered in the original site investigation are shown in Fig. 4*a* for the Photovac device and Fig. 4*b* for the Ecoprobe. The profile line was selected to cross the anomaly identified in 1988.

Measurements were made at a depth of 0.5 m using both instruments; the measurements, approximately 0.6 m (2 ft) apart, were alternated between the two instruments. Measurements were not performed in the same hole to minimize sampling problems.

The measurement data were smoothed and the smoothed Ecoprobe relative values were converted to absolute values. The correlation between the Photovac and relative Ecoprobe measurements is excellent. Both devices yielded essentially the same results. The profile results indicate low levels of hydrocarbons still exist in the soil air in the area previously delineated with test holes.

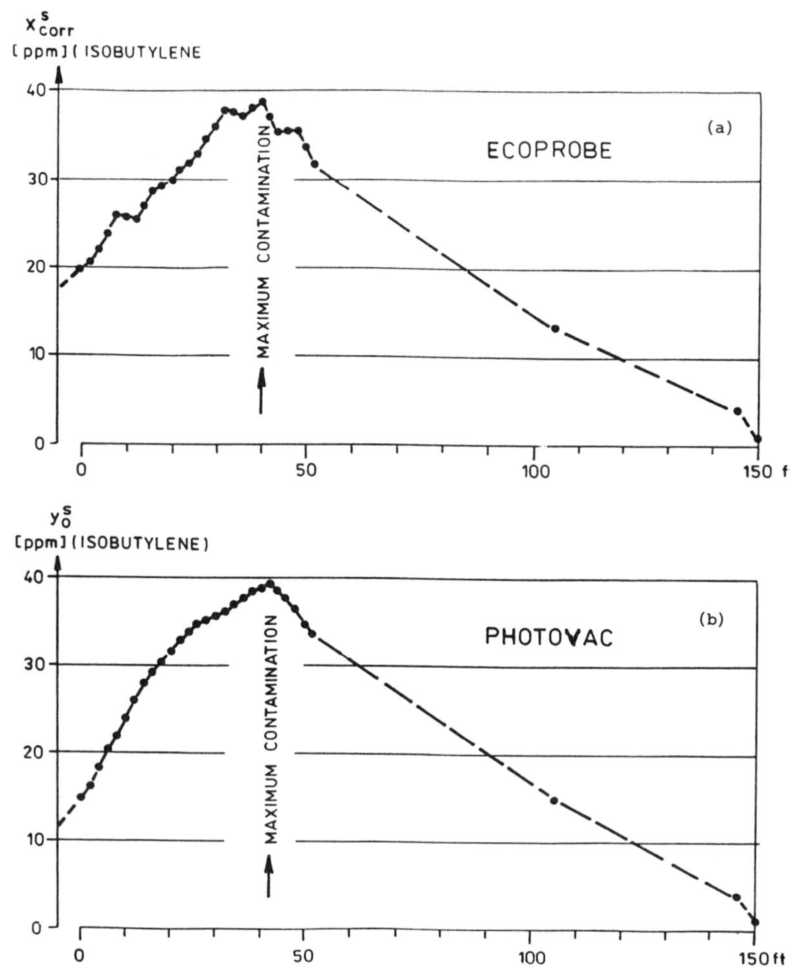

FIG. 4—*Soil gas contaminant concentration values along a profile of a petroleum contaminant plume;* (A) *Using a Photovac instrument, and* (B) *using the Ecoprobe.*

Rudna, Czech Republic

A soil gas survey was made at a state farm near Rudna, a small town about 20 km west of Prague. Geologically, the locality lies within the Barrandian Precambrian. The purpose of the survey was to determine the source of hydrocarbon pollution observed in ground water at the farm. The source of the pollution became controversial when the farm was transferred to its new owner.

Survey results, depicted in Fig. 5, showed the source of the pollution was within the boundaries of the farm with a waste dump and former fueling area being the only likely sources. Follow-up analysis of soil and soil gas revealed that concentration levels requiring remediation were exceeded several times. Based on these results, remediation was agreed to and the work is currently being conducted.

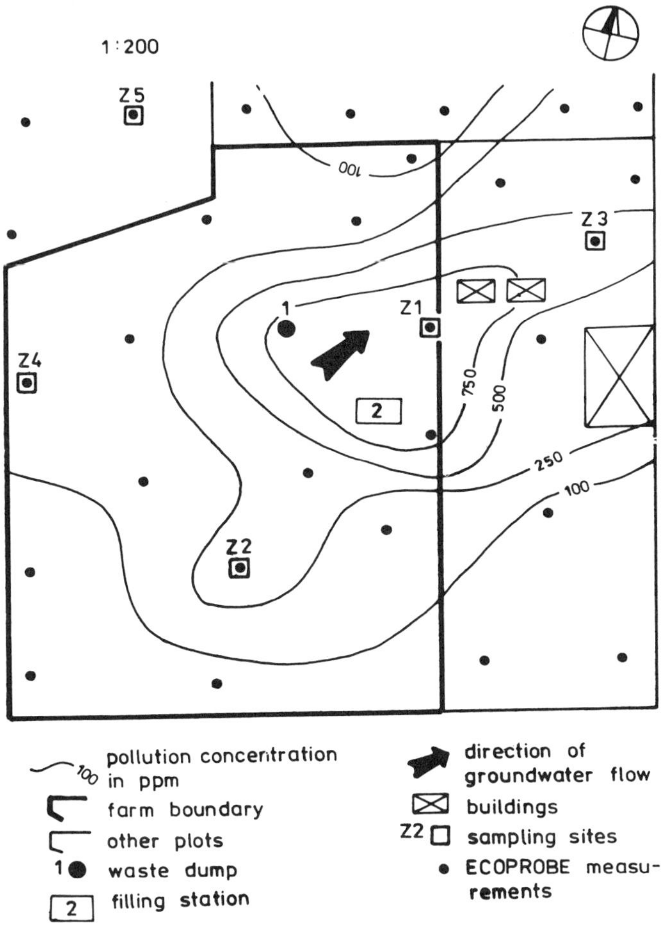

FIG. 5—*Soil gas survey results from Runda, Czech Republic, 1993.*

Conclusions

The Ecoprobe 30 has a number of advantages which makes it very convenient for soil gas surveys. Some advantages include: pollutant concentrations are determined in situ; no operating gas is required; measurements are independent of the temperature of the medium within the temperature range from −20 to 60°C; measurements can be made rapidly with the average sensor response time being 5 to 10 s and the sensor recovery time of 10 to 50 s depending on the previous concentration value; and up to about 200 measured points are possible on one fully charged battery. The Ecoprobe produces, like many other field instruments, relative measurements of unknown hydrocarbons that are adequate for initial field screening of contaminant plumes.

Three case histories demonstrate the use of the device under field conditions. Two of the case histories are for locations in the Czech Republic near Prague; they demonstrate the use of the device in actual remediation decisions. The third case history is from a site in Lincoln, Nebraska in the United States. This case history demonstrates the use of the Ecoprobe 30 side-by-side with a commonly used PID device. The results obtained with the Ecoprobe correlate very well with those obtained with the PID device.

Acknowledgments

The authors are pleased to acknowledge the assistance of M. Hagenmeister and B. Simons, who assisted in the field survey and made the Photovac measurements; also A. Hottovy of HWS who provided results of the 1988 survey, and J. Schuckebier of the University of Nebraska–Lincoln who helped arrange the measurements in Lincoln.

Environmental Monitoring

Cynthia P. Ardito,[1] *Thomas A. Duval,*[1] *and Douglas W. Reaber*[1]

Characterizing a VOC Site Through On-Site Analysis of Soil Gas

REFERENCE: Ardito, C. P., Duval, T. A., and Reaber, D. W., **"Characterizing a VOC Site Through On-Site Analysis of Soil Gas,"** *Volatile Organic Compounds in the Environment, ASTM STP 1261,* W. Wang, J. Schnoor, and J. Doi, Eds., American Society for Testing and Materials, 1996, pp. 203–218

ABSTRACT: An alternative approach for characterization of sites contaminated by volatile organic compounds (VOCs) has been employed at Sandia National Laboratories, Albuquerque, New Mexico (SNL/NM), in order to characterize a plume from a chemical waste landfill (CWL). The characterization approach is designed to maximize the use of soil gas analysis for definition of the nature and extent of contamination and for validation of the conceptual model of contaminant transport. Comparison of the soil gas data analyzed at an on-site laboratory with those data generated by an EPA-approved analytical laboratory indicates that soil gas data provide a much more representative picture of site conditions. To assist in the validation of our conceptual model of contaminant transport for the CWL, we employed an approach developed at SNL/NM that allows for assessment of nonaqueous phase liquids (NAPLs) through equilibrium testing of the soil gas data.

Ground water beneath the CWL, approximately 146 m below ground surface (BGS), has been impacted by low levels of VOCs. Laboratory data from soil samples collected through split-spoon sampling techniques revealed detectable contaminants to a maximum depth of approximately 6 m BGS. However, soil gas concentrations revealed a contaminant plume at concentrations in the range of 100 ppm_v at depths greater than 60 m BGS. We demonstrate in this paper that soil gas data can provide a much more accurate picture of site conditions.

KEYWORDS: volatile organic compound (VOC), dense nonaqueous phase liquid (DNAPL), ground water, contamination, characterization, soil gas

To date, soil gas analysis is typically only used for screening a site to optimize the locations of soil borings where soil samples can be collected for subsequent analysis by an analytical laboratory. However, we believe analysis of soil gas samples, preferably using a field laboratory, provides a more realistic picture of site conditions and the basis for better decision making. Characterizing volatile organic compounds (VOCs) using data from soil samples analyzed by an off-site laboratory can be misleading. Volatilization of the contaminants from a soil (or water) sample during field handling is unavoidable. Perhaps even more problematic are the laboratory methods used for analysis of these chemicals.

Approved laboratory methods of analysis for the presence of VOCs [Environmental Protection Agency (EPA) Method 8240 or 8260] are inherently conducive to the loss of these compounds. In today's regulatory climate, the laboratory data are typically used for decision making; field analyses are generally not accepted. Therefore, it is possible that important decisions, such as sources and extent of contamination, are made through evaluation of biased data.

[1] Project manager, project hydrologist, and project hydrogeologist, respectively, INTERA Inc., 1650 University Blvd. NE, Suite 300, Albuquerque, NM 87102.

In response to the discovery of trichloroethylene (TCE) in ground water at the Sandia National Laboratories, New Mexico (SNL/NM) Chemical Waste Landfill (CWL), an alternative strategy for characterizing the contamination at this site has been developed. Through data interpretation and discussion we will demonstrate the following:

1. Reliable soil gas samples can be collected and analyzed in the field, providing for a high level of cost effectiveness.
2. Soil gas samples can provide crucial information for sites contaminated by VOCs.
3. Depending on the nature of contaminant transport at a given location, the value of collecting destructive soil samples for laboratory analysis should be questioned.

Site Background

The CWL was used for the disposal of hazardous and chemical waste through the years 1962 to 1985. Low levels of TCE, a dense nonaqueous phase liquid (DNAPL), were detected in ground-water samples from a water table aquifier approximately 146 m BGS. The CWL is approximately 0.8 hectares in areal extent and contains separate disposal pits known to have been used for the disposal of acids, oxidizers, reducers, organics, reactives, bulky materials, metal, neutral compounds, and salts (Fig. 1). Written records of disposal practices were not kept between 1962 and 1975, resulting in a large degree of uncertainty concerning the actual locations of waste pits at the site and the actual volumes of different types of wastes disposed there. It was inferred from preliminary reconnaissance studies of the CWL that the waste pits were uniformly distributed around the full extent of the CWL. Based on known disposal rates and volumes, a conservative estimate of the volume of waste disposed at the landfill is 600 m^3. Liquid organics in the form of solvents and waste oils comprised the largest percentage of wastes disposed at this site.

The average precipitation for the Albuquerque area is 22.4 cm per year. During the summer monsoon season (July to September) up to 2.5 cm of rain can be discharged in one event. Recent environmental isotope recharge studies conducted by the U.S. Geological Survey (USGS) suggest "negligible" recharge in this area.

The sediments beneath the CWL are heterogeneous, with discontinuous sedimentary layers of alluvial and possibly fluvial origin that are composed of varying percentages of sand, silt, and clay. Ground water is at a depth of approximately 146 m BGS, with a northwest local ground-water flow direction. Ten monitor wells were installed at the site; five of these wells have been approved by State of New Mexico regulators for assessment monitoring (Fig. 1). The contaminants of concern at the CWL are chlorinated solvents that were sometimes mixed with oils during the bulking process.

Conceptual Model Development/Implications for Site Investigation Strategy

The physical characteristics of DNAPLs provide for transport over a thick (146 m) unsaturated zone. To date, only minor impacts to ground-water quality [that is, concentrations near the EPA's drinking water standard of 5 parts per billion (ppb)] have been identified. For a compound such as TCE, the pathways are specifically the three fluid phases present in the unsaturated zone beneath the CWL: (1) the aqueous phase (that is, dissolved in soil water), (2) a separate organic liquid phase (that is, DNAPLs), and (3) the gas phase (that is, soil gas).

If contaminants were carried in the water as an aqueous phase, transport would be accomplished by the dissolution of a contaminant into precipitation infiltrating the CWL and mi-

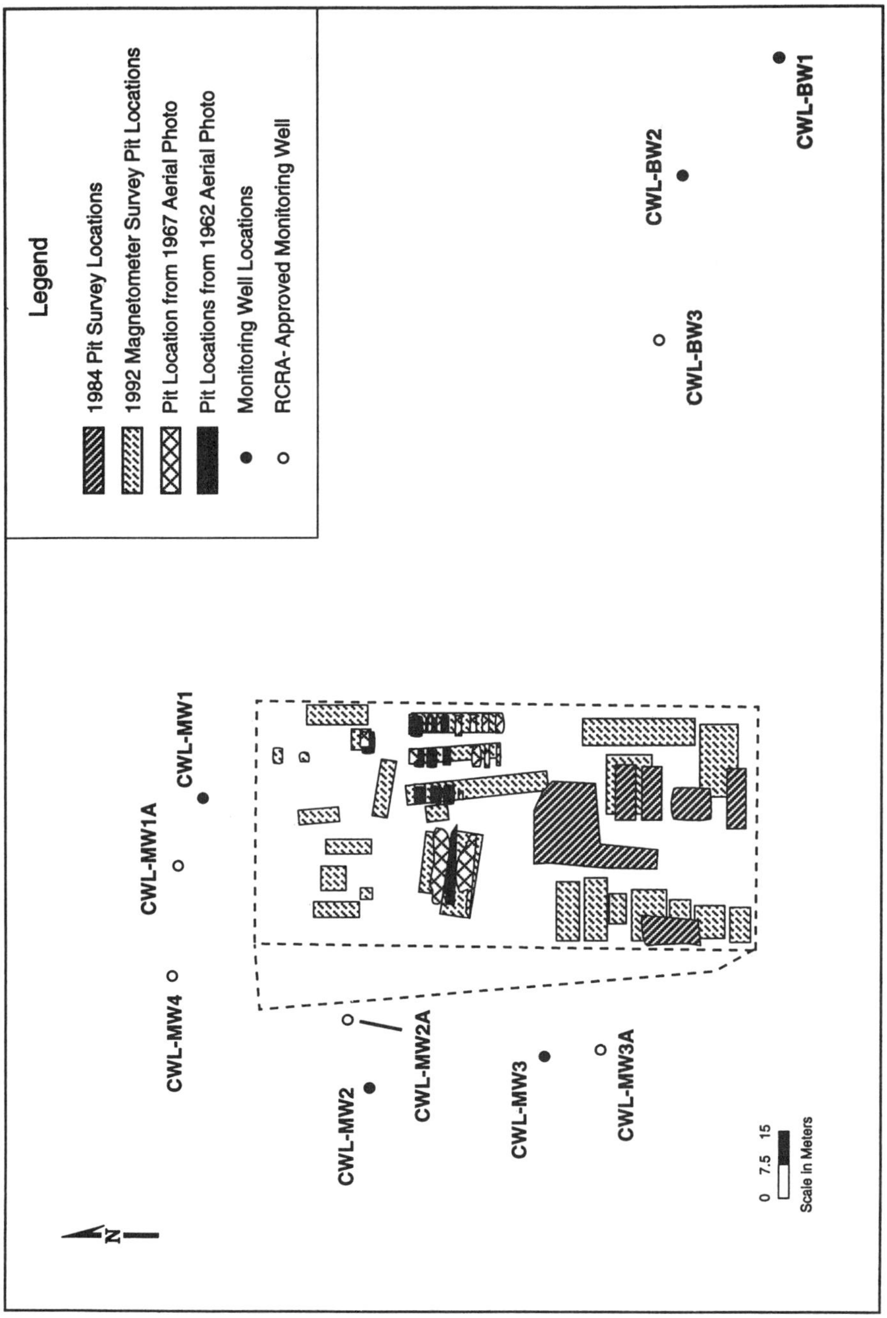

FIG. 1—*Chemical waste landfill base map.*

grating to the water table. Because of New Mexico's arid environment, only a small flux of rainwater infiltrates the CWL. Therefore, this is the least likely pathway for solute transport.

Transport of organic liquids as a separate DNAPL is the second possible migration pathway through the unsaturated zone. For a separate organic-liquid phase to move, disposal of large volumes of organic solvents and oils is necessary. Once the organic-liquid phase reaches the water table, the soluble constituents dissolve into the ground water and the dense liquid sinks. As a preliminary indication of site conditions, SNL/NM performed conservative calculations of contaminant transport based on estimates of organic liquid volumes disposed of in the CWL (SNL 1992). Assuming all of the organic liquid (estimated from inventory records) was disposed there at one time, and employing a simple one-dimensional model for transport with a high permeability medium, a separate liquid phase could only have migrated to approximately 90 m BGS.

The third potential migration pathway is gas-phase advection and diffusion through the unsaturated zone. As the gas phase reaches the water table, ground water would be impacted through dissolution from the gas phase to the liquid phase. As discussed in Mendoza and Frind (1990), TCE volatilizes to form a mixture with soil gas that is about 1.35 times as dense as uncontaminated soil gas. They showed that under favorable conditions it is possible to achieve density-driven transport of VOCs. Although sediments under the CWL are fairly coarse and dry, the highly layered soil structure makes density-driven transport more unlikely. SNL/NM (1992) conservatively calculated 11 years as the necessary time for travel of a gas plume to the water table through gaseous diffusion alone.

As confirmed through the investigation described in detail below, TCE migrated to the water table beneath the CWL through multiple pathways (that is, a combination of a DNAPL plume in the unsaturated zone with gas phase transport of contaminants that have volatilized from the DNAPL plume and migrated to the ground water). As the organic liquid phase moves through the unsaturated zone, a residual is left behind. Further migration of contaminants from the residual phase can only occur through evaporation of the contaminant from this sorbed phase and subsequent movement of the contaminant in the gas phase.

With this conceptual model in mind, the site investigation strategy described below was designed to focus on the collection of soil gas data to guide the investigation and optimize the sampling locations. Though required by state regulators to take destructive soil samples (that is, split-spoon samples), we realized at the start of the investigation that the soil samples would probably not provide very useful information because the contaminant would be present predominantly in the gas phase.

Sample Collection and Laboratory Analysis Methodology

Field operating procedures (FOPs) used to collect soil gas samples in the shallow subsurface (0 to 9.1 m) and in the deep subsurface (9.1–67.1 m) are described next. These procedures were developed at SNL/NM to ensure that acceptable and consistent soil gas samples were collected.

Shallow Soil Gas Sampling

We used a Model 8-MU Geoprobe to collect samples in the upper 9.1 m of the CWL (Fig. 2). Using this equipment, we found sampling gas at locations below 9.1 m was often unreliable, difficult, and time consuming. To collect samples, we advanced the Geoprobe to the desired depths by drive point (typically 3.1, 6.1, and 9.1 m) and withdrew the probe approximately 15 cm to create an airspace beneath the probe and to allow the drive head to

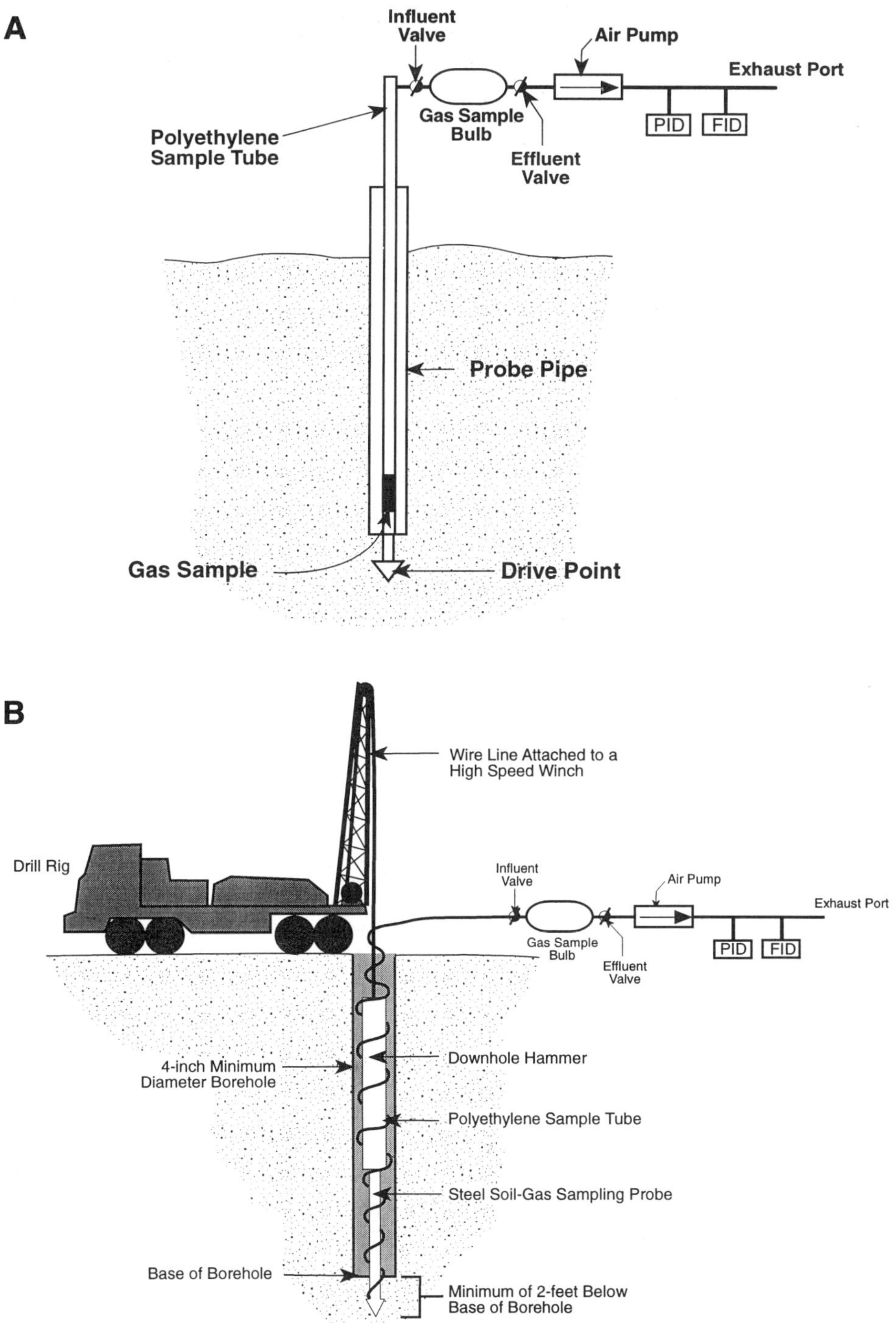

FIG. 2—*Schematic of equipment setup for shallow and deep soil gas sampling.*

slide out of the probe exposing the intake screen. Polyethylene tubing was attached to the Geoprobe drive head and was unrolled from a spool as the probe advanced.

At each sampling location, the tubing was connected to a 500 mL glass bulb and air was drawn through the bulb with an SKC constant-discharge air sampling pump. Using the SKC air pump allowed us to monitor the discharge through the tubing with an organic vapor meter (OVM). Figure 2*a* illustrates the sampling equipment configuration for shallow soil gas samples. The polyethylene tubing is connected to the glass bulb influent fitting and the effluent fitting is connected to the air pump. A "T" fitting was placed in the air pump discharge line to monitor the discharge with a Thermoenvironmental 580B photoionization detector (PID) and a Sensidyne flame ionization detector (FID). These data were used to optimize the sampling time (that is, when the instrument readings stabilized, indicating the soil gas being discharged through the lines was in equilibrium with the surrounding soils). At least three equivalent pore volumes of the sampling tubing were evacuated prior to sampling. Additionally, the PID/FID readings were used for data validation and to provide an indication of any leaks in the sampling system.

After the lines were purged and the PID/FID readings stabilized, the soil gas sample was collected by first closing the effluent fitting on the glass bulb, allowing the bulb to equilibrate to atmospheric pressure, and then closing the influent fitting. At approximately 20% of the sampling locations, SUMMA canister samples were collected and sent to an EPA-approved analytical laboratory for analysis.

Deep Soil Gas Sampling

Figure 2*b* illustrates the sampling equipment setup for deep soil gas sampling. Soil gas samples were collected from depths greater than 9.1 m BGS by using the techniques described above and adapting the Geoprobe sampling equipment to a dual-tube percussion drill rig. We found that retrieval of the soil gas probe was more efficient when the probe was connected to a wire line operated with a high-speed winch. Additionally, the technique required a borehole of at least 10.2 cm in diameter to allow the drill rig to cleanly drive the soil gas probe into the base of the borehole. A dual-tube percussion drill rig was used because this technology is efficient for drilling in heterogeneous sediments without the aid of drilling fluids to depths greater than 30 m.

After the borehole was advanced to the sampling location, the drill rig hammer was removed from the borehole and the Geoprobe sampling rod was threaded onto the end of the drill rod. Polyethylene tubing was attached to the Geoprobe tip. Plastic ties were used to keep the tubing from kinking in the borehole. The sample probe was advanced 5.1 to 7.6 cm into the undisturbed soil below the borehole and the drive mechanism was pulled back approximately 15.2 cm to expose the probe tip. Collection of the soil gas samples was accomplished in the same manner as described above for the shallow gas survey.

Soil Sample Collection

Soil samples were collected with a split-spoon sampler following SNL/NM FOPs. The sampler was advanced into the undisturbed soils in advance of the borehole. The soil samples were always taken prior to soil gas sample collection. When the soil samples were removed from the split-spoon sampler, they were immediately put into glass containers with Teflon (polytetrafluoroethylene resin)-coated screw caps in such a manner as to minimize the headspace in the jar and the loss of volatiles during shipment to the laboratory. The samples were shipped to the laboratory in a cooler and analyzed within the specified holding period of 14 days.

Laboratory Analysis

The following section describes the methods used to analyze soil gas samples in SNL/NM's Field Analytical Screening Laboratory (FASL) and split-spoon soil samples at the contract laboratory.

Field Analytical Screening Laboratory

Soil gas samples were analyzed at the FASL with a Viking 600 gas chromatograph/mass spectrometer (GC/MS). This instrument was designed to be durable enough for use in a mobile laboratory. The GC/MS was calibrated to quantify the following compounds to 10 ppm_v (the instrument detection limit was 1 ppm_v):

(*a*) trichlorotrifluoromethane
(*b*) dichloromethane (DCM)
(*c*) carbon tetrachloride
(*d*) 1,1,1-trichloroethane (TCA)
(*e*) trichloroethane (TCE)
(*f*) perchloroethylene (PCE)
(*g*) benzene
(*h*) toluene
(*i*) xylenes

The procedures for gas analysis by the field GC/MS were developed at SNL/NM and are described in detail in SNL/NM (1992). As discussed further here, the data produced from the FASL were validated in accordance with EPA procedures (EPA 1986). Duplicate samples, approximately 7% of the total number of samples, were collected in SUMMA canisters and sent to a contract laboratory for analysis by EPA Method TO14 (EPA 1988).

Contract Laboratory

All soil samples were sent to an EPA-approved laboratory under contract to SNL/NM and analyzed via EPA Method 8240 (EPA 1986). Approximately 5% of the soil samples were analyzed by the "high-level" Method 8240, which involves putting soil samples into a laboratory-prepared glass jar containing methanol. This was done to compare the results to the standard or "low-level" Method 8240 analysis in hopes that the field extraction would help to minimize the loss of volatiles during shipment to the laboratory.

Quality Assurance/Quality Control (QA/QC) Program

The quality of analytical data for the list of parameters specified above was assessed by the following data quality indicators: analytical method reporting limits, accuracy, precision, completeness, representativeness, and comparability (EPA 1986). In general, there were no reportable occurrences of QA/QC violations resulting from the analysis of this data set. All of the soil sample data described in this paper are acceptable as Level I or II data.

Approximately 95% of the soil gas samples were analyzed at SNL/NM's FASL. However, the data were validated in accordance with "Data Quality Objectives for Remedial Response Activities" (EPA 1987). The methodology was slightly modified to allow for a refinement in the screening data criteria. Data generated from the FASL fell into three categories of Level II data as defined below:

Level 1: Screening Data—Data classified as Level I typically were missing an acceptable QC sample at either the beginning or end of the analytical run, or were analyzed when the QC data were outside of the range of acceptable surrogate recovery, or both. Data that fell into this category correspond with EPA Level I data.

Level 2: Conditionally Acceptable Field Lab Data—Data classified as Level II typically were analyzed with one of the QC samples out of the acceptable range. Data that fell into this category correspond with EPA Level II data.

Level 3: Acceptable Field Lab Data—These data met all QC requirements as specified for EPA Level II data.

Approximately 95% of the data generated from the FASL for this investigation were classified as Level 2 or Level 3 data.

Site Investigation Strategy and Shallow Soil Gas Survey Results

Soil gas analysis was first used at the CWL in May of 1992 to define "hot spots" and the horizontal extent of the gas plume for optimization of soil boring locations. Approximately 165 soil gas samples were collected from 75 locations on an evenly spaced grid covering the landfill area. Geophysical survey data and historical information on pit locations were used to determine final sampling locations. Samples were collected to approximately 12 m BGS.

The results of analysis for the total target VOCs (predominately TCE) at 9 m BGS are illustrated in Fig. 3. As a result of this survey, three hot-spot locations were identified in the southern half of the landfill and the extent of the 100 ppm_v was defined. In general, VOC concentrations increased with depth, which is indicative of some migration of the source from the landfill. The highest gas concentrations were observed in the southwest corner of the CWL. The results of the shallow soil gas survey were then used to optimally locate the soil borings. Figure 3 also illustrates the locations of the 21 soil borings used to confirm the horizontal event of the plume and to characterize the vertical extent of the plume.

Due to a lack of confidence in the quality of data, field screening methods were considered unacceptable by state regulators for final delineation of the contamination and compliance with the regulations governing closure of this site. However, realizing the importance of collecting this data to validate our conceptual model, we collected gas samples at greater than 90% of the soil sampling locations. During this second phase of the investigation, 178 soil gas samples and 230 soil samples were collected and analyzed for DNAPLs. The soil borings were installed using dual-tube percussion methods to approximately 65 m BGS. Soil samples were taken at 1.5-to-6.1-m intervals, depending on a sample's location relative to the source area. Soil gas samples were initially taken at 3.1-m intervals during the first phase of the investigation and then at all locations where soil samples were collected.

Vapor Plume Characterization/Results and Discussion

TCE is the dominant compound present in the subsurface at the CWL. TCA also was detected at higher concentrations in many of the contaminated samples. The combined concentrations of TCE and TCA in the gas samples comprised the majority (80 to 90%) of the total VOCs in most gas samples. This is consistent with the fact that these compounds are the only VOCs that have been verified as ground-water contaminants at the site. The following discussion is focused on the TCE data to illustrate that on-site soil gas analysis provided the best data for site characterization.

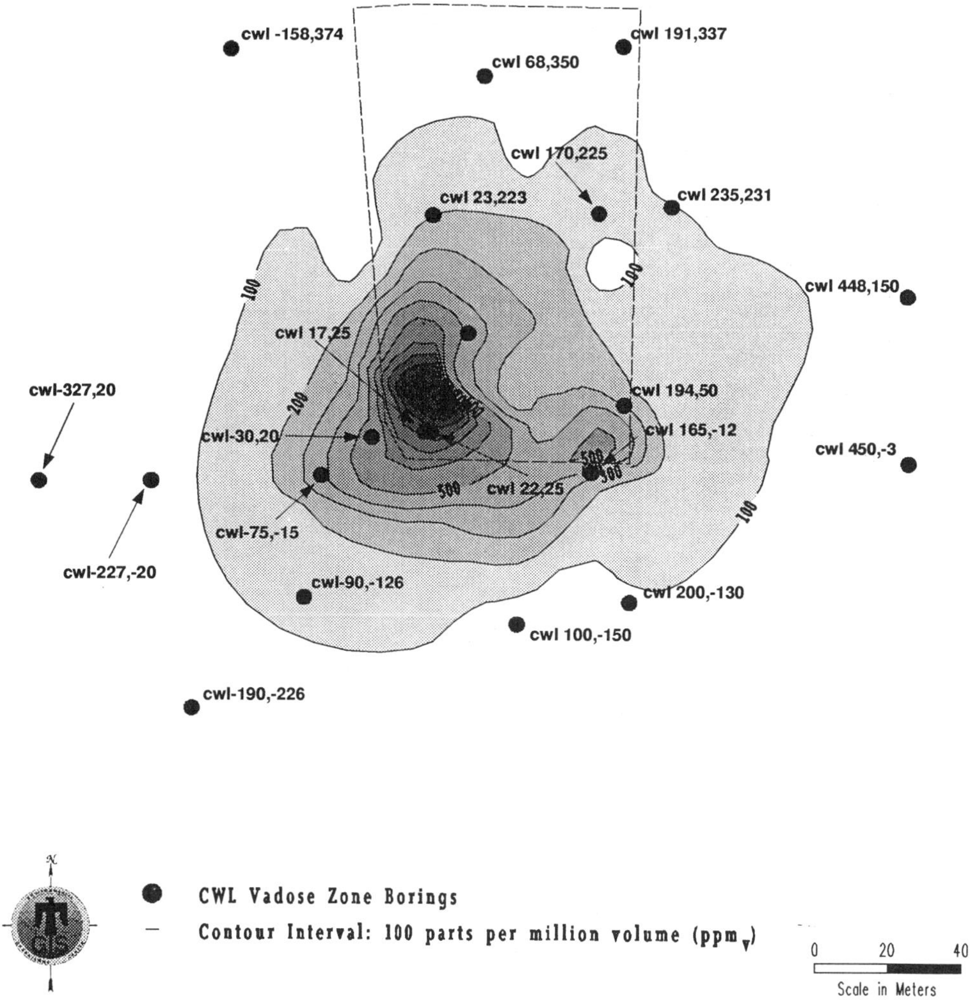

FIG. 3—*Total target VOCs at 9 m BGS and soil boring locations.*

Comparison of Soil Gas and Soil Sample Data for TCE

Figure 4 is a composite three-dimensional view of the 100 ppm$_v$ TCE plume in the subsurface at the CWL. The data used for this figure were produced at SNL/NM's FASL for both the shallow and the deep soil gas surveys. The figure was created using a program called Spyglass (with the subprogram Dicer to create the cross sections). The data are contained in layers and composited to create a three-dimensional image. Cross sections through the data illustrate the hot spot in the southwest corner of the landfill, which is most likely the source of the soil gas migrating to the water table. This figure clearly illustrates the conceptual model described previously; lateral and vertical movement of NAPL with the development of a gas phase plume that diffuses to the water table. Assuming a homogeneous distribution of TCE contamination throughout the volume illustrated in Fig. 4, we estimated 256 000 m^3 of soil are contaminated above the specified action levels.

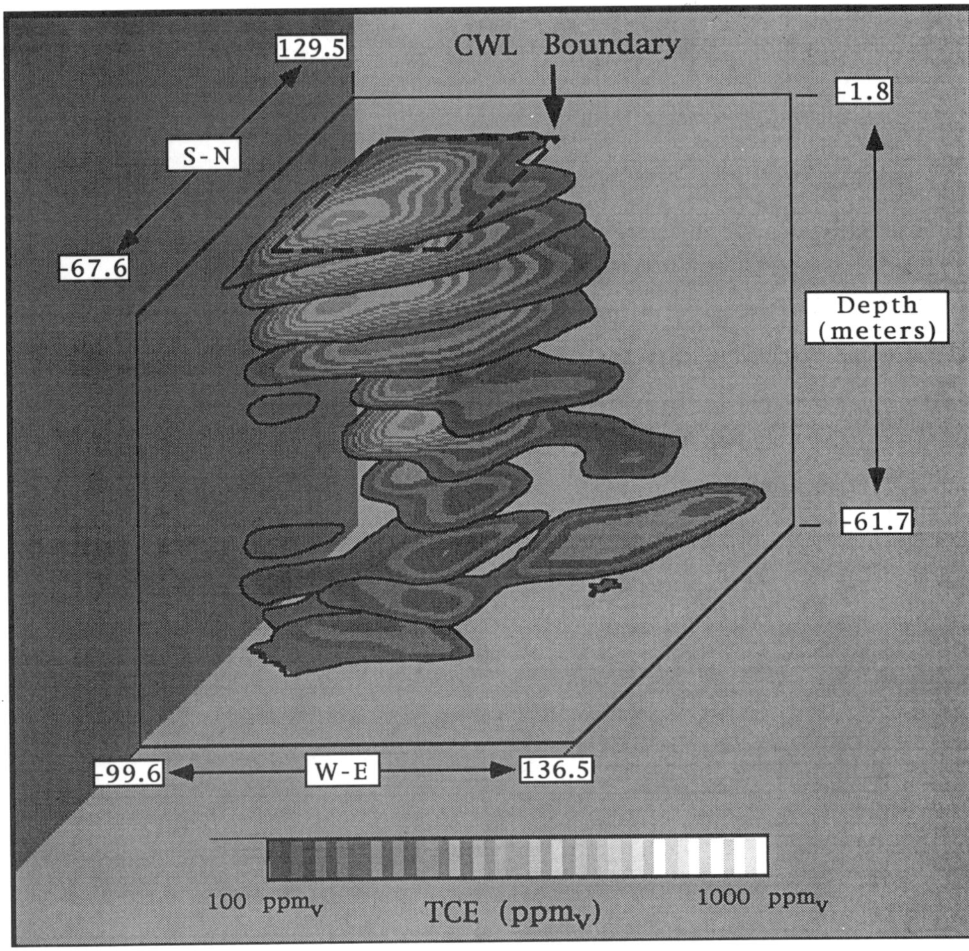

FIG. 4—*Composite TCE gas plume beneath the chemical waste landfill.*

In contrast, Table 1 provides a summary of all soil samples that contained detectable levels of TCE analyzed by the contract laboratory. Only 20 samples of the 250 samples taken revealed any TCE impact. The detection limit for these samples varied but generally fell within the range of 5 to 10 μg/g. A review of all laboratory data revealed detectable soil contamination in the hot-spot area illustrated in Fig. 3 (CWL-22, 25, and CWL-17, 25). A very conservative estimate of the extent of soil contamination defined by this data set indicates a volume of approximately 1400 m^3 in size, distinctly different from the volume estimated through soil gas analysis.

We believe there are two main reasons for the discrepancy in contaminant volumes as calculated from these two separate data sets. As described in EPA (1986a), a relatively small aliquot of soil, approximately 550 g, was removed from the split-spoon sampler and placed into the sample container. During this transfer, volatilization of VOCs is inevitable. Subsequently, the laboratory analyzed approximately 5 g of soil from each sample and extracted VOCs from the sample for analysis by EPA Method 8240. Volatilization of the contaminants

TABLE 1—*Comparison of TCE concentrations in soil samples (high level versus low level EPA Method 8240).*

Borehole Location	Sample Depth, (m)	Field Methanol-Extraction, (m) High Level	Soil TCE Concentration, (mg/kg) Low Level	Detection Limits, (mg/kg)
CWL-30, 20	1.52	0.091	...	0.005
	3.05	...	0.040	0.005
	4.57	0.056	0.002	0.005
	6.10	...	0.007	0.005
	7.62	0.087	...	0.005
CWL-75, -15	24.08	0.016	0.098	0.005
	33.53	...	0.2	0.005
CWL 22, 25	1.52	100	11	3.1
	3.05	...	51	1.2
	3.05	...	14	0.62
	4.27	110	77	3.1
	4.27	130	140	4.2
	4.72	160	180	6.2
	6.10	...	110	2.5
	7.62	...	30	3.1
CWL 17, 25	1.52	...	1300	62
	1.98	...	9500	310
	2.59	...	210	6.2
CWL 55, 115	3.05	...	0.002	0.005
CWL 170, 225	24.38	...	0.008	0.0085
CWL 165, -12	6.10	0.48	0.002	0.005
	6.10	...	0.008	0.005
CWL-194, 50	3.05	0.026	...	ND
	6.10	0.048	...	ND

from the sample most likely occurs at this stage and has an even greater impact on the results than field handling. Additionally, the size of the sample analyzed is small relative to the mass initially collected and may not be representative of the sampling location. In comparison, the soil gas sample is collected from a larger undisturbed zone of contamination, and may be more representative of VOC levels in the soil at any given sampling location.

Methanol-Extracted Volatile Organic Compounds

In addition to the soil samples analyzed for VOCs through a standard Method 8240 analysis (low level), approximately 5% of the total soil samples were analyzed for VOCs following a field extraction with methanol (high level) as discussed previously. The purpose of these additional analyses was to evaluate whether the potential loss of contaminants via volatilization during field handling and laboratory preparation could be minimized. The results of these analyses are presented in Table 1.

Though the sample size may be too small to conclude that one method is better than another, at concentrations less than 0.1 mg/kg, five of the six samples analyzed using the field methanol extraction method revealed higher levels of TCE than the EPA Method 8240 samples. In general, for TCE, the methanol-extracted samples revealed a higher level of the compound than the standard Method 8240 samples, particularly in areas where contaminant levels were low (<0.1 mg/g). In areas of higher TCE levels (>100 mg/g), the difference between the results from these two methods is not conclusive; however, in most cases, the methanol results are higher. Additional evaluation of this method would be necessary before any definitive conclusions could be drawn. Further, there are two inherent problems with this approach; it does not remove the problem of volatilization during field handling and there is often a problem with laboratory contamination of the methanol used for the field extraction.

Conceptual Model Validation Using Indicator Gas Concentrations

Because we believe soil gas data provide a more accurate representation of the extent of the contaminant plume, we decided that a mechanism was needed to establish the relationship between a soil gas concentration and the presence of DNAPLs. In order to validate the conceptual model of gas-phase transport to the water table and to define the limits of the source area for remedial action design, it was necessary to establish a soil gas concentration which would be indicative of the presence of DNAPL.

At this time, the presence of DNAPL is usually assessed through visual inspection and off-site laboratory analysis of samples from soil borings. Visual identification is effective if residual DNAPL contents are high and contaminant zones are thick. If residual concentrations are low or the DNAPL is distributed heterogeneously in the sample, or both, visual inspection of the samples may be misleading. To date, analytical methods that can directly confirm the presence of residual DNAPL are not available (Feenstra et al. 1991). A chemical analysis cannot indicate directly how the VOC is distributed between the three phases present in the soil sample: (1) dissolved in the pore water of the sample, (2) sorbed onto the soil solids (DNAPL), or (3) distributed as soil gas in the pores of the soil sample.

Feenstra (1991) developed an indirect method to calculate whether DNAPL may be present in a soil sample analyzed by a laboratory. The calculations, described in detail in his paper, are based on the assumption that DNAPL is not present in a soil sample. The apparent pore water concentration of the contaminant of interest is calculated from the measured total soil concentration by assuming equilibrium partitioning between the solid, pore water, and soil gas phases. If DNAPL is not present, then there is a hypothetical maximum mass of chemical that can be present in the pore water and the soil gas. If DNAPL is present, then the maximum concentration would be exceeded and the calculated pore water concentration would exceed the solubility of the chemical.

Applying the Feenstra analysis to the soil sample results from CWL-22, 25 where field observations revealed free-phase organics (see Table 1), we found that residual DNAPL was present at 7.6 m BGS. However, in consideration of our lack of faith in the laboratory data, we followed Feenstra's lead of using indirect methods to test for the presence of DNAPL and developed a method of determining the gas concentration that would be indicative of the presence of DNAPL (that is, an indicator vapor concentration) (SNL/NM 1993).

Indicator Vapor Concentration (IVC) Analysis

The IVC approach was developed at SNL/NM and the following procedures for completing this analysis are briefly summarized below. We made the following assumptions to perform the analysis:

1. The NAPL in the NAPL/soil sample approximates an ideal liquid mixture. Feenstra et al. (1991) stated that this assumption may be plausible for mixtures of structurally

related, hydrophobic organic liquids. Except for the possible presence of acetone, this assumption appears to be valid for our purposes.

2. The laboratory analysis of the soil sample containing DNAPLs identified all the volatile and semivolatile compounds present in the sample. This assumption appears to be valid in that the laboratory did not identify the presence of any unknown compounds.
3. The NAPL composition remains proportionately the same at other locations in the CWL. This assumption is difficult to assess because NAPL was only observed at CWL-17, 25 and CWL-22, 25. In addition, very few of the soil samples collected during this investigation contained levels of volatile and semivolatile compounds above their respective method detection limits (MDLs).
4. VOCs in the gas phase occur primarily in the soil gas and not in the water vapor that may also be present. Preliminary calculations indicate that as much as 98% of the VOCs could be expected to be contained within the soil gas. Given the conservatism of the other assumptions (including the use of 70% of the indicator gas concentration as the threshold gas concentration), we concluded that the influence of water vapor was relatively insignificant.

We initially calculated the soil gas concentrations indicative of a DNAPL phase by comparing the laboratory analysis of a soil sample collected from boring CWL-17, 25 at a depth of 1.8 m BGS. This sample was chosen because free-phase liquids could be observed in the field, unlike any other sample obtained from the landfill. Table 2 provides the results of the laboratory analysis for the dominant volatile components in the sample and an evaluation of these results. There were many compounds other than TCE present in the sample, including: TCA, PCB, benzene, toluene, and naphthalenes, and these data were taken into consideration (as described below).

We calculated the partial pressure of TCE in the mixture using Raoult's law for ideal mixtures and the mole fraction of each compound identified in the NAPL/soil sample. The compounds detected in the laboratory method blanks and the sample were not included in the analysis. The partial pressures were converted to microlitres per litre (corrected for ambient atmospheric pressure) for comparison with the observed field results. These calculated values are the saturated vapor concentrations of the compound in equilibrium with the NAPL mixture. The calculated IVC is the concentration limit above which DNAPL would be expected to be present in the sample. Because the accuracy of the soil gas sample analysis ranged between 70 and 130% of the quantified values, we conservatively assumed that gas concentrations 70% of the IVC were indicative of NAPL. This 70% value is listed in Table 2 as the threshold gas concentration (TGC).

We empirically tested this analysis by comparing the results of the soil gas sample collected from CWL-22, 25 at a depth of 5.5 m with the calculated IVC derived from the soil sample collected at CWL-17, 25 where free phase was present. Even though these samples were collected at two separate locations, the boreholes were only a few feet apart and can therefore be related. We were unable to collect gas samples from the CWL-17, 25 area due to logistical problems. Visually observable DNAPL was present in the soil samples between 1.5 and 4.6 m BGS CWL-17, 25. The bar graph shown in Fig. 5 illustrates the comparison between the IVC calculated for a sample where we observed free-phase liquids in the field, CWL-17, 25, and the observed gas concentration in a gas sample from CWL-22, 25 for TCE as well as several other dominant compounds. The observed gas concentration for each VOC is greater than the IVC. This implies that NAPL is present in this area.

As shown in Table 2, the TGS for TCE is approximately 1500 ppm_v. Only samples from CWL-22, 25 contained gas concentrations at or above this threshold value. Figure 6 depicts

TABLE 2—*Indicator vapor concentration calculations for CWL-17, 25 at 1.8 m BGS.*

Compound	Concentration (mg/kg)[a]	Mass in Soil (g)[b]	Molecular Weight, (g/m)[c]	No. Moles[d]	Mole Fraction[e]	Vapor Pressure (mm Hg)[f]	Partial Pressure (mm Hg)[g]	IVC (ppm_v)[h]	Threshold Vapor Concentration (ppm_v)[i]
PCE	75	0.075	166	0.0005	0.0001	18	0.0026	4	3
DCM	110	0.11	99	0.0011	0.0004	182	0.0643	103	72
TCA	2300	2.3	133	0.0173	0.0055	123	0.6765	1085	760
TCE	9500	9.5	131	0.0725	0.0231	58	1.3377	2146	1502
Chloroform	98	0.098	246	0.0004	0.0001	246	0.0312	50	35
Benzene	130	0.13	78	0.0017	0.0005	95	0.0504	81	57
Toluene	680	0.68	92	0.0074	0.0024	28	0.0661	106	74
Ethylbenzene	270	0.27	106	0.0025	0.0081	7	0.0057	9	6
Xylene	810	0.81	106	0.0076	0.0024	10	0.0243	39	27

[a] Concentration determined by laboratory analysis of CWL-17, 25, 06 sample.
[b] Mass of each compound present in 1 kg of soil.
[c] Molecular weight of each compound from Cohen and Mercer (1993) and EPA Superfund Public Health Manu (1986).
[d] Number of moles of each compound present in the sample (Column [b]/Column [c]).
[e] Mole fraction of each compound present in the NAPL (Column [d]/total moles).
[f] Saturated vapor pressure of the pure compound from the Superfund Public Health Manual and Cohen and Mercer (1993).
[g] Partial pressure of each compound calculated with Raoult's Law (Column [f] * Column [e]).
[h] Indicator Vapor Concentration {[Column [g]/(760 * 0.82)]*1,000,000}.
[i] Threshold Vapor Concentration (Column [i] * 0.70).

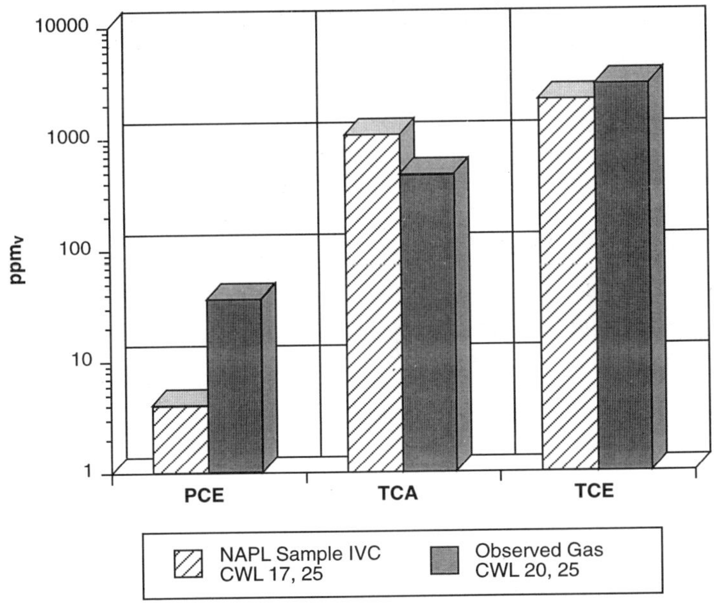

FIG. 5—*Comparison of the NAPL IVC and observed soil gas at CWL-22, 25.*

the gas concentration versus depth profile for TCE at CWL-22, 25. The broken vertical line represents the TCE-NAPL TGC of 1500 ppm_v. Data points to the right of the threshold are indicative of NAPL. The IVC gas analysis suggests that the TCE-NAPL has penetrated to a total depth of approximately 15.2 m beneath the CWL.

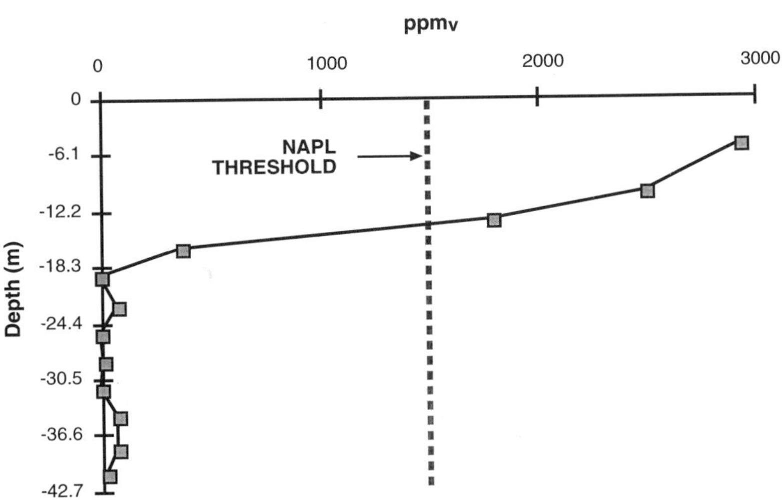

FIG. 6—*TCE vapor concentrations in soil boring CWL-22, 25.*

Conclusions

The results of the shallow and deep soil gas surveys at the CWL provided the most reliable information for evaluating the distribution of VOCs in the subsurface at the CWL. As stated earlier, only 20 of the 250 soil samples collected for laboratory analysis revealed the presence of laboratory contamination. All soil samples collected below 24 m BGS did not contain detectable levels of VOCs. Soil gas analysis revealed the presence of a 100 ppm_v plume to depths greater than 60 m and the presence of DNAPL to depths greater than 15 m. Without the benefit of analysis of the gas data, we would have underestimated the volume of contaminated media for remedial design. We would have been unable to explain the transport mechanism by which ground water is being impacted at this site, and, without a valid conceptual model, we would more than likely have been asked to complete additional investigation of this site because the regulator would have been forced to believe we had not defined the source area. Further, the collection and analysis of soil gas samples can be performed relatively inexpensively when compared with the high costs of drilling, sampling, handling, and expensive laboratory analysis.

Acknowledgment

This work was funded by the Environmental Impact and Restoration Department 7585 and performed at Sandia National Laboratories, which is operated for the U.S. Department of Energy under Contract No. DE-AC04-76DP00789.

References

Cohen, R. M. and Mercer, J. W., 1993, *DNAPL Site Evaluation,* CRC Press, Inc., Boca Raton, FL.

Feenstra, S., Mackay, D. M., and Cherry, J. A., 1991, "A Method for Assessing Residual NAPL Based on Organic Chemical Concentrations in Soil Samples," *Ground Water Monitoring Review,* Spring, pp. 128–135.

Mendoza, C. A. and Frind, E. O., 1990, "Model Development," *Water Resources Research,* Vol. 26, No. 3.

Sandia National Laboratories/New Mexico, 1993, *The Chemical Waste Landfill Unsaturated Zone Contaminant Characterization Report,* Environmental Impact and Restoration Department, Albuquerque, NM, Nov.

Sandia National Laboratories/New Mexico, 1992, *The Chemical Waste Landfill Closure Plan and Post-Closure Permit Application,* Environmental Impact and Restoration Department, Albuquerque, NM.

U.S. Environmental Protection Agency, 1988, *Compendium of Methods for Determination of Toxic Organic Compounds in Ambient Air,* U.S. EPA, Research Triangle Park, NC, EPA/600/4-89/017.

U.S. Environmental Protection Agency, 1988a, Office of Solid Waste and Emergency Response, "RCRA Corrective Action Plan," Interim Final Report, EPA/530/SW-88-028, Washington, DC.

U.S. Environmental Protection Agency, 1987, *Data Quality Objectives for Remedial Response Activities,* PB90-272634.

U.S. Environmental Protection Agency, 1986, *RCRA Ground-Water Monitoring Technical Enforcement Guidance Document,* OSWER-9550.1, Washington, DC.

U.S. Environmental Protection Agency, 1986a, *Test Methods for Evaluating Solid Waste, Physical and Chemical Properties,* 3rd ed., Revision 0, Sept.

Anthony D. Daus[1] *and Manoel M. Nobre*[2]

A DNAPL Emergency Remedial Action Program in a Developing Country

REFERENCE: Daus, A. D. and Nobre, M. M., **"A DNAPL Emergency Remedial Action Program in a Developing Country,"** *Volatile Organic Compounds in the Environment, ASTM STP 1261,* W. Wang, J. Schnoor, and J. Doi, Eds., American Society for Testing and Materials, 1996, pp. 219–234.

ABSTRACT: A chlorinated hydrocarbon (CHC) manufacturing facility in northeastern Brazil inadvertently released chlorinated hydrocarbons to the subsurface. The underlying ground water lies 60 to 70 m below the ground surface and is part of a large drinking water aquifer (Barreiras aquifer). Downgradient potential receptors include local inhabitants who draw their potable water from hand-dug wells (cacimbas), wetlands, and a large estuary.

The Barreiras aquifer is alluvial in origin and consists of interbedded clays, silts, sands, and gravels. A linear geologic structure was identified at the base of the Barreiras aquifer that resembles a paleo-channel. This channel appears to correlate with the zone of highest CHC contamination.

CHCs were subsequently detected in nearby production wells 500 m downgradient. An accelerated ground-water investigation and interim remedial action plan were implemented to contain the highest CHC concentrations and provide hydraulic control downgradient of the CHC source. Ground-water production wells were installed along the paleo-channel in the alluvial sediments. Ground-water level and quality monitoring suggest that the program has been successful although additional monitoring is required to confirm the initial results. Between April 1992 and December 1993, over 1.2 billion gallons of ground water have been pumped and over 52 tons of dissolved solvents have been removed. Future plans for the project include evaluation and optimization of the ground-water capture system and evaluation of dense nonaqueous phase liquid (DNAPL) recovery.

KEYWORDS: ground-water remediation, Brazil, dense nonaqueous phase liquid, emergency response, paleo-channel, extraction wells, 1,2,3-trichloropropane, 1,2-dichloropropane, 1,2-dichloroethane

Geomatrix Consultants, Inc. (Geomatrix) was retained to conduct a subsurface investigation and provide assistance in implementing a soil and ground-water remediation program at the Polo Chloroquimico de Alagoas (a chemical complex) and surrounding area (referred to in this paper as the Polo) located near Maceio, Alagoas, Brazil (Fig. 1). The investigation was initiated in response to a request by the state regulatory agency (IMA) to investigate and remediate the sources and extent of chlorinated hydrocarbons (CHCs) including 1,2-dichloroethane (1,2-DCA), 1,2-dichloropropane (1,2-DCP), and 1,2,3-trichloropropane (1,2,3-TCP) in the soil and ground water beneath the Polo.

[1] Vice president and principal hydrogeologist, Geomatrix Consultants, Inc., Newport Beach, CA 92660.
[2] Associate professor, Federal University of Alagoas, Brazil.

FIG. 1—*Site location.*

Background

The Polo is located about 5 km from Brazil's coast on a plateau which rises sharply from near sea level to an elevation of approximately 80 to 85 m (as shown in Fig. 2). The plateau is bounded to the east by the Rio dos Remedios Valley, to the west by a valley, and to the south by the edge of the plateau as it drops toward Mundau Lagoon (Lagoa Mundau). Sugar cane fields lie to the north of the Polo. The southern, eastern, and western forested areas surrounding the plateau and the lagoon are considered an environmentally sensitive area. Villages and residences at the base of the plateau obtain their drinking water from shallow hand-dug wells (cacimbas).

Construction of the Polo was completed in 1988. Two companies on the Polo manufactured CHCs. In 1989, a leak of waste CHCs (unknown quantity) as a dense nonaqueous phase liquid (DNAPL) was discovered at one of the facilities and the source of the leak was repaired. In early 1990, CHCs were discovered in one of the drinking water supply wells at the Polo and state environmental agencies and law enforcement officials became involved. The chemical plants and the downgradient population were supplied with treated water. Subsequent preliminary investigations indicated that migration of the CHCs off the Polo was likely unless immediate action was undertaken. The potential for off-site migration made timely hydraulic control of the CHCs in ground water the key initial priority. The company that had the accidental leak of CHCs terminated operations in 1991.

The project strategy included investigating subsurface conditions and initiating hydraulic control measures on an accelerated basis, educating the client and regulators, and the transfer of appropriate technology. We anticipated that this technology transfer must consider the available resources and infrastructure and use local labor, vendors, and expertise as much as possible to accomplish the goals of the project. This resulted in a stronger local interest in the project, minimized costs, and developed local expertise (critical in operating and maintaining long-term projects).

Geology

The stratigraphy of the Barreiras Formation and underlying shale was evaluated by drilling exploratory soil borings, installing ground-water monitoring and extraction wells, and reviewing geophysical borehole logs.

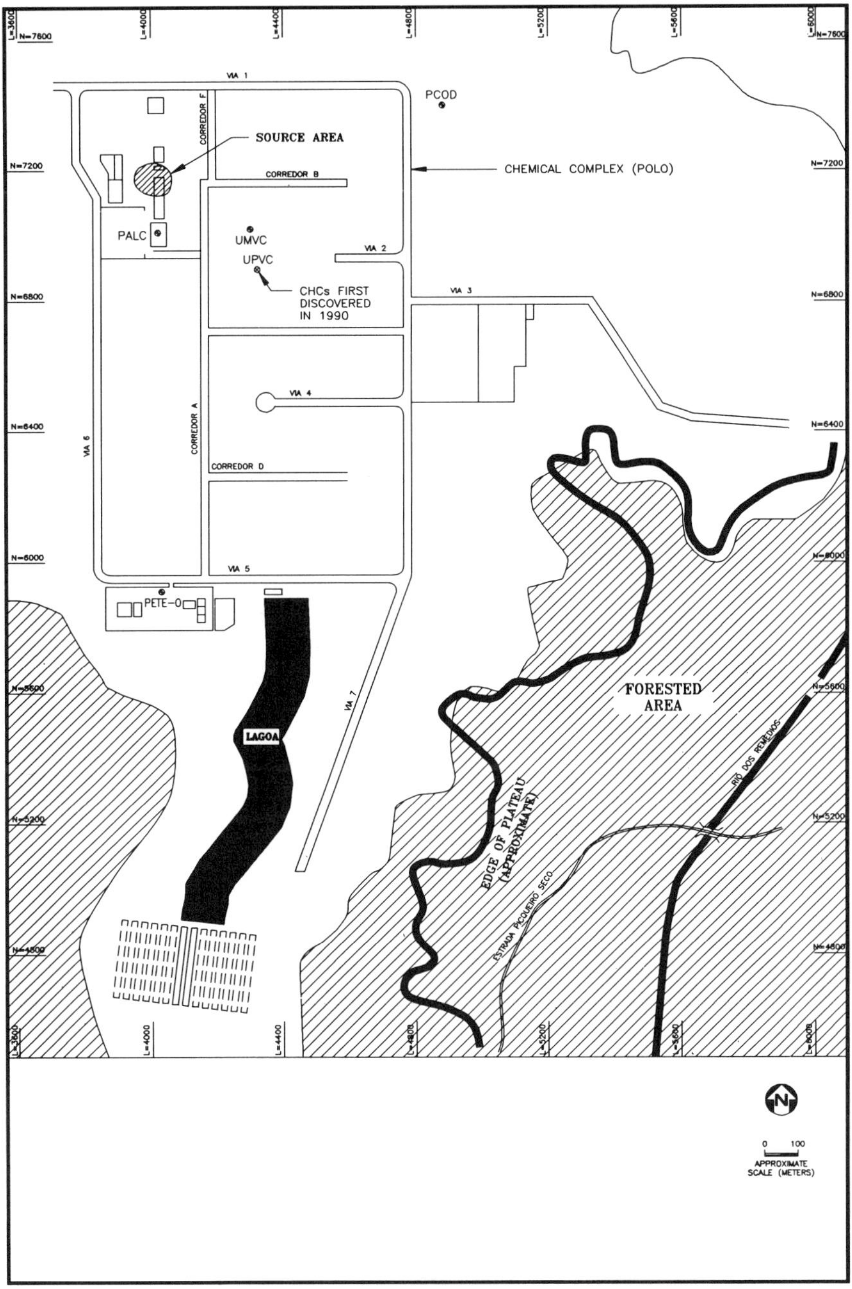

FIG. 2—*Site map.*

The Polo is underlain by approximately 85 to 95 m of interbedded gravel, sand, silt, and clay deposits that comprise the Barreiras Formation. The sediments that comprise the Barreiras Formation are primarily of alluvial origin, having been derived from the central Brazilian granitic craton northwest of the Polo. The deposits lie unconformably on shale units that represent the upper section of the Mesozoic/Cenozoic Sergipe Group basement complex. The shale units appear weathered into a clayey interval at the contact with the Barreiras Formation. The shale contains coarse-grained intervals in the southern portion of the Polo and is referred to as the Coquéiro Seco shale. In the northern part of the Polo, the shale unit is more massive, has few, if any, course-grained intervals and is referred to as the Ponte Verde shale.

The contact between the Barreiras Formation and the underlying shale is unconformable, showing as much as 9 m of total relief across the Polo. As shown in Fig. 3, the elevation of the shale unit is greatest in the northwest part of the Polo, decreasing to the south and east. After correlating the borehole geophysical and geologic data, a northwest trending trough is interpreted in the shale through the central portion of the Polo. The elevation of the shale appears to drop off significantly in the southwest portion of the Polo. This configuration of the top of the shale appears to influence ground-water flow and CHC plume migration across the Polo.

Hydrogeology

The hydrogeology at the Polo was evaluated based on lithologic descriptions of aquifer materials including the geophysical logs, measurements of ground-water elevations, and results of aquifer tests. Monitoring well locations are depicted in Fig. 4.

The recharge area for the Barreiras aquifer is primarily in the highlands north and west of the Polo. Recharge is also expected to occur in farmland adjacent to the Polo, particularly where irrigated, and minimally on the Polo proper where unpaved or where precipitation, which averages approximately 2 m/year and primarily falls during the winter, is allowed to infiltrate. Ground-water recharge is also likely through the bottom of the unlined lagoon (lagoa). This lagoa currently only receives runoff from precipitation events (Fig. 2).

Regionally, ground-water discharges to the estuary system located on the coastal plain west and south of the Polo. Springs and marshes located at the base of the plateau adjacent to the Polo confirm these ground-water discharge areas.

In general, the unsaturated zone at the Polo, particularly in the northern portion of the facility, is dominated by finer-grained deposits of silts, clays, and silty/clay mixtures interbedded with discontinuous sandy deposits. Where ground-water recharge is sufficient, more continuous clay zones appear to result in local perched ground-water conditions. Evidence for these perched conditions is found in the geophysical logs (neutron logs) and historically in the occurrence of springs and seeps along the plateau escarpment. Springs and seeps have been noted approximately midway down the escarpments immediately west of the lagoa (Fig. 2). The dense jungle that occupies most of the plateau escarpments limits access and a more thorough identification of these perched ground-water discharge zones.

The lower 20 to 25 m of the Barreiras Formation is saturated and represents a regional aquifer referred to as the Barreiras aquifer. The Barreiras aquifer is bounded on the bottom by the Ponte Verde or Coqueiro Seco shale. Depth to water under the Polo ranges from approximately 60 to 75 m below ground surface under nonpumping conditions and generally increases from north to south across the Polo. The depth to water has increased two to three meters since ground-water extraction was initiated in April 1992. Under nonpumping conditions, the direction of ground-water flow is regionally toward the southeast across the Polo. The horizontal hydraulic gradient, which remains generally uniform across the Polo, was

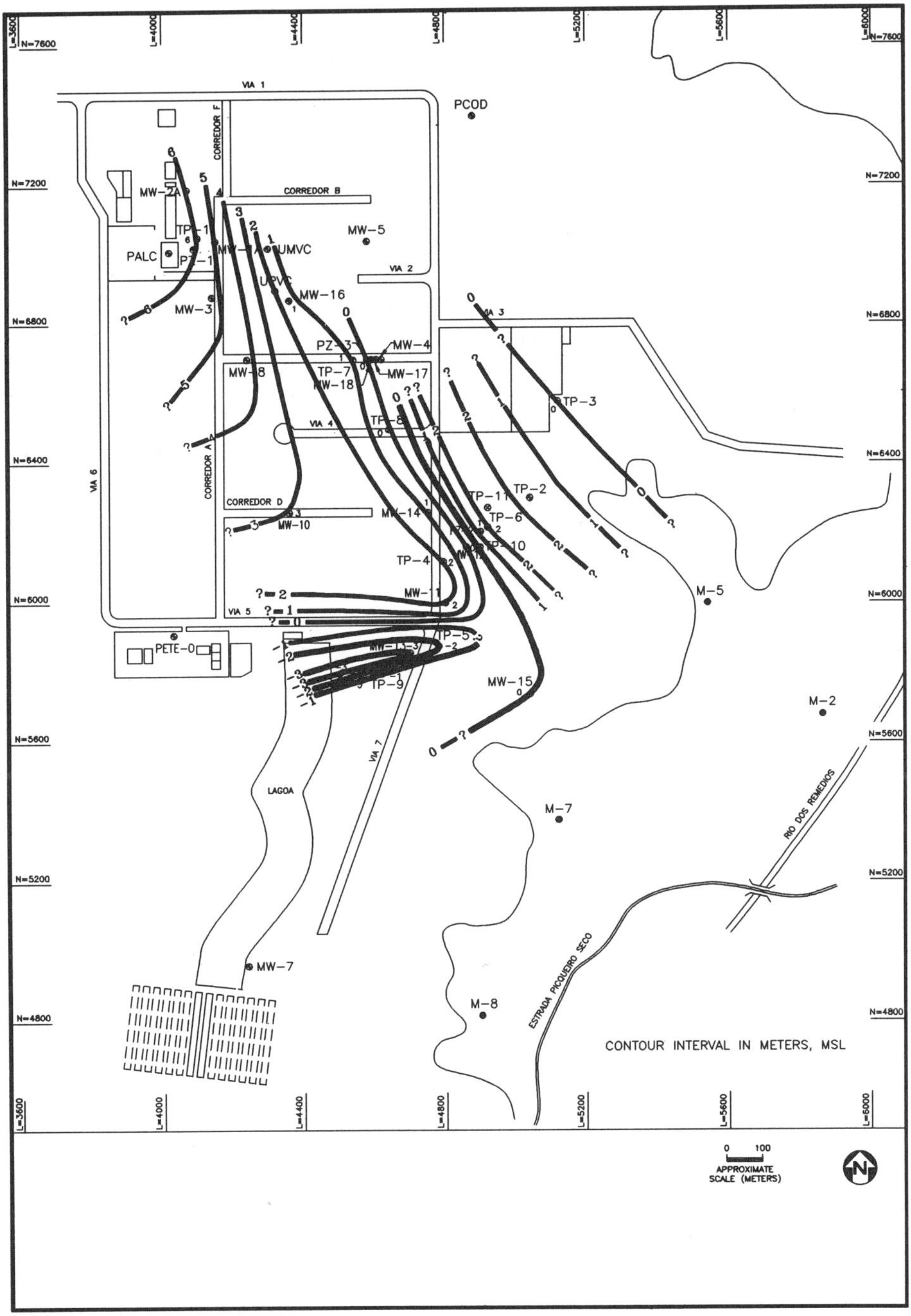

FIG. 3—*Structure contour of top of "weathered shale" contact.*

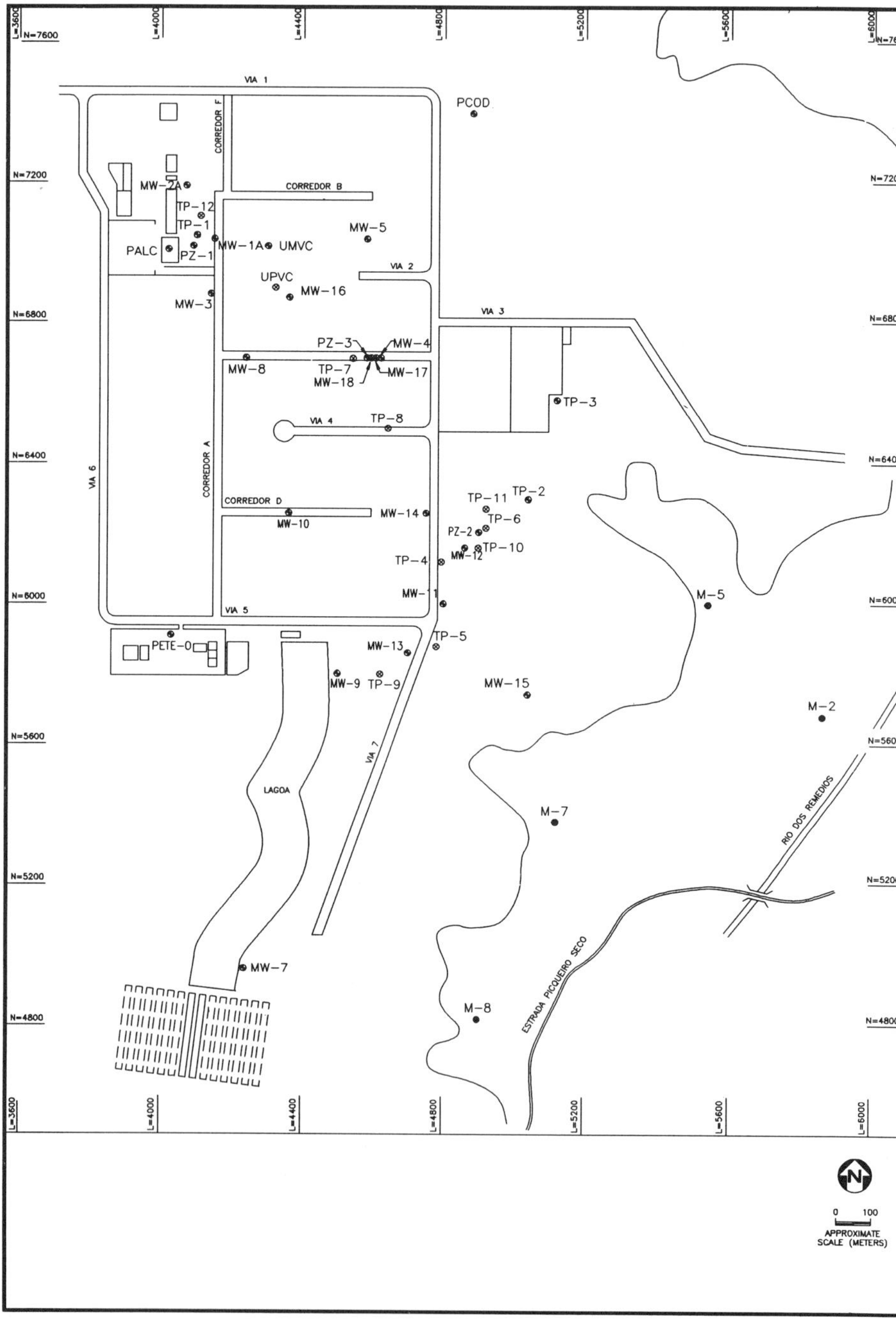

FIG. 4—*Ground-water monitoring and extraction well locations.*

approximately 0.007 m/m before pumping started. The gradient appears to have increased slightly since the onset of ground-water extraction (Fig. 5).

Constant rate and stepped discharge rate aquifer tests were performed in production wells at the Polo and have provided information about the type of aquifer and hydraulic properties. The stepped discharge rate aquifer test results were also utilized to assess the efficiency of the ground-water extraction wells. In general, the aquifer tests indicate that hydraulic conductivities increase to the south across the Polo. The hydraulic conductivities range from 13 m/day (m/d) at well TP-01, near the CHC source, to 170 m/d at well TP-06, in the southern part of the Polo. The aquifer appears to exhibit an unconfined hydraulic response in the central and southern portion of the Polo. A semiconfined response was observed in data obtained from a constant rate pumping test in TP-01.

A numerical ground-water model of the Barreiras aquifer was developed as a sensitivity analysis tool to assess various ground-water extraction wellfields and assist in placing key ground-water monitoring wells (Franz and Guiguer 1991). The conceptual hydrogeologic model consists of a single ground-water unit (the Barreiras aquifer) with a variable geometry and hydraulic conductivity. Model development and application is dynamic in that it incorporates recent data thereby refining the ground-water flow system interpretation. We anticipate continued use of this modeling tool to assess extraction wellfield performance.

Ground-water Quality

The distribution of chemical constituents in soil and ground water was assessed based on the results of the soil borings and monitoring well installation and sampling. Although high concentrations of CHCs have been found in both soils and ground water, and remediation activities are being implemented for both media; only ground-water remediation activities are discussed in this paper.

Based on the ground-water sampling results from the monitoring and extraction wells, 1,2-DCA, 1,2-DCP, and 1,2,3-TCP represent the principal CHCs found in ground water at the Polo. Ground-water quality discussions in this paper focus on these three compounds. Other organic compounds, including chloroform, tetrachloroethene (PCE), and monovinyl chloride (MVC), were detected at lower concentrations and less frequently than 1,2-DCA, 1,2-DCP, and 1,2,3-TCP (target compounds).

In August 1991, a site-wide ground-water sampling program was initiated at the Polo for all monitoring wells (that is, MW-series), extraction wells (TP-series), and existing Polo production wells (that is, UPVC, UMVC, etc). This regular monitoring program has been expanded on an as-needed basis to include new monitoring and extraction wells, cacimbas, and surface water sampling points, including springs.

Ground-water samples collected for this project were obtained initially under the direction of Geomatrix staff. On-site chemical plant personnel were trained by Geomatrix to obtain the samples and have been conducting the sampling since August 1991. Ground-water samples are analyzed at the Polo using an on-site gas chromatograph. Split samples are regularly sent to a U.S. laboratory as part of the quality assurance/quality control (QA/QC) program. In general, there has been acceptable reproducibility between the Polo and the U.S. laboratory. The typical relative percent difference for split samples analyzed by the two laboratories has been less than 25%. These steps were taken in accordance with the overall project strategy which stressed local involvement to reduce cost and develop local expertise.

The distribution of CHCs in the ground water beneath and downgradient from the Polo was evaluated by constructing chemical plume concentration (isocon) maps for 1,2-DCA, 1,2-DCP and 1,2,3-TCP. Since 1,2-DCA is the most widespread CHC and its distribution is

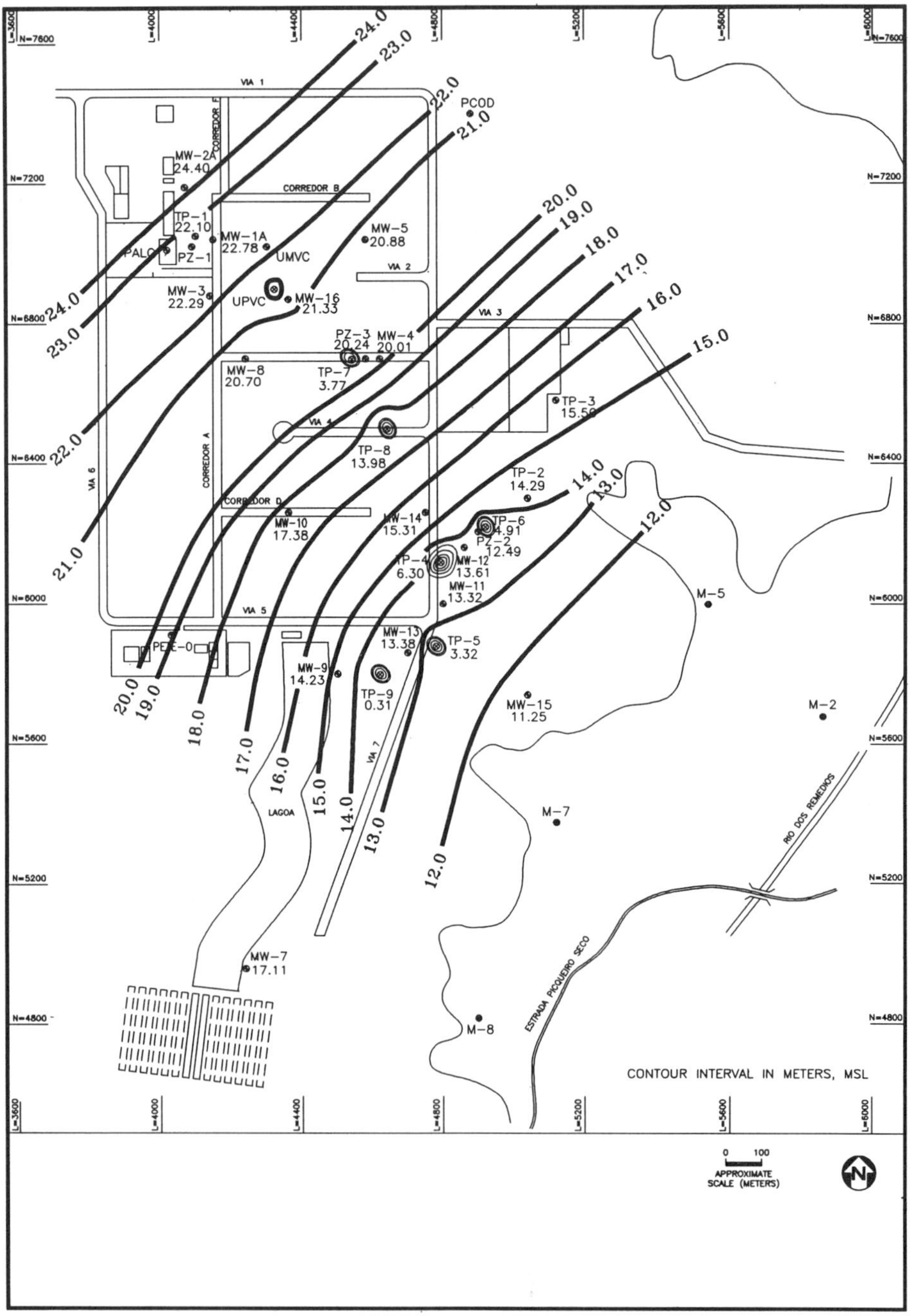

FIG. 5—*Ground-water elevation contour map, December 1992.*

similar to that of 1,2-DCP and 1,2,3-TCP, this discussion of ground-water quality will primarily center around 1,2-DCA.

Overall, the lateral distribution of 1,2-DCA can be described as a long and narrow zone of relatively high concentrations, greater than 1.0 mg/L, surrounded by a dispersed plume of lower concentrations, less than 1.0 mg/L (Fig. 6). All three target CHCs appear to follow this same trend. In August 1992 (three months after ground-water pumping was initiated) the 1,2-DCA plume length was on the order of 2 km down the plume axis and approximately 1 km wide at its widest point.

The vertical distribution of CHCs in the aquifer was evaluated through the installation of wells screened at various intervals within the Barreiras aquifer and using a monitoring well (MW-17) screened solely within the underlying Ponte Verde shale near TP-07 near the central portion of the plume. CHCs were not detected in the monitoring well completed in the underlying shale. Higher concentrations of CHCs were found in monitoring wells completed at the base of the Barreiras aquifer than in monitoring wells screened shallower. These data indicate that the CHCs are migrating principally within the alluvial Barreiras aquifer, apparently within the paleo-channel identified from geologic information.

During the sampling program, specific inorganic constituents have also been analyzed. Inorganic parameters have included alkalinity, calcium, magnesium, aluminum, copper, manganese, silica, and zinc. Selected ground-water samples have also been analyzed for acidity, biological oxygen demand (BOD), chemical oxygen demand (COD), and total dissolved solids (TDS). These inorganic parameters have been analyzed to (1) assess the scaling/biofouling potential for extraction wells and ground-water treatment equipment, (2) evaluate the relationship between the CHC plume and inorganic parameters, and (3) assist in assessing the CHC migration pathways.

Of critical importance to the implementation of this project has been the elevated iron concentrations found within the CHC plume. The concentration of total dissolved iron in the ground water collected from monitoring and production wells at the Polo ranges from approximately 1 mg/L to greater than 123 mg/L. The iron concentration has been observed to vary by one order of magnitude over time. The source of variability is being evaluated. The Barreiras aquifer matrix contains significant iron oxides as evidenced by the rust-colored outcrops and soil samples. The elevated iron appears to be associated with high CHC concentrations and may be due to changes in the ground-water chemistry (redox and/or pH) associated with the DNAPL source. Although field data do not indicate a substantial decrease in pH downgradient of the DNAPL source, additional data are being collected for further evaluation (Stumm and Morgan 1981). As expected, the high iron concentrations have significantly impacted the operation of the ground-water extraction system through the rapid growth of iron bacteria and iron-oxide scale within and near the extraction wells. Control of well clogging through well rehabilitation has become a significant cost in operating the ground-water extraction wells.

Ground-water Extraction and Treatment

A ground-water extraction well network consisting of plume axis wells and downgradient barrier wells was calculated to provide an effective means to accomplish the hydraulic control of the CHC plume. Ultimately, the number of barrier extraction wells was decreased due to the inability to treat and discharge the large volumes of ground water likely required to maintain complete plume cutoff (1200–1500 m^3/h). The focus of the extraction system was subsequently directed toward installation and operation of plume axis wells and cut-off of the CHC source in the northwest corner of the Polo. CHCs beyond the extraction wells were thus allowed to continue to migrate downgradient. Ground-water quality monitoring data

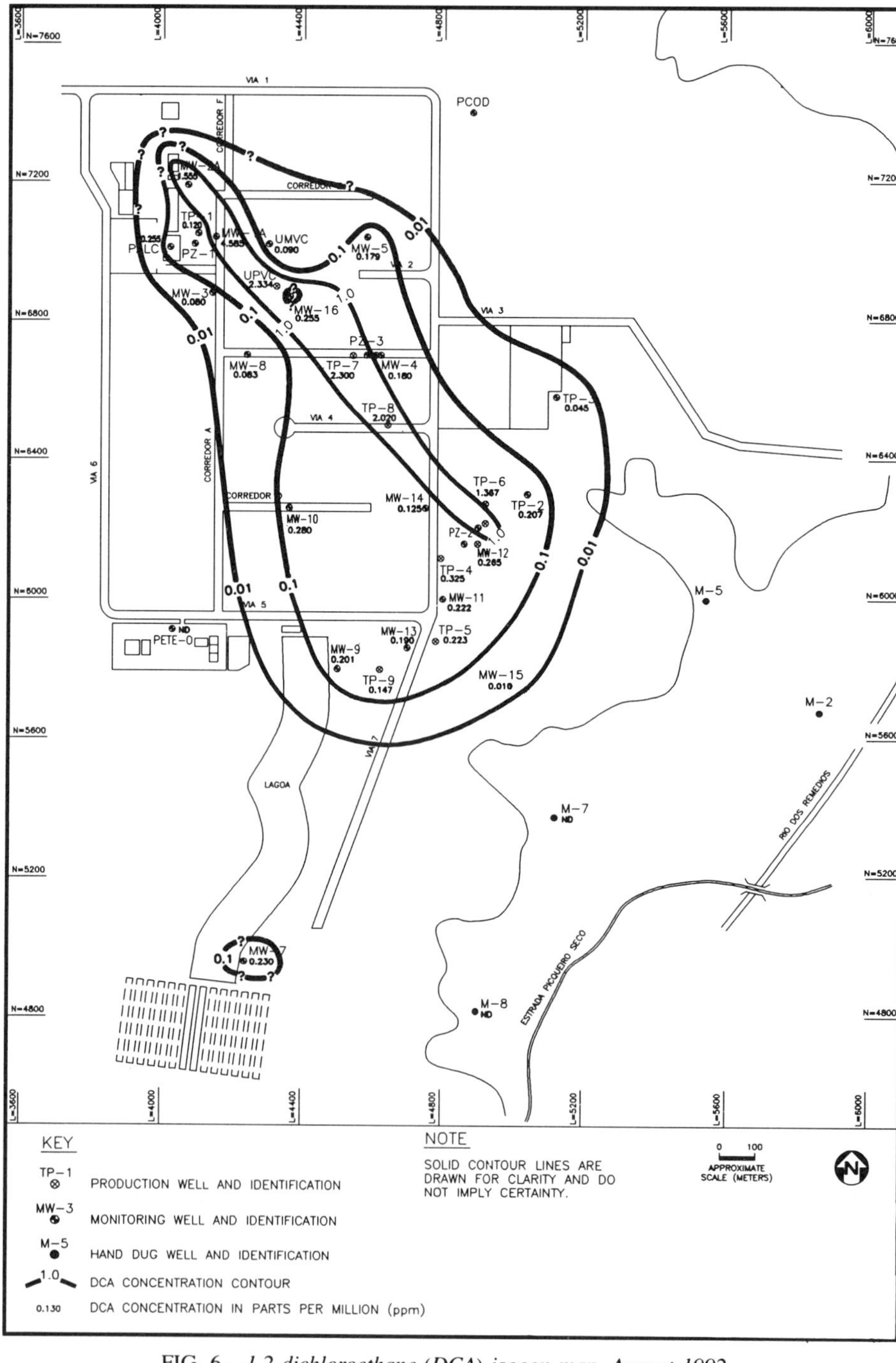

FIG. 6—*1,2-dichloroethane (DCA) isocon map, August 1992.*

downgradient of the extraction wells was used to assess the effectiveness of this pumping strategy. This strategy is consistent with recent U.S. EPA guidance documents regarding the design of ground-water hydraulic control systems (U.S. EPA 1993).

Fully penetrating extraction wells along the plume axis include TP-06, TP-07, TP-08, TP-10, TP-11, TP-12, and UPVC (Fig. 4). These wells have been installed over the course of the first 18 months of operation as ground-water monitoring data were utilized to assess ground-water extraction effectiveness. Additional wells (TP-03, TP-04, TP-05, and TP-09) were installed as part of a barrier system. TP-04 is currently being pumped and TP-05 and TP-09 were pumped until the CHC concentrations in these two wells declined and remained very low. As stated previously, completion of the barrier wellfield was terminated due to the high volumes of water that would likely have to be extracted and treated in order to maintain capture of the entire CHC plume. TP-12 and UPVC are the initial source cut-off extraction wells. Additional source cut-off wells are anticipated in the future.

Ground-water treatment is being accomplished by converting portions of the closed chemical manufacturing facility into a ground-water treatment facility. This includes oxidation ponds, cooling towers, and distillation columns. After the ground water is treated to regulatory limits (0.05 mg/L), the water is discharged to an ocean outfall pipeline where it is mixed with other treated wastewater. The goal of the treatment facility design was to minimize capital costs and maximize the use of local technology and expertise. Existing on-site equipment was utilized to reduce the cost of buying, importing, and maintaining specialty equipment.

In total, approximately 47 000 m^3 of water have been extracted and treated at the Polo since April 1992. Approximately 52 metric tons of dissolved phase CHCs have been removed from the treated ground water (Fig. 7).

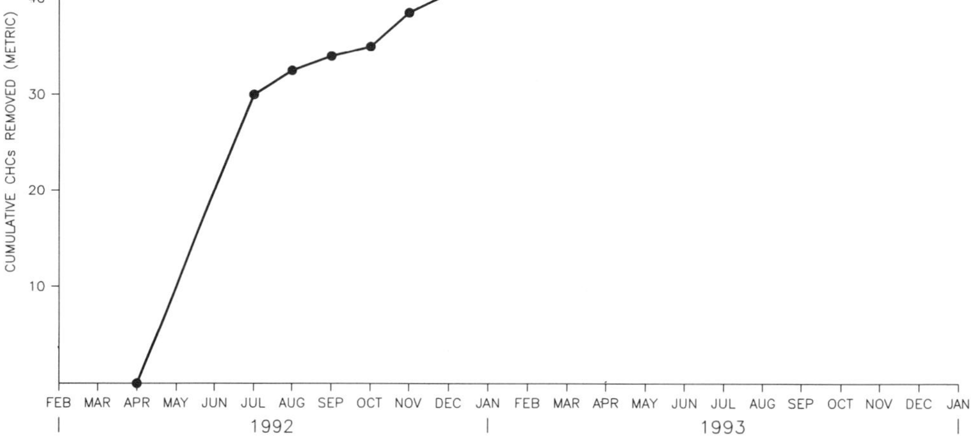

FIG. 7—*Cumulative dissolved CHCs removed since April 1992.*

Evaluation of Ground-water Extraction System Effectiveness

The effectiveness of the ground-water extraction system was assessed by evaluating the overall plume evolution and the temporal trends in ground-water CHC concentrations from individual extraction and monitoring wells. The following discussion focuses on the distribution and temporal trends in 1,2-DCA concentrations for the same reasons as previously presented.

The evolution of the 1,2-DCA plume was evaluated by plotting the 0.01 mg/L CHC isocon for August 1992 and January 1993 (Fig. 8). The overall size of the 1,2-DCA plume (as defined by the 0.01 mg/L isocan) has decreased by approximately 30% since ground-water extraction was initiated. As expected, this decrease in plume is particularly evident when evaluating the higher CHC concentrations due primarily to extraction from UPVC which is closest to the DNAPL source area.

These plume evolution observations are supported by the total CHC versus time plots for extraction well TP-08, which lies along the plume axis, and for TP-05, which was installed originally as a barrier well (Figs. 9 and 10). These plots suggest a general decrease in concentration along the plume axis and along the plume edges since ground-water extraction was initiated. TP-05 has decreased to below detection limits, however, additional sampling is being performed to further assess this trend. Total CHC concentrations in TP-08 have decreased rapidly and continue to decrease with time although at a slower rate than initially observed. Other axis wells such as TP-06 and TP-07 have displayed similar trends.

In the eastern portion of the Polo near TP-03, the plume appears to be relatively unchanged or slightly expanded (Fig. 11). Barrier wells have not been installed and operated in this area due to the limited capacity to treat ground water, and dissolved CHCs are allowed to continue to migrate downgradient of the Polo, TP-03, along with other monitoring wells, is currently used to help assess the effectiveness of the axis extraction wells by evaluating the temporal changes in ground-water quality (Fig. 11). CHC concentrations in this well are sensitive to the operation of the upgradient extraction wells. If the extraction wells are not effectively providing hydraulic containment and therefore capturing CHCs, the concentrations of CHCs in TP-03 are anticipated to increase and/or stabilize with time until the upgradient source of CHCs is removed or isolated. If the extraction wells are effective, the concentrations in TP-03 would be expected to fall after a short rise in concentrations, similar to the trend observed in a slug-type CHC source (Freeze and Cherry 1979). The concentration trend initially suggested that the extraction wells may be effective (Fig. 11). Although limited data currently are available, the recent slight rise in CHC concentrations in TP-03, as well as in TP-08, may reflect reductions in the pumping rates from selected extraction wells, particularly UPVC, which were being rehabilitated due to iron bacteria clogging. Well rehabilitation requires that purging from the extraction well is temporarily stopped in order to chemically and mechanically remove the bacteria mass and iron scale. Additional data is being collected to confirm this CHC trend and the effectiveness of the hydraulic containment system.

One observation that was made regarding concentration versus time trends for extraction wells is the relationship between total CHC concentration and initial pumping from the extraction wells. In most of the extraction wells, CHC concentrations were observed to slowly rise during sustained pumping and fall if pumping was terminated. These data suggest that the CHCs are being pulled either vertically or horizontally into the extraction well from another part of the aquifer. Monitoring well MW-17, which was installed within the Ponte Verde shale below the Barreiras aquifer near TP-07 to assess the vertical distribution of CHCs, indicated that the CHCs are not migrating within the underlying shale. The CHCs are therefore likely migrating within the Barreiras aquifer. These data further support the

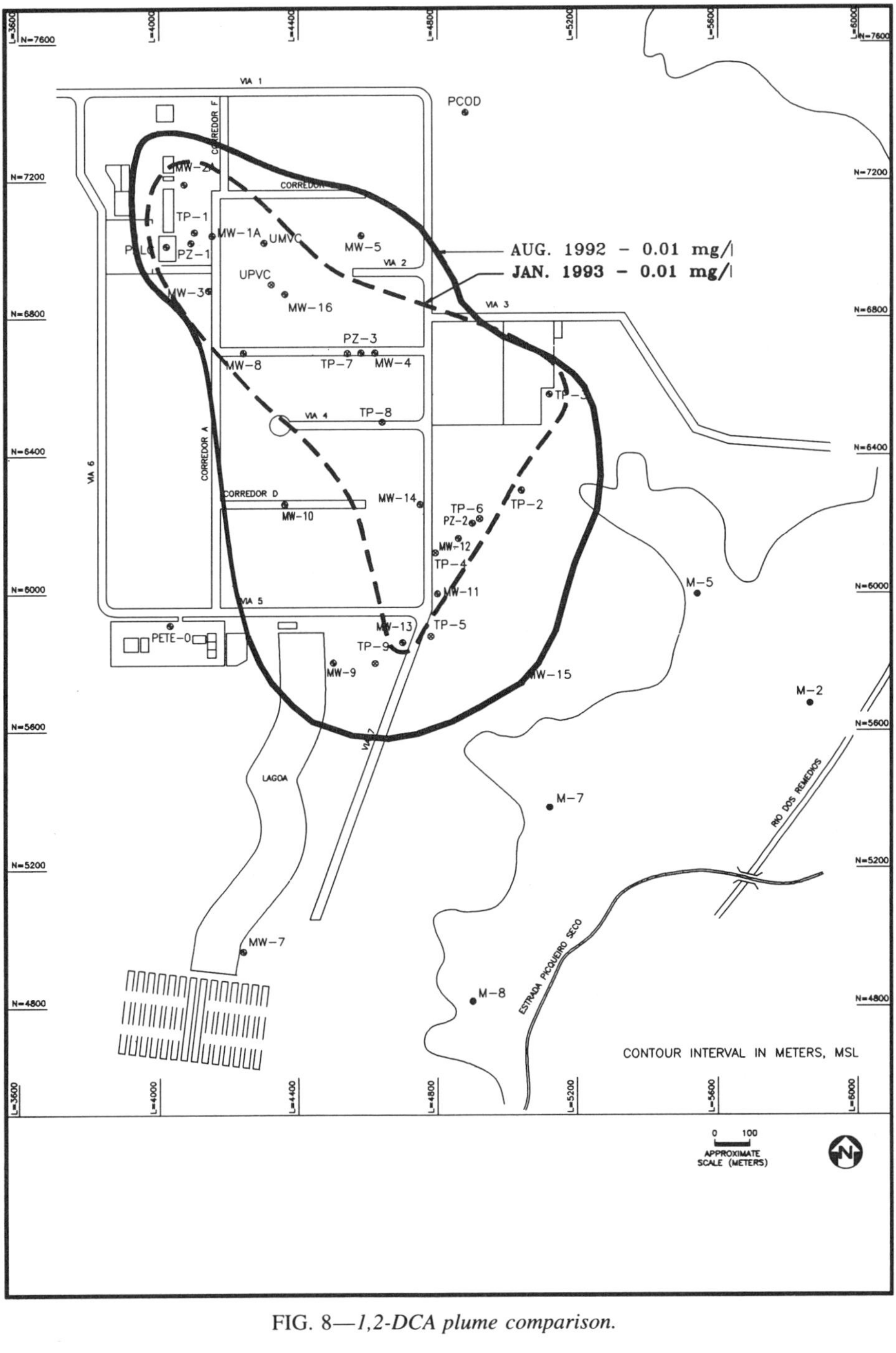

FIG. 8—*1,2-DCA plume comparison.*

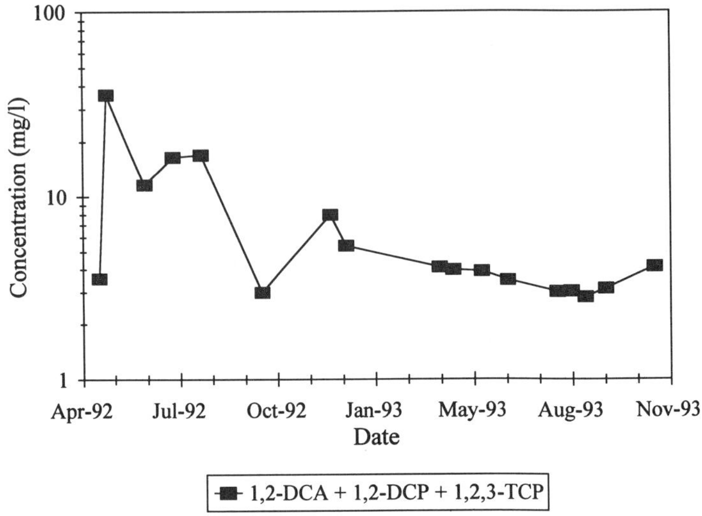

FIG. 9—*Total CHC concentration versus time for well TP-8.*

interpretation of a preferred CHC migration pathway within the Barreiras aquifer near the axis extraction wells. The interpreted paleo-channel and shallower monitoring wells suggests that the CHC migration is near the base of the Barreiras aquifer.

Summary

Depending on the volume of the source and rate of release, CHCs can migrate quickly through thick sections of unsaturated zone and significantly impact ground-water resources. These impacts are a growing concern in developing countries, where regulators and law enforcement officials are beginning to require the investigation and cleanup of CHCs in the subsurface.

Under certain conditions such as the Polo, time and money are not available to perform a detailed hydrogeologic investigation prior to initiating ground-water extraction. However, limited geologic information in combination with extraction wellfield data can provide information to further evaluate hydrogeologic conditions. It should be cautioned that care should be exercised while utilizing this approach to avoid exacerbating the spread of DNAPL (that is, drilling within a known source area).

The transfer of appropriate technology is important to complete a project successfully in a developing country. The technology transfer must consider the local limitations, if any, regarding investigation methods and the installation and operation of treatment equipment. These limitations include importation costs, material availability, and labor costs. For example, available information suggests that local Brazilian engineers with comparable expertise levels earn 50% of their American counterparts. Local vendors and specialists should be used as much as possible so that a timely response can be provided for unanticipated conditions.

Solicitation of local expertise is critical to developing an effective project team, particularly given the extensive time typically required on ground-water remediation projects. This transfer of technology should include educating the client and regulators regarding realistic cleanup objectives and the time required to meet those objectives.

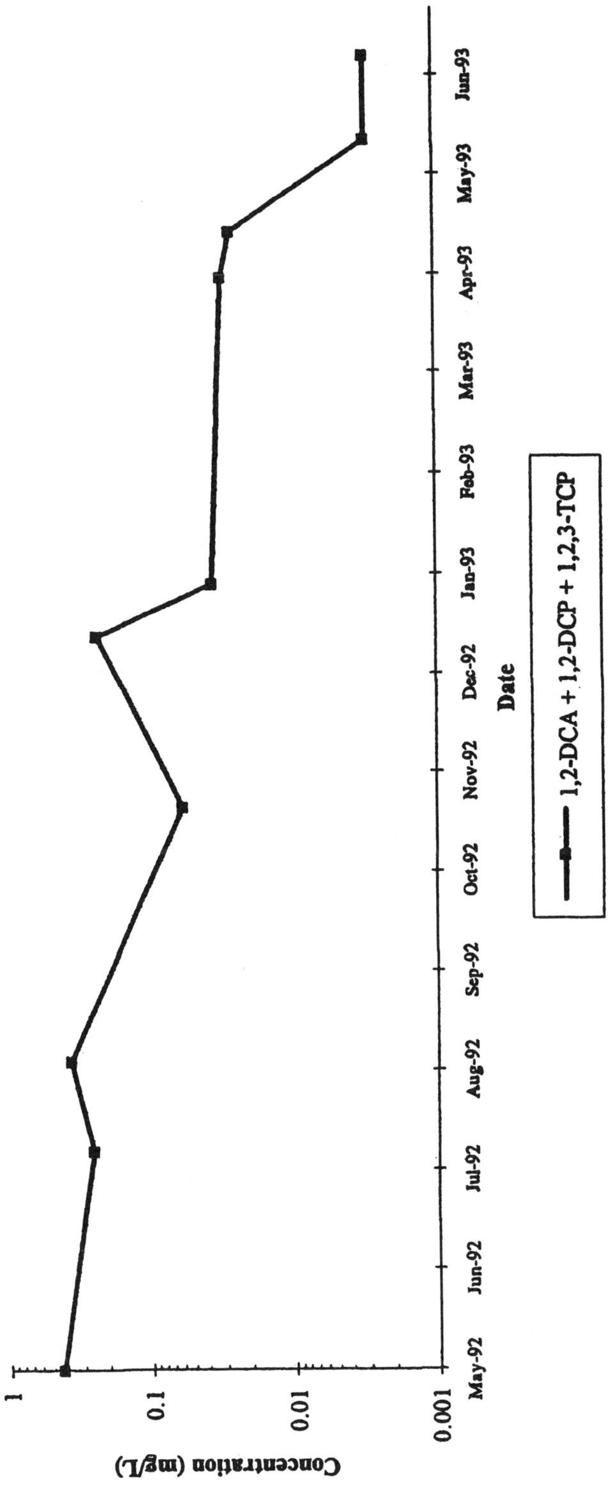

FIG. 10—*Total concentration of 1,2-DCA, 1,2-DCP, and 1,2,3-TCP versus time for well TP-5.*

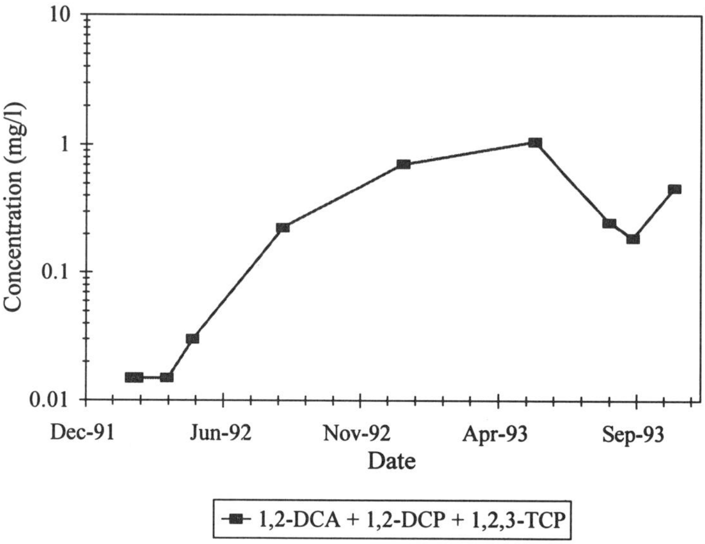

FIG. 11—*Total concentrations versus time for well TP-03.*

Additional monitoring and analysis are required to determine the long-term effectiveness of the hydraulic control program currently in place at the Polo. After a dependable hydraulic control system has been installed, operated, and verified, DNAPL source removal technologies can be developed and evaluated.

References

Franz, T. and Guiguer, N., 1991, "Steady-State Two-Dimensional Horizontal Aquifer Simulation Model," Waterloo Hydrogeologic Software, Waterloo, ON, Canada.

Freeze, A. R. and Cherry, J. A., 1979, *Groundwater,* pp. 394–395.

Stumm, W. and Morgan, J. J., 1981, "An Introduction Emphasizing Chemical Equilibria in Natural Waters," *Aquatic Chemistry,* 2nd edition, pp. 447–448.

U.S. Environmental Protection Agency, Sept. 1993, "Guidance for Evaluating the Technical Impracticability of Ground-Water Restoration," Publication 9234.2-25.

Control and Remediation

Jae Y. Kim,[1] *Jae K. Park,*[1] *Tuncer B. Edil,*[1] *and Jhung K. Jhung*[2]

Sorption Capacity of Ground Tires for Vapor-Phase Volatile Organic Compounds

REFERENCE: Kim, J. Y., Park, J. K., Edil, T. B., and Jhung, J. K., **"Sorption Capacity of Ground Tires for Vapor-Phase Volatile Organic Compounds,"** *Volatile Organic Compounds in the Environment, ASTM STP 1261,* W. Wang, J. Schnoor, and J. Doi, Eds., American Society for Testing and Materials, 1996, pp. 237–244.

ABSTRACT: Batch sorption tests were conducted to determine the partition coefficient of volatile organic compounds (VOCs) by ground tires. The partition coefficient in the vapor phase was estimated by dividing the partition coefficient in the aqueous phase by Henry's law constant. Under a diluted condition (<0.6 mM), VOCs are sorbed onto ground tires noncompetitively regardless of the existence of other VOCs. Polar compounds such as methylene chloride were less sorbed onto ground tires than nonpolar compounds. The vapor-phase partition coefficient was found to have a logarithmic relationship with the saturation vapor concentration. VOCs emitted from waste-water treatment facilities can be effectively retarded by the ground tires.

KEYWORDS: ground tires, off-gas, partition coefficient, sorption, volatile organic compounds

The waste water from commercial, industrial, and agricultural activities collected in publicly owned treatment works (POTWs) contains volatile organic compounds (VOCs) (Witherspoon et al. 1992). In most cases, VOC concentrations are low and vary greatly with time. VOCs which enter into POTWs are removed by volatilization, biodegradation, chemical oxidation, sorption onto biomass, and so on. Volatilization has been considered to be a dominant mechanism of VOC removal from waste water.

The Clean Air Act Amendments of 1990 require new technology-based regulations to meet stringent ambient air quality and risk standards. Many POTWs will be required by local and state regulatory agencies to reduce emissions to meet air quality or health risk standards. A number of technologies are available for controlling VOC emission: adsorption (50 to 95% removal), absorption (>90% removal), condensation (50 to 90% removal), flares (>98% removal), thermal incineration (>98% removal), and catalytic incineration (>95% removal). The use of granular activated carbon (GAC) seems one of the most effective treatment technologies for removing VOCs from the off-gas; however, it is a relatively expensive process.

It was found that scrap tires have a significant VOC sorption capacity (Park and Jhung 1993). The capacity of ground tires to sorb VOCs was found to be 1 to 6% of GAC on a volume basis. In the United States, approximately 240 million tires are being discarded in a

[1] Graduate student, associate professor, and professor, respectively, Department of Civil and Environmental Engineering, University of Wisconsin-Madison, Madison, WI 53706.

[2] Associate professor, Department of Urban Planning Engineering, Won-Kwang University, Iri City, Korea.

year. Since most states have strict disposal regulations, over two billion tires have been stockpiled or illegally dumped (EPA 1991).

The objectives of this study were to determine the vapor phase organic compound sorption capacity of ground tires, to evaluate the desorption rate, and to assess the feasibility of using ground tires as a sorbent for VOCs emitted during waste-water treatment processes.

Theories

In two immiscible phases, the concentration of a compound in one phase can be linearly related to the concentration in the other phase. The parameter for this linear relationship is the partition coefficient (or distribution coefficient). In order to estimate the partition coefficient between two different phases, the concentrations in both phases are required. In this study, the concentration in the aqueous (water) phase could be directly measured but the concentration in the solid phase (ground tire) could not be directly measured. However, the concentration in ground tires can be indirectly estimated because the initial concentration of a target VOC in ground tires can be assumed to be zero and the initial and equilibrium concentrations in the aqueous phase are known. Consequently, the partition coefficient in the aqueous phase, K_p, in litres per kilogram (L/kg), can be obtained as follows

$$K_p = \frac{C_{\text{tire},e}}{C_{l,e}} \tag{1}$$

$$C_{\text{tire},e} = \frac{(C_{l,o} - C_{l,e}) \cdot V_l}{M_{\text{tire}}} \tag{2}$$

where

$C_{\text{tire},e}$ = equilibrium concentration of target compound in ground tire, mg/kg,
$C_{l,o}$ = initial concentration of target compound in aqueous phase, mg/L,
$C_{l,e}$ = equilibrium concentration of target compound in aqueous phase, mg/L,
V_l = volume of solution, L, and
M_{tire} = mass of applied ground tire, kg.

Thus, the tire-VOC partition coefficient in the aqueous phase can be directly estimated. The partition coefficient in the vapor phase, on the other hand, can be indirectly estimated by the extension of the linear partitioning between tire and water. The partition coefficient in the vapor phase, K_v (L/kg), can be expressed as follows

$$K_v = \frac{C_{\text{tire},e}}{C_{v,e}} \tag{3}$$

where $C_{v,e}$ is the equilibrium concentration of the target compound in the vapor phase (mg/L).

At equilibrium, the concentration in tire can be expressed using Eqs 1 and 3 as follows

$$C_{\text{tire},e} = K_p \cdot C_{l,e} = K_v \cdot C_{v,e} \tag{4}$$

Henry's law constant is expressed as follows

$$H_c = \frac{C_{v,e}}{C_{l,e}} \tag{5}$$

where H_c is Henry's law constant, dimensionless.

Thus, from Eqs 4 and 5, the partition coefficient in the vapor phase can be determined from the following relationship (Park et al. 1991):

$$K_v = \frac{K_p}{H_c} \tag{6}$$

Experimental Materials and Methods

Materials

Ground tires were obtained from Tire Technology, Inc. (Rockton, Illinois). No attempt was made to separate tires based on different manufacturers or sizes. The diameter of the ground tire granules ranged from 0.18 to 1.67 cm. The diameter of the tire granules was estimated by the sieve analysis. The surface area of ground tire was measured by a Surface Area Analyzer (Micromeritics Co., ACCQUSORB 2100B). The surface areas ranged from 0.16 to 0.56 m^2/g. The pore size was estimated by a porosity meter. The pore diameter ranged from 0.003 to 3.0 μm and the dominant value was 0.0385 μm. The apparent particle density was 1.204 g/cm^3.

Before conducting tests, ground tires were rinsed with deionized water, dried in a 40°C dry oven for a day, and stored in a desiccator. Since tire is composed of several different organic materials, a high-temperature oven was not used to dry the ground tires.

Chloroform (CF), ethylbenzene (EB), methylene chloride (MC), toluene (TOL), 1,1,1-trichloroethane (TCA), trichloroethylene (TCE), and *m*-xylene (XYL) were selected for testing. These organic compounds are the most frequently detected volatile organic compounds in POTWs (Witherspoon et al. 1992). The physical/chemical properties of the VOCs tested are summarized in Table 1.

TABLE 1—*Properties of volatile organic compounds tested (Verschueren 1983).*

VOCs	Molecular Weight, g/mol	Specific Gravity	Solubility, mg/L	Vapor Pressure, mm	Henry's Constant[a]	Saturation Vapor Concentration, mg/L
CF	119.38	1.489	8000	160	0.154	1027
EB	106.17	0.867	152	7	0.324	40
MC	84.93	1.327[b]	20000	349	0.105	1549
TOL	92.10	0.867	515	22	0.271	110
TCA	133.41	1.350	4400	100	0.170[c]	726
TCE	131.50	1.460	1100[b]	60	0.366	415
XYL	106.16	0.864	200[c]	6	0.283	35

NOTE: CF = $CHCl_3$, EB = $C_6H_5C_2H_5$, MC = CH_2Cl_2, TOL = $C_6H_5CH_3$, TCA = CCl_3CH_3, TCE = CCl_2 = CHCl, and XYL = $C_6H_4(CH)_2$.

[a] Dimensionless (Mackay and Shiu 1981).

[b] At 25°C.

[c] Dostal (1990).

Batch Isotherm Sorption Tests

The batch sorption tests were conducted to estimate the partition coefficients of VOCs in the aqueous phase. The batch isotherm test consists of submerging a measured amount of ground tire into a glass tube containing the organic compounds solution. The equilibrium concentration of VOCs in the aqueous phase was measured to estimate the partition coefficient. Before starting the batch isotherm tests, the required time to reach the equilibrium condition was estimated. In less than one day, the equilibrium condition was reached.

Two types of screw-capped glass tubes with a polytetrafluoroethylene-coated septum, approximately 43 or 25 mL, were also used for the batch isotherm tests. Glass tubes were tumbled end-over-end by the Millipore® Rotary Agitator (Millipore Co., Bedford, Massachusetts) for five days. All isotherm tests were conducted at 20°C. The initial aqueous concentrations of organic compounds were around 50 mg/L. The dosage of ground tires ranged from 0.3 to 4% by volume. To prevent bacterial growth, sodium azide was added during each test at a concentration of 0.05% (by weight). An aliquot of 1% sodium azide stock solution was made to be diluted further.

The target compounds easily vaporized during the tests. Thus, the headspace within the test tubes was kept as small as possible. The initial VOC concentration in the aqueous phase was corrected by using control tubes since the VOC concentration could not be held constant while the test tubes were being filled. The control tubes, which contained no ground tire, were filled with the test solution in every six test tubes. The initial concentration was recalculated proportionally based on the concentrations of the control tubes. For better accuracy, the volume of the solution added in the test vials was estimated by weighing the vials.

Analytical Methods

Gas chromatographic analyses were performed on a Varian 3600 gas chromatograph with 60 m × 0.25-mm-inside-diameter Supelcowax-10 megabore column and a flame ionization detector (FID). An aliquot of 1.0 μL liquid sample was directly injected into the column using an autosampler. Helium was used as the carrier gas.

The column temperature was programmed to initially hold at 50°C for five minutes, then to climb to 85°C at a rate of 5.0°C/min, to climb again immediately to 230°C at a rate of 40°C/min, and finally to hold for two minutes. A Varian® 1093 Septum Equipped Programmable Injector (SPI) was programmed to initially hold at 50°C for 0.5 min, then to immediately climb to 230°C at a rate of 225°C/min, and then hold for 5 min.

Calibration curves were produced using standards at 0, 10, 20, 30, 40, and 50 mg/L. The calculated concentrations were not reliable below some values of peak area because of the recording noise of the gas chromatograph system. Thus, detection limits were approximately 0.5 mg/L for chloroform, methylene chloride, 1,1,1-TCA, and TCE, and 0.1 mg/L for ethylbenzene, toluene and *m*-xylene, respectively. All weighings were made on a scale sensitive to 10^{-4} g order.

Results and Discussion

Batch Isotherm Sorption Tests

The partition coefficient can be estimated by finding the slope of the best fitting line by linear regression to the experimentally obtained concentrations in each phase at equilibrium (Mackay 1991). Alternatively, the ratio between each pair of concentrations of the two phases is calculated and the partition coefficient can be estimated by averaging all the ratios. The

TABLE 2—*Comparison between two different partition coefficient estimation methods in the single-component system.*

VOCs	Vapor Partition Coefficient, L/kg	
	Averaged Ratio	Linear Regression (R^2)
CF	331	331 (0.38)
EB	3123	3022 (0.98)
MC	105	105 (0.57)
TOL	1096	1085 (0.98)
TCA	1000	959 (0.64)
TCE	713	719 (0.88)
XYL	3534	3696 (0.98)

partition coefficients in the aqueous phase are converted to those in the vapor phase using Eq 6.

Table 2 lists the estimated vapor partition coefficients based on both of these methods of computation. The aqueous batch isotherm tests indicated strong linearity for four of the organic compounds tested with high coefficients of correlation. Chloroform, methylene chloride, and 1,1,1-trichloroethane had lower coefficients of correlation. This reflects the relative significance of measurement errors in the case of these three organic compounds because of the small degree of aqueous sorption onto ground tires. The difference between the partition coefficients estimated by the two methods was within ±5%.

The presence of other VOCs may affect the sorption of organic compounds onto tire. Theoretically, the partition coefficient is the distribution ratio of activities between two different phases. When two VOCs are mixed together and exist in aqueous or in gas phase, the activities of the two VOCs are affected. The activity of VOCs in a mixed condition can be estimated by the UNIQUAC equation (Prausnitz 1986). Due to the complexity and impracticability of this approach, the activity coefficient was assumed to be unity in this study.

Batch isotherm tests were conducted to evaluate the effect of the presence of other organic compounds on the sorption of a given organic compound onto tire. The results are summarized in Table 3. The *t*-test was conducted to determine whether the partition coefficients

TABLE 3—*Comparison between single-component and multi-component systems.*

VOCs	Vapor Partition Coefficient, L/kg					
	Single-Component			Multicomponent		
	Average	S.D.[a]	No. of Tests	Average	S.D.	No. of Tests
CF	331	73.4	13	351	81.8	13
EB	3123	211.7	15	2630	920.7	11
MC	105	17.1	14	114	17.1	11
TOL	1096	79.0	15	1070	304.8	13
TCA	1000	162.4	14	1170	316.5	13
TCE	713	209.0	14	648	189.6	12
XYL	3534	419.1	14	3378	1101.8	11

[a] S.D. = standard deviation.

TABLE 4—*Results of* t-*test between single-component and multicomponent sorption tests.* H_o = K_v*(single)* = K_v*(multi).*

VOCs	ν^a	$t_{0.025,n}$	$\lvert t \rvert$	Decision
CF	24	2.064	0.604	do not reject H_o
EB	24	2.064	2.019	do not reject H_o
MC	23	2.069	2.044	do not reject H_o
TOL	26	2.056	0.350	do not reject H_o
TCA	25	2.060	1.786	do not reject H_o
TCE	24	2.064	0.557	do not reject H_o
XYL	23	2.069	0.481	do not reject H_o

[a] Degree of freedom.

are different between the single-component and the multicomponent systems. The results are summarized in Table 4. The hypothesis that the mean value of the partition coefficient measured in the single-component system is equal to that in the multicomponent system is not rejected. Therefore, within the experimental accuracy and for the range of concentrations tested (that is, in the diluted condition), VOCs do not seem to affect each other's sorption onto tires significantly.

Figure 1 shows the relationship between the partition coefficients in the vapor phase and the saturation vapor concentration of the seven VOCs tested. Polar compounds which have higher saturation vapor concentration were less sorbed onto ground tires than nonpolar compounds. The logarithmic value of the partition coefficients of VOCs onto ground tires was proportional to the logarithmic value of the saturation vapor concentration. This relationship can be extended to estimate the vapor partition coefficient over a variety of VOCs. Equation 7 was derived from both the single and the multicomponent system data.

$$\log K_v = 4.5876 + 0.6930 \cdot \log C_s \tag{7}$$

where C_s is the saturation vapor concentration (mg/L).

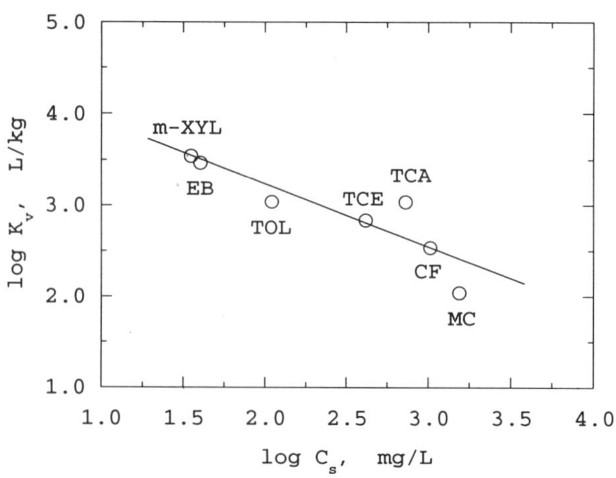

FIG. 1—*Relationship between vapor partition coefficient and saturation vapor concentration.*

Desorption Tests

Three batch desorption tests in the vapor phase were conducted. An aliquot of 10 mL of m-xylene was dropped into the 25 mL glass tube used in the batch tests and a piece of ground tire was inserted in each tube. The sorption of m-xylene vapor onto ground tire was completed within 24 h. After one day, the ground tire was exposed to air and the weight of the ground tire was monitored under well-ventilated conditions.

Figure 2 shows the desorption rate of *m*-xylene. M_{tire} is the weight of the ground tire before conducting the sorption tests. M_t/M_o is the ratio of the weight of the ground tire during different times since the beginning of the desorption test to the initial weight. After 2 h, less than 30% of the sorbed *m*-xylene remained in the ground tire. After 12 h, the weight of the ground tire returned to the initial weight. Thus, the vapor-phase sorption of VOCs onto ground tire is reversible and the desorption is a fast reaction.

Sorption/desorption tests were conducted with tire chips in the aqueous phase (Park et al. 1993). After sorption reached equilibrium, the solution containing VOCs was replaced with deionized water and the desorption test was performed for seven days. Only 3.4 to 7.9% of VOCs sorbed onto tire chips were desorbed, which means that the VOCs sorbed onto tires will not be leached out completely under nonzero external boundary conditions.

Potential Application

A potential application for ground tires is to install a ground tire layer above the aeration basin or channels with a simple supporting structure such as a geotextile. VOCs which are stripped from the waste-water treatment plants can be removed by a ground tire layer.

A laboratory-scale column test showed that the VOC removal efficiencies by a ground tire layer were over 95% for nonpolar organic compounds and 85% for polar organic compounds under typical stripping conditions occurring in a full-scale activated sludge aeration basin (Park and Jhung 1993).

Conclusions

The removal of VOCs in the off-gas appears to be feasible with a sorption layer of ground tires. From the batch isotherm sorption tests, the following conclusions can be drawn:

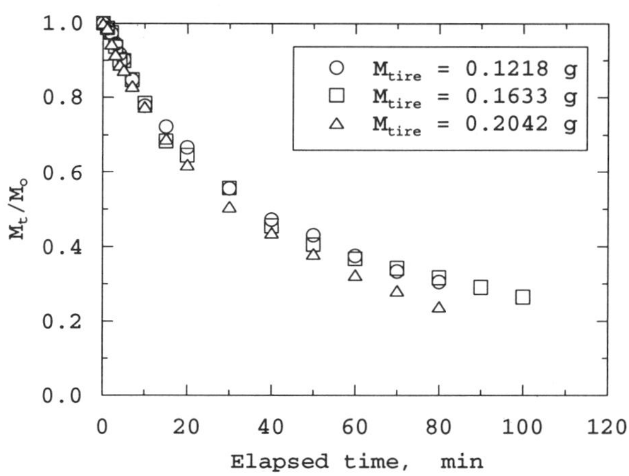

FIG. 2—*Vapor phase desorption rate of m-xylene.*

1. The partition coefficients in the vapor phase for the seven VOCs tested were 2.7 to 9.5 times greater than those in the aqueous phase since the Henry's law constants are less than 1.
2. Polar organic compounds were less sorbed onto ground tire than nonpolar organic compounds.
3. Under a diluted system (<0.6 mM), the sorption of VOCs onto ground tires was not affected by the presence of other VOCs.
4. The vapor-phase partition coefficients of organic compounds are logarithmically proportional to the saturation vapor concentration.
5. Due to the relatively high capacity of ground tires to sorb VOCs, it may be possible to use ground tires for control of the VOCs emitted from waste treatment processes.

References

Dostal, K. A., 1990, *WERL Treatability Data Base,* Risk Reduction Engineering Laboratory, U.S. EPA, Cincinnati, OH.

EPA, 1991, *Markets for Scrap Tires,* EPA/530-SW-90-074, U.S. EPA, Office of Solid Waste, Washington, DC. Oct.

Mackay, D., 1991, *Multimedia Environmental Models—The Fugacity Approach,* Lewis Publishers, Chelsea, MI, pp. 69–70.

Mackay, D. and Shiu, W. Y., 1981, "A Critical Review of Henry's Law Constants for Chemicals of Environmental Interest," *Journal of Physical Chemical References Data,* Vol. 10, No. 4, pp. 1175–1199.

Park, J. K., Bontoux, L., Holsen, T. M., Jenkins, D., and Selleck, R. E., 1991, "Permeation of Polybutylene Pipe and Gasket Material by Organic Chemicals," *Journal of the American Water Works Association,* Vol. 83, No. 10, pp. 71–78.

Park, J. K. and Jhung, J. K., 1993, "Removal of Volatile Organic Compounds Emitted During Wastewater Treatment by Ground Tires," *Proceedings,* 66th Annual Conference of the Water Environment Federation, Anaheim, CA, 3–7 Oct.

Park, J. K., Kim, J. Y., and Edil, T. B., 1993, "Mitigation of Organic Compound Movement in Landfills by a Layer of Shredded Tires," *Proceedings,* 66th Annual Conference of the Water Environment Federation, Anaheim, CA, 3–7 Oct.

Prausnitz, J. M., Lichtenthaler, R. N., and de Azevedo, E. G., 1986, *Molecular Thermodynamics of Fluid-Phase Equilibria,* 2nd ed., Prentice-Hall, Englewood Cliffs, NJ.

Verschueren, K., 1983, *Handbook of Environmental Data on Organic Chemicals,* Van Nostrand Reinhold, New York.

Witherspoon, J. R., Bishop, W. J., and Wallis, M. J., 1992, "Emission Control Options for POTWs," *Environmental Protection,* Vol. 4, No. 5, pp. 59–66.

Jae K. Park,[1] Joni P. Sakti,[2] and John A. Hoopes[1]

Determination of Volatile Organic Compound Permeation Through Geomembranes

REFERENCE: Park, J. K., Sakti, J. P., and Hoopes, J. A., "**Determination of Volatile Organic Compound Permeation Through Geomembranes,**" *Volatile Organic Compounds in the Environment, ASTM STP 1261,* W. Wang, J. Schnoor, and J. Doi, Eds., American Society for Testing and Materials, 1996, pp. 245–258.

ABSTRACT: Synthetic geomembranes are commonly used as barriers or covers in hazardous waste management facilities. A confined double-compartment method (CDCM) was developed to evaluate the sorption and diffusion mechanisms of volatile organic compounds (VOCs) through high-density polyethylene (HDPE), very low-density polyethylene (VLDPE), and polyvinyl chloride (PVC) geomembranes. The loss of VOCs during testing was measured and its effect on mass transport parameters assessed. The capacity of geomembranes to adsorb VOCs, indicated by weight gain, was large, ranging from 21% to 52% of the initial geomembrane weight. In general, the total amount of VOCs lost decreased as the aqueous concentration decreased and increased as the test duration increased. The VOC loss in the CDCM test was within the maximum gas chromatograph analytical error of about 3% to 4%; thus, the CDCM enables sorption onto and diffusion through a geomembrane to be evaluated accurately along with the influence of concentrations above and below the geomembrane on the transport behavior.

KEYWORDS: geomembrane, landfill, loss, organic chemicals, permeability test, high-density polyethylene, very-low-density polyethylene, polyvinyl chloride, transport

U.S. Environmental Protection Agency (EPA) Method 9090 (1986) is currently used for chemical resistance testing of geomembranes with leachates and other liquids. However, this test method does not address the transport of volatile organic compounds (VOCs) through an intact geomembrane.

Many VOCs have been found in landfills, which even at extremely low concentrations are hazardous to human health (Friedman 1987, Dimmick et al. 1990, Gibbons et al. 1992, Lee and Jones 1992). About 75% of the landfills in the United States are polluting ground water and significant VOCs have been detected in leachates and ground waters (Friedman 1987, Gibbons et al. 1992).

The general chemical resistance guidelines of commonly used geomembranes, summarized by Koerner (1990), state that polyethylene (PE) geomembranes passed all physical and mechanical property tests for all the chemicals used. It is possible that the physical and mechanical properties of a geomembrane are practically unaltered after a long period of direct contact with VOCs; however, this does not guarantee that VOCs will not permeate a PE

[1] Associate professor and professor, respectively, Department of Civil and Environmental Engineering, University of Wisconsin-Madison, Madison, WI 53706.
[2] Engineer, Indonesia.

geomembrane in amounts that may pose a serious threat to the environment surrounding a containment facility. Many studies acknowledge this problem but experiments are not adequate to address VOC permeation.

August and Tatzky (1984) and August et al. (1984) measured permeation rates of concentrated (pure) organic solvents, mixtures of hydrocarbons, and aqueous solutions of VOCs through geomembranes of various types and thicknesses using a permeation cell. Their permeation cell was composed of two compartments separated by a sheet of a geomembrane. The concentration in the top compartment was assumed constant, while the concentration in the bottom compartment was assumed to be zero, as the permeated substances were removed from the bottom compartment. The removed substances were frozen in a cooling trap and their concentrations were determined by a gas chromatograph (GC). The permeation rates were determined from the concentration change with time.

August et al. (1984) found that the permeation rates were dependent on compound type and on geomembrane thickness and type. August et al. postulated that VOC transport in/through geomembranes may involve physical phenomena such as diffusion, absorption, and desorption. The permeation rate was estimated solely on the amount of VOC that passed through the geomembrane surface area in a unit time. This estimation is incorrect for it does not account for partition and diffusion and assumes a constant concentration above the geomembrane.

Haxo et al. (1984) and Haxo and Pierson (1991) measured permeation rates using a two-compartment (separated by a geomembrane) solvent vapor transmission apparatus. The bottom compartment was partially filled with a VOC so that a saturated atmosphere within the compartment might be attained. The top compartment was open to the atmosphere to approximate zero concentration. The apparatus was weighed periodically. The permeation rate was determined from the weight loss. Similar to the work by August and Tatzky (1984) and August et al. (1984), this kind of permeation rate has no physicochemical transport basis.

Haxo et al. (1987) measured the distribution and permeation of a mixture of nine VOCs (equal weight) in a dilute aqueous solution through a linear low-density polyethylene (LLDPE) geomembrane. They used a two-compartment apparatus separated by the LLDPE geomembrane. Approximately 85% of the top compartment was filled with the solution and the rest was air space. The bottom compartment was all air space. After 190 h they measured the weight (concentration) of each compound in the geomembrane and the compartments. This experiment gives the amount of VOCs retained in the geomembrane and permeated to the bottom compartment.

In order to simulate the service conditions of a geomembrane in a dry tomb landfill, Haxo et al. (1987) and Haxo and Lahey (1988) extended their two-compartment method to a three-compartment method. Each compartment was separated by a LLDPE geomembrane. The middle compartment was filled with an aqueous solution of VOCs up to 62% of its volume (the rest was air space), the top compartment was air space, and the bottom compartment was initially filled with deionized water up to about 37% of its volume (the rest was air space). In these experiments, significant amounts of trichloroethylene (TCE), toluene, and o-xylene migrated to the air spaces in the top and bottom compartments in 24 h. Though VOCs may eventually reach equilibrium in each phase, these experiments determined the amount of each VOC distributed in each phase of the compartments and in the geomembranes but not the rate of permeation. Prasad et al. (1994) adopted a similar setup to Haxo and Lahey (1988) to determine the diffusion coefficient of geomembranes.

Britton et al. (1989) estimated the partition (called solubility) and diffusion coefficients of eight compounds (included toluene and methylene chloride) in an HDPE geomembrane using the Standard ASTM Test Method for Resistance of Protective Clothing Materials to Permeation by Liquids or Gases under Conditions of Continuous Contact (F 739-81) permeation cell. The solution used was an equimolar mixture (liquid nonaqueous) of eight compounds (31%) and 2-methylnaphthalene (50%) in dimethylformamide. Britton et al. used the immersion/weight gain method to estimate the solubility (partition) coefficient of the penetrants. The diffusion coefficients were estimated from the weight gain data and a mathematical solution of the partition-diffusion process. Although it was possible to estimate diffusion coefficients, these values were obtained from nonaqueous solutions where the concentration of each compound was much higher than its maximum solubility in water. This kind of solution is not typical in landfills or other containment systems.

Park and Nibras (1993) investigated diffusion and partition coefficients in dilute aqueous solutions of VOCs with a HDPE geomembrane. Each geomembrane was placed in a bottle filled with a particular solution. The weight gain of the geomembrane was measured and used to calculate the partition coefficient. The diffusion coefficient was estimated, assuming a constant concentration (average concentration of initial and final stages) and employing a partition-diffusion model. Park and Nibras estimated the partition and diffusion coefficients of each VOC for various aqueous concentrations. Their work could not determine a relationship between partition and diffusion coefficients with concentration. Furthermore, the weight gain method does not show concentration changes (with time) in each compartment or the breakthrough time, which are needed to understand the transport process and to assess the ability of geomembranes to contain VOCs.

Other researchers (Giroud 1984, Haxo et al. 1984, Lord et al. 1988, Koerner 1990) in an attempt to compare the permeability of a geomembrane and a soil, investigated the potential amount of water (vapor or liquid) that could pass through a geomembrane. The apparatus utilizes two compartments separated by a geomembrane. One compartment (upstream) was filled with water vapor or liquid water under constant pressure. The other compartment (downstream) was open to the atmosphere or connected to a pressure device. By measuring the amount of water lost from the upstream compartment, they estimated the permeability coefficient (equivalent hydraulic conductivity) and the water vapor transmission rate employing Darcy's equation. For an intact geomembrane this procedure is not correct, since substances move through an intact geomembrane as a result of a concentration gradient following Fick's law of diffusion. In an aqueous solution, each substance diffuses separately, except water, which moves as a result of hydraulic gradient.

VOC transport through a geomembrane can be expressed by a three-step physicochemical process: partition (adsorption) to the surface of the geomembrane (upstream interface); diffusion through the geomembrane; and partition (desorption) to an external phase in contact with the geomembrane (downstream interface). If the phases of the upstream and downstream media in contact with a geomembrane are different (that is, liquid in the upstream and vapor in the downstream), a single partition coefficient is not correct, since the partition coefficient in the vapor phase is equal to the partition coefficient in the aqueous phase divided by the Henry's law constant (Park et al. 1991). If VOC flux rates are predicted using measured values under different conditions on each side of the geomembrane, a mathematical model must be used with appropriate partition and diffusion coefficients.

The weight gain method does not provide the concentration change in the compartment or the breakthrough time. The equivalent hydraulic conductivity (permeability coefficient) and the water vapor transmission rate, acquired from Darcy's equation, are not appropriate ways to represent the permeability of a geomembrane. The permeability coefficient should be based on the diffusion and partition coefficients. Data on transport properties are indeed scanty, and there are no investigation/data in the available literature regarding the effect of geomembrane tension on transport properties, although geomembranes can experience large tensions in the field.

The main objectives of this study were to develop an experimental method for studies of VOC sorption and diffusion mechanisms in various geomembranes and to assess the effect of VOC loss during a test on VOC transport. A mixture of VOCs in dilute aqueous solutions, normally encountered in landfills, was used. The experimental procedure employs a confined double-compartment method (CDCM), giving concentration changes in both compartments. Incorporating these two sets of concentration changes, the partition coefficient is calculated directly and the diffusion coefficient estimated using the partition-diffusion model. The effects on partition and diffusion coefficients of aqueous concentration and of geomembrane thickness, tension, and type can be investigated with the test method developed in this study. These results will be discussed in subsequent papers along with a mathematical model developed specifically for CDCM.

Experimental Materials and Methods

Organic Compounds Tested

Each aqueous solution was prepared by mixing equal (target) concentrations of methylene chloride (MC), toluene, trichloroethylene (TCE), and m-xylene in deionized (DI) water in a 2-L glass flask with a glass stopper. The solutions were continuously stirred for several hours to give a well-mixed solution. These volatile VOCs were chosen because they are frequently detected pollutants in leachates (Friedman 1987, Dimmick et al. 1990, Gibbons et al. 1992, Lee and Jones 1992) and cover a wide range of vapor pressures and solubilities in water. The initial target aqueous concentrations used were 10, 50, and 100 mg/L.

Geomembranes

Three most commonly used geomembranes—HDPE (high-density polyethylene—Gundle), VLDPE (very-low-density polyethylene—Gundle), and PVC (polyvinyl chloride—OXY)—were investigated. Both VLDPE and HDPE contained 97.5% polymer and 2.5% of carbon black, antioxidants, and heat stabilizers. The PVC contained plasticizers (mainly phthalates), approximately 30% by weight. These plasticizers can leach out because of heat, soil chemicals, and stresses in the liner, causing the liner to become brittle with time. The thicknesses of geomembranes tested were 0.76, 1.52, and 2.54 mm.

Confined Double-Compartment Apparatus

The laboratory experiments were conducted using a confined double-compartment apparatus, consisting of two aluminum columns separated by a geomembrane sheet (Fig. 1). The sheet is clamped between grooved flanges of the two columns and may be placed in tension. The diameter of both the top and bottom columns are the same. However, the height of the bottom column is 1/4 that of the top column in order to obtain and detect higher penetrant concentrations and achieve overall equilibrium faster. O-rings were placed in both grooves to further minimize leakage of liquid along the interfaces of the geomembrane. These O-ring-and-groove systems were designed such that when the top and bottom flanges were clamped tightly, the surfaces of the flanges would be in direct (firm) contact with the geomembrane. Therefore, the geomembrane surface area in direct contact with the aqueous solution was known. This surface area is an important transport parameter. The bottom (downstream) compartment was initially filled with DI water, containing no VOCs. The geomembrane was placed carefully on top of the bottom compartment such that no air bubbles were trapped; the top column then was placed on the geomembrane and clamped

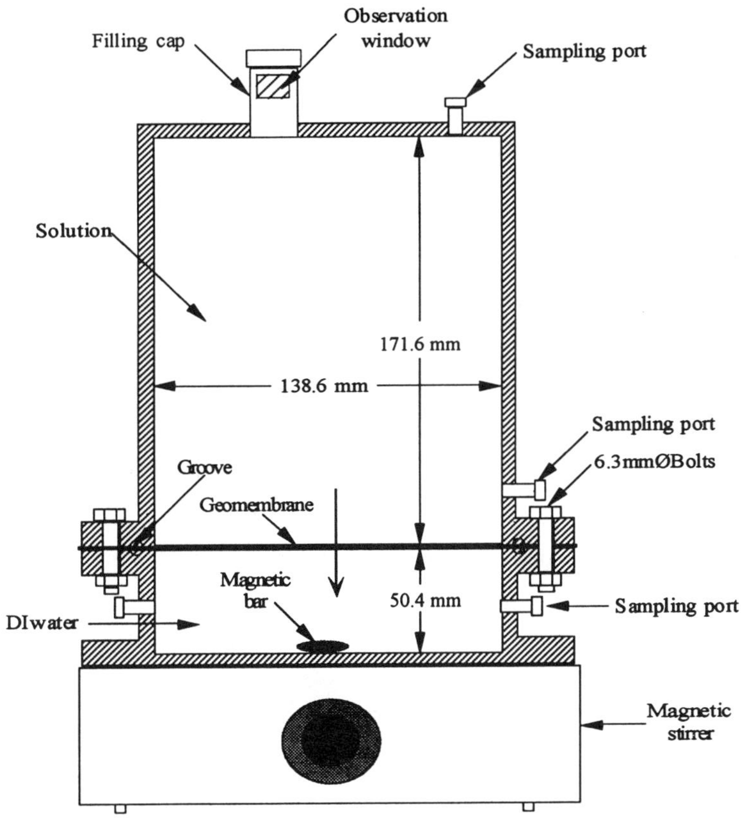

FIG. 1—*Confined double-compartment apparatus.*

tightly with eight 0.63-mm (inside diameter) bolts, evenly distributed around the perimeter. The top (upstream) compartment was filled with the well-mixed aqueous solution up to the observation window. During an experiment, the solution in the downstream compartment was stirred continuously with a magnetic stirrer so that a well-mixed solution was obtained.

Samples were withdrawn with a 5 μL gastight syringe at various times through sampling ports, equipped with trifluoroethylene-lined septa, and immediately injected to a GC, minimizing VOC loss due to volatilization. Samples from both compartments were taken periodically until an approximate steady-state condition was achieved. The volume of each sample was about 3 μL. Though samples were taken as many as 30 times from each compartment in a test, the total sampled volume (approximately 180 μL) was negligible compared with the volumes of the upstream and downstream compartments, 3.1146 L and 0.779 L, respectively.

In both compartments, samples were obtained from two sampling ports (Fig. 1) at each time. Samples from two ports provided a check on concentration uniformity and minimized analytical errors of the GC. The average value of the two samples was considered the true concentration in each compartment.

All the apparatus components and water used to make a VOC solution were autoclaved at 121°C and 103 kPa for 20 min to prevent potential microbial activity during the test.

Analytical Methods

A Varian Model 3600 gas chromatograph, equipped with a flame ionization detector (FID), was employed for sample analysis. An aliquot of 1.5 μL liquid sample was directly injected into a capillary column using a gastight syringe. The column was a 60-m-long, SupelcoWax® 10 with 0.25 mm inside diameter and 0.25 m film thickness. Helium was used as the carrier gas through the column (6 mL/min flow rate) and as makeup gas through the FID (24 mL/min flow rate). Hydrogen and air were used for the flame with flow rates of 30 and 300 mL/min, respectively. The detector temperature was 250°C and the initial injector temperature was 60°C with a hold time of 0.5 min. The temperature program employed for the GC column was as follows: initially the column temperature was set at 70°C for 13 min, and then the temperature was changed at a rate of 25°C/min to the final column temperature of 220°C.

A calibration curve of each VOC tested was determined by plotting standard concentrations (0.5, 20, 60, and 100 mg/L) versus their specific GC area values. Injections of each standard concentration were conducted five times to determine the GC reproducibility. The coefficients of variation (standard deviation divided by expected value) were less than 3% for m-xylene, 3.5% for TCE and toluene, and 4% for MC. Utilizing the calibration curves which were linear for all VOCs tested, the concentrations of the VOCs in the samples were readily determined. The FID detection limits were approximately 0.5 mg/L for all the target compounds.

Mass Balance

In order to assess the accuracy of the experimental results, the amount of VOCs in the various regions of the closed experimental system must be determined (that is, upstream compartment, downstream compartment, and geomembrane in direct contact with the aqueous solutions in both compartments). If the decreased amount of the VOC in the upstream compartment at any time is not equal to the increased amount in the geomembrane and downstream compartment, some VOC loss has occurred through the geomembrane clamped by the flanges, into other parts of the CDCM apparatus, or into the air due to volatilization. The step-by-step procedure used to perform the mass balance is explained below.

Step 1

Estimate total amount of VOCs in geomembrane in direct contact with the solution and clamped by the flanges. The CDCM apparatus was dismantled after each experiment. The geomembrane in direct contact with the solution was cut into three pieces (approximately 50 × 40 mm): one in the middle and one near each of the two flanges. Two 50 × 40 mm (approximate width of the flange) pieces were cut from the geomembrane clamped in the flanges. The three pieces of geomembrane in direct contact with the solution were used to determine sorption uniformity across the geomembrane, while the two pieces from the clamped portion of the geomembrane were used to determine chemical loss through that area.

All of the geomembrane pieces were quickly blotted to remove surface liquid and weighed on an analytical balance. Next, the pieces were heated in an oven at 80°C (two to three days) to remove the compounds from the geomembrane quickly without destroying its internal structure, and then removed from the oven, cooled to room temperature, and weighed again. This process was repeated several times until the weight of each piece was practically constant.

In order to determine if the original geomembranes were losing weight due to heating (especially important for PVC which contains many plasticizers), following each experiment two pieces of untested geomembrane (about 5 × 4 cm) were tested in the same manner as the tested geomembrane. The O-rings used were also tested following the same procedures as the geomembrane.

Step 2

Determine the amount of water sorbed in the geomembrane using the immersion/weight gain method (same temperature and pressure as in the CDCM experiments). This series of tests was performed because water moves through the geomembrane although at a slower rate than for VOCs. The 0.76- and 2.54-mm-thick HDPE, the 0.76-mm-thick PVC, and the 0.76-mm-thick VLDPE geomembranes were cut into 50 by 120 mm pieces and soaked in a DI water-filled mason jar. Regularly, the geomembranes were withdrawn, quickly blotted to remove surface liquid, and weighed. Since the initial weight of each geomembrane was known, the weight gain versus time plot of each geomembrane was obtained. The water uptake experiments were performed in duplicate for each geomembrane type and thickness. The water uptake for the 1.52-mm-thick HDPE was not determined but was interpolated from the 0.76- and 2.54-mm-thick HDPE results.

Step 3

Perform mass balance for the VOCs tested. The weight gain of each geomembrane in a test is the sum of the four VOCs, whose weight contribution cannot be separated, and the water; hence, the following formula is used:

$$\sum_{i=1}^{i=N} (M_{i,0}^{u} - M_{i,f}^{u}) = (W_{d,f} - W_{w,f}) + W_{c,f} + \sum_{i=1}^{i=N} M_{i,f}^{d} - Q \tag{1}$$

where

M = amount (mass) of a VOC in the compartments (superscripts u and d refer to the upstream and downstream compartments, respectively, subscripts 0 and f denote initial and final conditions, respectively, and subscript i denotes a VOC), mg,
N = number of VOCs involved (in this study $N = 4$),
W = weight gain of the geomembrane (subscripts d and c refer to geomembrane in direct contact with the solution and clamped, respectively, and subscript w refers to water alone), mg, and
Q = total amount of the VOCs lost from the closed system, except into the clamped geomembrane, mg

In Eq (1), the mass of each VOC in both compartments is estimated from:

$$M_0^u = C_{s,0}^u V^u \tag{2}$$

$$M_f^u = C_{s,f}^u V^u \tag{3}$$

$$M_f^d = C_{s,f}^d V^d \tag{4}$$

where C_s is the VOC concentration in solution, mg/L, and V is the volume of the compartment, L.

From experimental data, Q can be calculated from Eq 1. W_c may be excluded from Q because this term can be incorporated in the mathematical modeling.

Results and Discussion

Replicate runs were conducted for several types and thicknesses of geomembranes and initial aqueous concentrations to assess the reproducibility of the experiments using the CDCM apparatus. The concentration changes of VOCs for 0.76-mm-thick PVC at 100 mg/L initial aqueous concentration are shown in Fig. 2. The dashed lines shown in Fig. 2 depict the duplicate run. The replicate runs for all cases were within the GC analytical error (less than 4%). Therefore, results from this test method are highly reproducible.

The experimental conditions and breakthrough times are summarized in Table 1. The concentration changes in both upstream and downstream compartments for 0.76-mm-thick HDPE, VLDPE, and PVC at 100 mg/L initial aqueous concentration are shown in Fig. 3. Small concentrations of the VOCs appeared in 20 to 200 h for 0.76 to 2.54-mm-thick HDPE geomembranes (breakthrough time varied with the square of thickness for each VOC), in 8 h for the 0.76-mm-thick VLDPE, and in 9 h for the 0.76-mm-thick PVC.

The total amount of VOCs lost from the closed system except into the clamped geomembrane, Q, must be known to assess the accuracy of the experiments. Rewriting Eq 1, Q is given by

$$Q = \sum_{i=1}^{i=N} (M^u_{i,0} - M^u_{i,f}) - \sum_{i=1}^{i=N} M^d_{i,f} - (W_{d,f} - W_{w,f}) - W_{c,f} \quad (5)$$

Since $W_{d,f}$ and $W_{c,f}$ can be estimated only at the end of each experiment (by dismantling the CDCM apparatus), Q can be estimated only at that time. Table 2 shows detailed estimates of each parameter in Eq 5 for 2.54-mm-thick HDPE at 100 mg/L initial aqueous concentration.

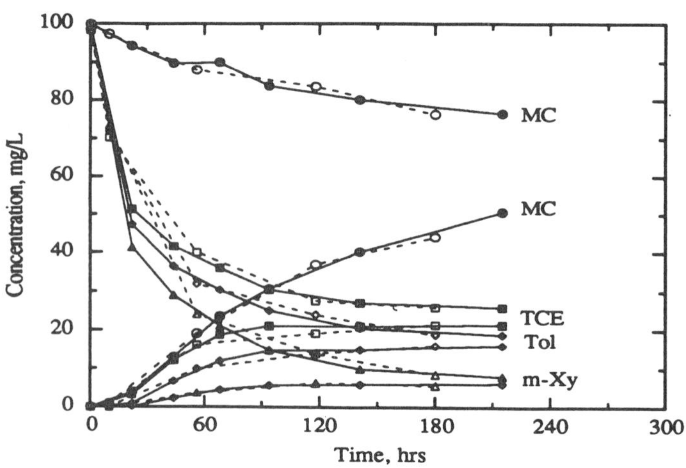

FIG. 2—*Concentration changes of VOCs in upstream and downstream compartments for 0.76-mm-thick PVC at 100 mg/L initial aqueous concentrations for duplicate runs.*

TABLE 1—*Summary of experimental conditions, breakthrough times, and mass balance parameters.*

Geomembrane type	HDPE	HDPE	HDPE	HDPE	HDPE	VLDPE	PVC
Thickness, mm	0.76	0.76	0.76	1.52	2.54	0.76	0.76
Initial aqueous concentration, mg/L	100	50	10	100	100	100	100
Breakthrough time (h)	20	25	28	80	200	8	9
Test duration (h)	696	624	656	1578	3545	181	215
$M^u_{T,0}$ (mg)	1231	611	123	1221	1216	1235	1245
$W_{d,f}/M^u_{T,0}$ (%)	20.5	22.0	23.7	32.3	37.4	31.4	52.2
$W_{c,f}/M^u_{T,0}$ (%)	0.94	0.71	0.67	1.82	3.30	1.05	1.25
$W_{w,f}/W_{d,f}$ (%)	0.41	0.77	3.53	0.65	1.31	0.52	2.31
$Q/M^u_{T,0}$ (%)	7.93	6.24	4.59	8.20	9.49	7.02	9.38

NOTES:
$M^u_{T,0}$ = Initial total mass of the four VOCs tested.
$W_{d,f}$ = Weight gain of the geomembrane in direct contact with the solution at the end of the experiment.
$W_{c,f}$ = Weight gain of the clamped geomembrane at the end of the experiment.
$W_{w,f}$ = Mass of water in the geomembrane at the end of the experiment.
Q = Total losses of the four VOCs from the closed system at the end of the experiment.

$W_{d,f}$ (weight gain of the geomembrane in direct contact with the solution) for each experiment was taken as the average value of the three pieces. The differences among these three values were within the balance limit (0.05 mg). Thus, it can be said that the transport processes occurred uniformly across the geomembrane. The ratios of $W_{d,f}$ to the sum of the initial mass of the four VOCs tested (written as $M^u_{T,0}$ in Table 1) show that the total mass of VOCs retained in each geomembrane varied from 20.5% (0.76-mm-thick HDPE at 100 mg/L initial aqueous concentration) to 52.2% (0.76-mm-thick PVC). It is clear that significant amounts of VOCs were held by the geomembranes, especially for m-xylene, toluene, and TCE, although the volume ratio of the geomembrane to the upstream compartment was only 0.0037 (0.76-mm-thick geomembrane). Hence, a substantial error can occur in determining mass transport parameters if it is assumed that VOC concentration in the upper compartment is constant during data analysis.

$W_{c,f}$ (weight gain of the clamped geomembrane) was taken as the average of the two pieces. The difference between these two values was within the balance limit. Table 1 summarizes the ratios of $W_{c,f}$ to $M^u_{T,0}$ for each experiment and shows that this ratio had a maximum value of 3.3% for a 2.54-mm-thick HDPE at 100 mg/L initial aqueous concentration. Since this test took more than 3500 h, the weight gain of the geomembrane surfaces clamped by the flanges is small. It is possible to incorporate this loss into the mathematical model for determining partition and diffusion coefficients. This model will be discussed in a subsequent paper.

The weight gains of both O-rings (located in the top and bottom flanges) for each experiment were also within the balance limit, suggesting that the geomembrane surfaces were in good contact with the flange surfaces.

The weight gain of each geomembrane due to water penetration alone, W_w, normalized by its initial weight (W_0) is shown in Fig. 4. The initial weights, W_0, of 0.76-mm-thick HDPE, 2.54-mm-thick HDPE, 0.76-mm-thick VLDPE, and 0.76-mm-thick PVC were 4889.3 mg, 15 819.6 mg, 3978.5 mg, and 6434.9 mg, respectively. Thus, W_w for each geomembrane at any time can be estimated readily. For the 1.52-mm-thick HDPE, its W_w value was directly interpolated from the W_w values of the 0.76-mm-thick HDPE and 2.54-mm-thick HDPE. The ratio of W_w to W_d for each experiment is given in Table 1.

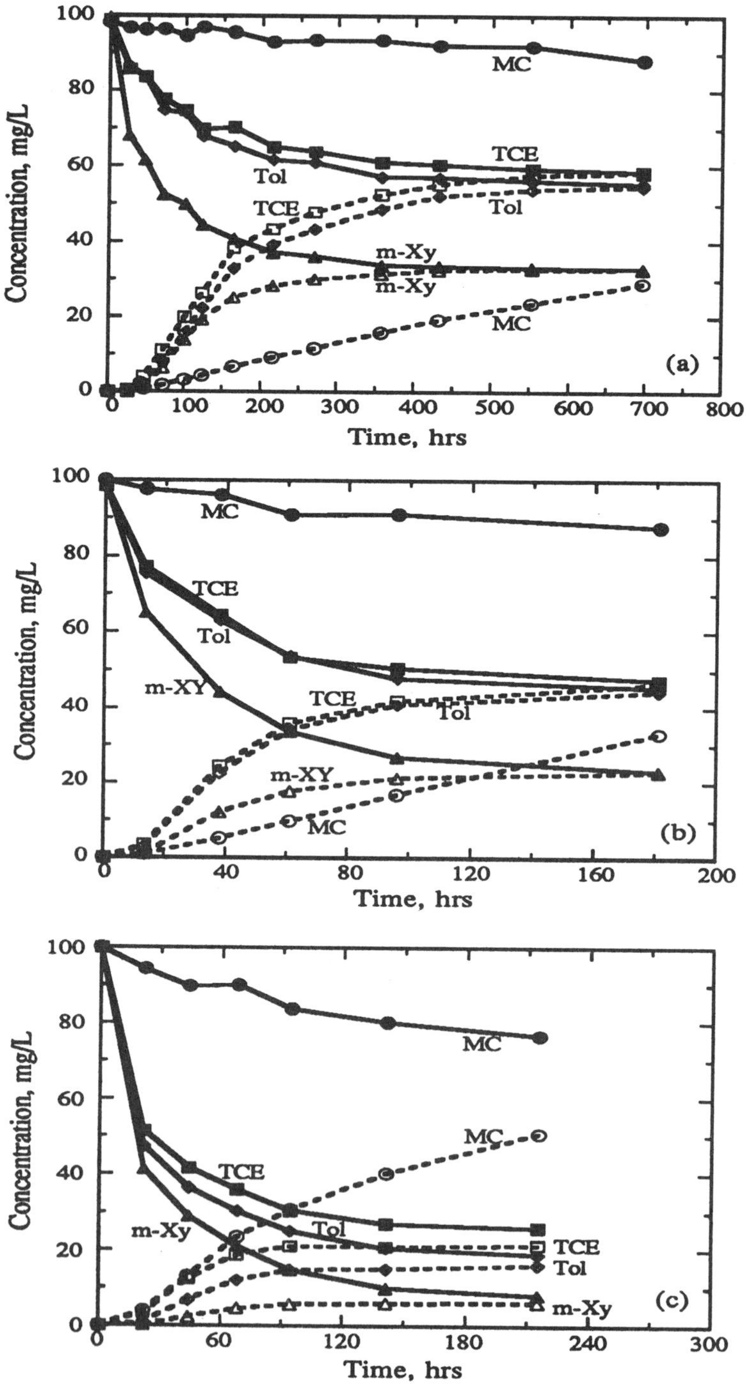

FIG. 3—*Concentration changes of VOCs in upstream and downstream compartments for 0.76-mm-thick HDPE* (a), *VLDPE* (b), *and PVC* (c) *at 100 mg/L initial aqueous concentrations.*

TABLE 2—*Mass balance parameters for 2.54-mm-thick HDPE at 100 mg/L initial aqueous concentration.*

Parameters	MC	TCE	Toluene	m-Xylene
$C^u_{s,0}$ (mg/L)	97.03	98.50	97.01	97.94
C^u_s (mg/L)	88.01	38.18	32.01	13.89
C^d_s (mg/L)	22.73	33.81	28.19	12.45
M^u_0 (mg)	302.21	306.79	302.15	305.04
M^u (mg)	274.16	118.92	99.70	43.26
$M^u_0 - M^u$ (mg)	28.00	187.87	202.45	261.78
M^d (mg)	17.71	26.34	21.96	9.70

(1) $\sum_{i=1}^{i=4} M^u_{i,0} = 302.21 + 306.79 + 302.15 + 305.04 = 1216.19$ mg

(2) $\sum_{i=1}^{i=4} M^u_i = 274.16 + 118.92 + 99.70 + 43.26 = 536.04$ mg

(3) $\sum_{i=1}^{i=4} (M^u_{i,0} - M^u_i) = 1216.19 - 536.04 = 680.15$ mg

(4) $\sum_{i=1}^{i=4} M^u_{i,0} = 17.71 + 26.34 + 21.96 + 9.70 = 75.71$ mg

(5) W_d (sum of the four VOCs) = 454.81 mg

(6) W_c (sum of the four VOCs) = 40.17 mg

(7) W_w (only DI water) = 5.98 mg

Q (Eq 5) = (3) − (4) − (5) − (6) + (7) = 115.44 mg

$Q/M^u_{T,0}$ (%) = 115.44/1216.19 · 100 = 9.49%

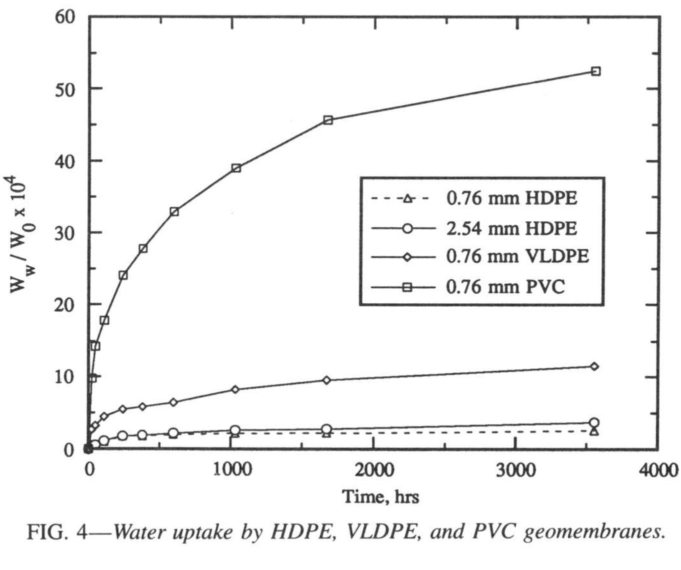

FIG. 4—*Water uptake by HDPE, VLDPE, and PVC geomembranes.*

Figure 4 shows that 0.76-mm-thick PVC had a significantly greater water uptake rate than 0.76-mm-thick VLDPE, which had a greater water uptake rate than 0.76-mm-thick HDPE. On the other hand, 2.54-mm-thick HDPE had only a slightly higher water uptake rate than 0.76-mm-thick HDPE. The extremely high water uptake rate by PVC compared with the PE geomembranes is probably due to the fact that PVC is a more polar polymer than PE and has no crystalline fractions. The significant difference in the water uptake rate between VLDPE and HDPE may be due mainly to crystallinity and density differences between these polymers. The small difference between 2.54-mm-thick HDPE and 0.76-mm-thick HDPE seems to be the result of thickness differences alone.

Based on the ratios of $W_{w,f}/W_{d,f}$ as shown in Table 1, it is evident that each geomembrane retained relatively small amounts of water compared to VOCs. This result suggests that VOCs penetrate geomembranes readily while water (solvent) does not. Moreover, with the distinct penetration rate of each VOC (except toluene and TCE) in each geomembrane (Fig. 3), it may be concluded that each VOC diffuses through the geomembrane as a molecule not as a bulk solution. For toluene and TCE, they may also diffuse separately; however, because they have similar diffusion and partition coefficients, they are not separable.

From the ratios of $Q/M^u_{T,0}$ in Table 1, it is seen that Q decreased as the aqueous solution decreased, implying that VOC losses from the closed system became smaller as the aqueous concentrations decreased. From the same ratios as shown in Table 1, Q increased as the test duration increased. However, the difference in Q between the tests that required 696 h (0.76 mm-thick HDPE and 100 mg/L initial aqueous concentration) and 3545 h (2.54 mm-thick HDPE and 100 mg/L initial aqueous concentration) was only 1.6%, indicating that most of the loss occurred at early times.

Table 1 shows that for each experiment, $Q/M^u_{T,0}$ values ranged from 4.6 to 9.5%. Approximately 4% of the weight loss occurred during the weighing processes (cutting the geomembranes after dismantling, blotting, and weighing). If this loss during sample preparation and weighing is included, the actual W_d value would be 473.00 mg, resulting in a total loss of 8.0% instead of 9.5%, shown in Tables 1 and 2. Since 8.0% is the sum of the four VOCs, the loss of each VOC is expected to be within the GC analytical error of 3 to 4% achieved in this study. Therefore, it may be inferred that VOC loss from the closed system does not significantly affect the diffusion and partition processes.

The mass balance properties may also be used to estimate the maximum swell of the geomembranes from penetration/plasticization by the VOCs and water. Assuming all losses are accumulated in the geomembrane in direct contact with the solvent, the volume of each VOC retained in the geomembrane at the end of each experiment is:

$$\forall_{i,f} = \frac{M^u_{i,0} - M^u_{i,f} - M^d_{i,f}}{\rho_i} \tag{6}$$

where $\forall_i$ is the volume of the VOC, L, and ρ_i is the density of the VOC, mg/L.

The PVC geomembrane retained more VOCs than the other geomembranes. Figure 4 also shows that the PVC geomembrane held much more water than the other geomembranes. Because the density of each VOC is known and its mass held in the PVC geomembrane can be obtained, its volume in the PVC geomembrane can be calculated from Eq 6. The computed volumes are 0.325, 0.276, 0.145, and 0.025 cm^3 for m-xylene, toluene, TCE, and MC, respectively. For water, its volume in the PVC geomembrane was estimated to be 0.015 cm^3.

The total volume of the VOCs and water in the PVC geomembrane is then approximately 0.786 cm^3. Dividing 0.786 cm^3 by the geomembrane surface area in direct contact with the solution (153.28 cm^2) gives a value of about 50 μm. Therefore, if all substances retained in the PVC geomembrane were equally distributed within the geomembrane in direct contact with the solvent, the increase in the initial geomembrane thickness (swell) would be about 6.6%. Considering the molecular size of each compound, 0.786 cm^3 is a large number. This swell should be considered as maximum swell only because the penetrant molecules may also occupy cavities and microvoids available within the geomembrane. The above scenario is for the case when losses are neglected. If the losses (Q and W_c) are considered, the total volumes of the VOCs and water retained in the PVC geomembrane would be smaller than 0.786 cm^3. Since HDPE and VLDPE geomembranes held smaller total masses of the VOCs and water than PVC, the total volumes of the VOCs and water present in PE geomembranes are expected to be smaller than 0.786 cm^3.

Summary and Conclusions

Previous studies have used different phases of VOCs in compartments separated by a geomembrane sheet (that is, liquid phase on one interface and vapor phase on the other interface). Since partition coefficients in the aqueous and vapor phases are different, it is not possible to evaluate directly mass transport for other conditions. A mathematical mass transport model with proper partition and diffusion coefficients should be used for prediction instead of the permeability coefficient (the product of the partition and diffusion coefficients). The CDCM developed in this study has the same phase in both upstream and downstream compartments and little loss during the test. The CDCM enables concentration change patterns to be tracked in the upstream and downstream compartments. In addition, the concentration change patterns provide important information in understanding the transport behavior of VOCs in aqueous solutions through intact geomembranes.

Small concentrations of the VOCs appeared in 20 to 200 h for the 0.76, 1.52, to 2.54-mm-thick HDPE geomembranes (breakthrough time varied with the square of thickness for each VOC), in 8 h for the 0.76-mm-thick VLDPE, and in 9 h for the 0.76-mm-thick PVC. The weight gain of HDPE geomembranes by sorption of VOCs ranged from 21% to 37% of the initial geomembrane weight. VLDPE had more than a 50% greater weight gain than HDPE under the same conditions. The weight gain was greatest for m-xylene, followed by toluene and trichloroethylene. The weight gain of the section of geomembrane clamped by the flanges was negligible ($<$1% for experiments $<$700 h and 3.3% for tests of 3545 h). In general, the total amount of VOCs lost decreased as the aqueous concentration decreased and increased as the test duration increased. The maximum total loss of 9.5%, for 100 mg/L initial aqueous concentration, 2.54-mm-thick HDPE, was considered small since this value was the sum of the four VOCs tested. Therefore, the effect of the total loss on transport behavior of individual VOCs for this test method is negligible.

In conclusion, the CDCM enables sorption onto and diffusion through a geomembrane to be evaluated along with the influence of concentrations above and below the geomembrane on the transport behavior. Second, the sorptive capacity, indicated by weight gain, for HDPE geomembranes is large, up to 37% of the initial weight. Finally, it appears that geomembranes may not be a suitable containment barrier, as VOCs passed through in 1 to 8 days (increased with thickness).

Acknowledgment

This research was supported by the National Science Foundation (NSF) (Grant No. MSS-9122501).

References

August, H. and Tatzky, R., 1984, "Permeabilities of Commercially Available Polymeric Liners for Hazardous Landfill Leachate Organic Constituents," *Proceedings,* International Conference on Geomembranes, Denver, CO, Vol. 1, Industrial Fabrics Association International, St. Paul, MN, pp. 163–168.

August, H., Tatzky, R., Pastuska, G., and Win, T., 1984, "Study of the Permeation Behavior of Commercial Plastic Sealing Sheets as a Bottom Liner for Dumps Against Leachate, Organic Solvent, and Their Aqueous Solutions," *Environmental Research Plan of the Federal Minister of Interior,* West Germany, Research Report No. 103-02-208.

Britton, L. N., Ashman, R. B., Aminabhavi, T. M., and Cassidy, P. E., 1989, "Permeation and Diffusion of Environmental Pollutants Through Flexible Polymers," *Journal of Applied Polymer Science,* Vol. 38, pp. 227–236.

Dimmick, D., Chow, A., Kuo, I., Howle, R., and Crenshaw, J., 1990, "U.S. EPA Clean Air Act Regulation for Landfill Gas Emission," *Proceedings,* Government Refuse Collection and Disposal Association, 13th Annual International Landfills Gas Symposium, pp. 243–255.

EPA Method 9090, 1986, "Compatibility Test for Wastes and Membrane Liners," Test Methods for Evaluating Solids Wastes, Vol. 1A, Laboratory Manual, *Physical/Chemical Methods,* 3rd ed., SW-846, U.S. EPA.

Friedman, M. A., 1987, "Volatile Organic Compounds in Groundwater and Leachate at Wisconsin Landfills," *Wisconsin Department of Natural Resources,* PUBL-WR-192.

Gibbons, R. D., Dolan, D., Keough, H., O'Leary, K., and O'Hara, R., 1992, "A Comparison of Chemical Constituents in Leachate from Industrial Hazardous Waste & Municipal Solid Waste Landfills," *Proceedings,* 15th Annual Madison Waste Conference, Municipal and Industrial Waste, Department of Engineering Professional Development, University of Wisconsin-Madison, pp. 251–276.

Giroud, J. P., 1984, "Impermeability: the Myth and Rational Approach," *Proceedings,* International Conference on Geomembrane, Vol. 1, Denver, CO, pp. 62–157.

Haxo, H. E. and Lahey, T. P., 1988, "Transport of Dissolved Organics from Dilute Aqueous Solutions Through Flexible Membrane Liners," *Journal of Hazardous Waste & Hazardous Materials,* Vol. 5, No. 4, pp. 275–294.

Haxo, H. E., Lahey, T. P., and Rosenberg, M. L., 1987, "Factors in Assessing the Compatibility of FMLs and Waste Liquids," U.S. EPA, *EPA/600/12,* Hazardous Waste Engineering Research Laboratory, Cincinnati, OH.

Haxo, H. E., Miedema, J. A., and Nelson, N. A., 1984, "Permeability of Polymer Membrane Lining Materials for Waste Management Facilities," *Proceedings,* Education Symposium on Migration of Gases, Liquids, and Solids in Elastomers, Denver, CO, pp. 151–156.

Haxo, H. E. and Pierson, P., 1991, "Permeability Testing: Geomembranes" in *Identification and Performance Testing,* A. Rollin and J-M. Rigo, Eds., Chapman and Hall, University Press, Cambridge, U.K., pp. 219–239.

Koerner, R. M., 1990, *Designing with Geosynthetic,* 2nd ed., Prentice-Hall, Englewood Cliffs, NJ.

Lee, G. F. and Jones, R. A., 1992, *Municipal Solid Waste Management in Lined "Dry Tomb" Landfills, A Technologically Flawed Approach for Protection of Groundwater Quality,* G. Fred Lee & Associates, El Macero, CA.

Lord, A. E., Koerner, R. M., and Swan, R. H., 1988, "Chemical Mass Transport Measurement to Determine Flexible Membrane Liner Lifetime," *Geotechnical Testing Journal,* Vol. 11, No. 2, pp. 83–91.

Park, J. K., Bontoux, L., Holsen, T. M., Jenkins, D., and Selleck, R. E., 1991, "Permeation of Polybutylene Pipe and Gasket Material by Organic Chemicals," *Journal of American Water Works Association,* Vol. 83, No. 10, pp. 71–78.

Park, J. K. and Nibras, M., 1993, "Mass Flux of Organic Chemicals Through Polyethylene Geomembranes," *Water Environment Research,* Vol. 65, No. 3, pp. 227–237.

Prasad, T. V., Brown, K. W., and Thomas, J. C., 1994, "Diffusion Coefficients of Organics in High Density Polyethylene (HDPE)," *Waste Management & Research,* Vol. 12, pp. 61–71.

Dennis Waslenchuk[1] *and Victoria Man*[1]

STERDI: An Innovation to Accelerate Remedial Investigations and Engineering

REFERENCE: Waslenchuk, D. and Man, V., **"STERDI: An Innovation to Accelerate Remedial Investigations and Engineering,"** *Volatile Organic Compounds in the Environment, ASTM STP 1261,* W. Wang, J. Schnoor, and J. Doi, Eds., American Society for Testing and Materials, 1996, pp. 259–271.

ABSTRACT: The purpose of this paper is to introduce a procedure for remedial investigations and for beginning cleanup more quickly than by traditional approaches. Simultaneous Testing, Exploration, Remedial Design, and Installation (STERDI) speeds the early and mid-phases of the remediation process, eliminates some traditional repetitive steps, and thereby results in lower project costs. The procedure involves the installation of in-ground components of an appropriate remedial system during the same exploration program that delineates the contaminant distribution. The method is built upon two capabilities: (1) planning for the multiple possible contaminant distributions, and (2) conducting "real-time" quantitative chemical analyses of soils during explorations in the vicinity of a release.

To implement STERDI, one has to recognize the various contaminant distributions which may have resulted from the release, prior to conducting subsurface explorations. Several spill "conceptual models" are developed, based on knowledge of the contaminant's geochemical behavior and on local geological characteristics. For each model, a "presumptive remedial design" (involving borehole-installed equipment) is prepared. An exploration program, incorporating in-field chemical analysis, is planned with sequencing which would allow early identification of which set of modeled conditions actually occurs. Once actual conditions are identified, the exploratory boreholes are completed by immediately installing remediation system components, guided by the in-field chemical testing results, according to the presumptive design which was developed for that model prior to the fieldwork.

At a gasoline release, the contaminant distribution was discovered and the in-ground portions of a vapor extraction system, including monitoring wells, were installed, all in a single three-day field effort. At a tetrachloroethylene spill, contamination was found through the unsaturated zone, and within the upper 20 ft (6.1 m) of the 50-ft-deep (15 m) overburden aquifer; a bubbler was installed in an exploratory borehole for remediation, accurately placed just beneath the contaminated interval, all in a two-day effort.

KEYWORDS: VOCs, low cost, remediation, innovative, presumptive remedy, accelerated, investigations, conceptual model, field screening

The elements of Simultaneous Testing, Exploration, Remedial Design, and Installation (STERDI), a procedure for accomplishing remedial investigations and remedial actions at contaminated sites, are explained. Also, two case studies are summarized to illustrate how the method has been applied successfully.

The traditional approach to investigating a contaminated site, and designing and installing subsurface remediation equipment, is logical and iterative, which, in the authors' experience,

[1] Vice president and senior scientist, respectively, Haley & Aldrich, Inc., 110 National Drive, Glastonbury, CT 06033; currently, president and vice president, respectively, ALTA Environmental Corporation, 7 South Main Street, Marlborough, CT 06447.

may be represented by the flow chart shown in Fig. 1. An initial exploration program is usually undertaken to begin to define the environmental conditions, followed by successively more focused iterations, until the extent of contamination is defined. Each iteration involves a work plan review and approval period, and an exploration equipment mobilization period, both of which can add considerably to the time required to complete the project.

Traditionally, once the extent of contamination has been defined, the remedial system is designed, installed, and started up. This too involves a work plan review and approval period, and another mobilization of drilling equipment.

The purpose of the STERDI procedure is to minimize the delays caused by repetitive reviews and approvals, and by repetitive drill-rig mobilizations. Only one comprehensive STERDI plan need be reviewed and approved, and only one drill rig mobilization is required (Fig. 2). All the elements of the STERDI procedure are in common practice today; the innovation is simply the arrangement of the elements, combined with an up-front commitment to take a project through to remedial component installation.

STERDI requires a good prior understanding of the local soil characteristics and hydrogeology, and of how the contaminants involved act when released to the environment (that

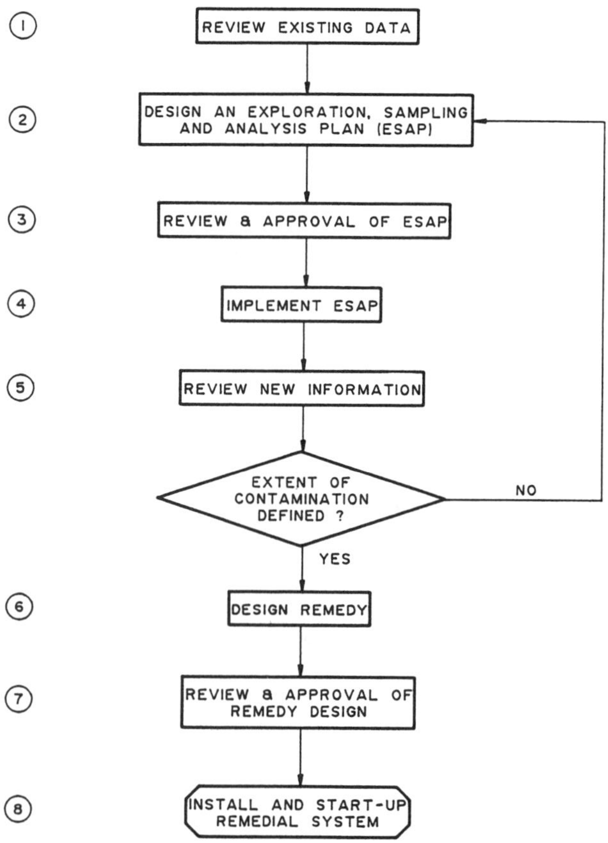

FIG. 1—*The traditional approach to spill characterization and remediation usually involves two or more iterations of Steps 2 through 5. Multiple reviews and approvals of planned work (Steps 3 and 7) often cause substantial delays. Multiple mobilizations of drilling equipment (Steps 4 and 8) are typical.*

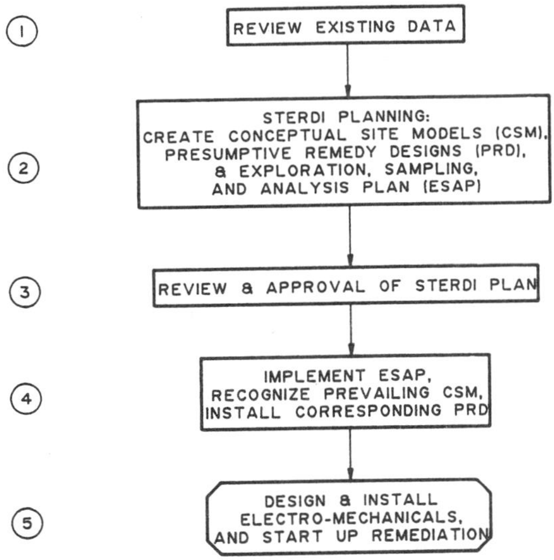

FIG. 2—*Unlike the traditional approach, the STERDI procedure does not incorporate a decision point that would trigger multiple iterations. Only one work plan approval and one drilling mobilization are required as long as the conceptual site model and exploratory layout and sequence are well though out.*

is, their transport, fate, and behavior characteristics). One must also have knowledge of the nature of the contaminant release (such as a subsurface gasoline leak, or a spill of solvent to the ground surface, also whether the release is recent or historic). In Connecticut, extensive mapping of surficial materials has been completed by the state and U.S. Geological Surveys. The maps provide sometimes-detailed information on local soil properties (for example, grain sizes, stratification, depth to bedrock) and hydrology (for example, depth to water table). Also, the state is densely developed and much subsurface information can be available from geotechnical (for example, foundation) studies. Where such prior knowledge is available, STERDI can be successfully applied.

With the aforementioned data in hand, one or more conceptual site models can be developed depicting subsurface conditions, including possible contaminant distributions. Then, drawing on the accumulated experience of the remedial engineering profession, one of the time-proven remediation systems can be specified and designed for each of the possible contaminant distributions hypothesized in the conceptual models. Only then is the exploration, sampling, and analysis plan developed to provide for testing the hypothesized contaminant distributions in a single program. The field program is conducted, being prepared to install down-borehole components of the appropriate remedial system in the exploratory boreholes.

The elements of the STERDI procedure are described in the following section.

Elements of STERDI

Conceptual Site Models (CSM)

It is often possible to accurately predict where contaminants have migrated, based on some knowledge of how and when the release occurred, and of the local soil characteristics and hydrogeology. The predictions can be shown in plan view and cross-section drawings; such

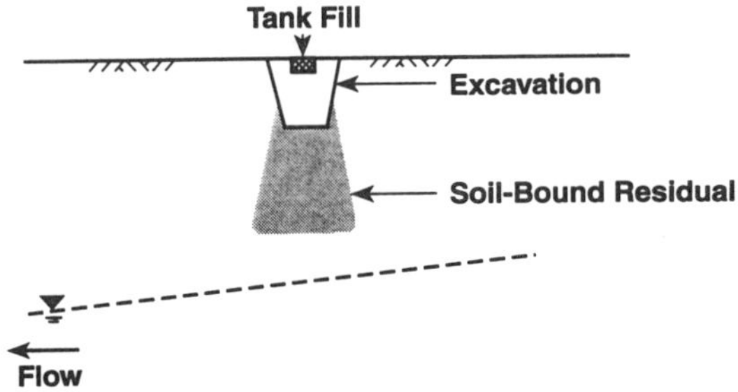

FIG. 3a—*A leaky underground gasoline line was observed to have contaminated shallow subsurface soil. This conceptual model hypothesized that only the vadose zone soils beneath the leak had been impacted and that soil-bound gasoline constituents remained below the pipe line excavation. The water table is depicted by the dashed line and inverted triangle symbol.*

conceptual models traditionally have been used to guide exploration and sampling programs, and recently have been used to guide presumptive remedy planning [for example, in accelerated Superfund remedial investigations/feasibility studies (RI/FS) and interim measures actions (U.S. EPA 1988)].

For illustration, consider a case where a leaky fill pipe running to an underground gasoline storage tank was discovered when the line was excavated to upgrade the system. Three conceptual models were developed which represented the range of contaminant distributions thought likely (Fig. 3). The hypothesized contaminant distributions recognized the environmental media which may have been impacted (such as vadose zone soil or ground water), and the various possible phases of the contaminants [for example, soil-bound, dissolved, and Low-density Non-Aqueous Phase Liquid (LNAPL)].

Information on soil characteristics and stratification, depth to ground water, and groundwater flow direction was available from geotechnical engineering test boring logs which had been prepared earlier, prior to construction of the facility. The transport mechanisms and physical properties of gasoline, and its constituents, are well-enough understood that the models were simple to develop.

Presumptive Remedy Design (PRD)

Many instances of contamination are amenable to remediation technologies which make use of everyday monitor well construction materials (for example, solid and slotted pipe,

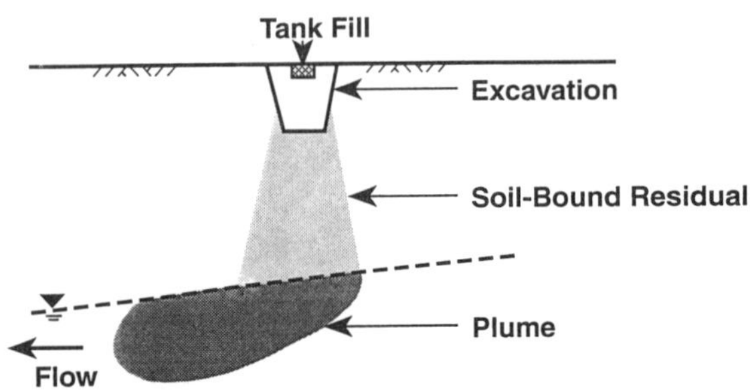

FIG. 3b—*This model represented a possible condition where gasoline constituents had migrated down to the water table, and dissolved gasoline constituents were migrating in a ground-water plume.*

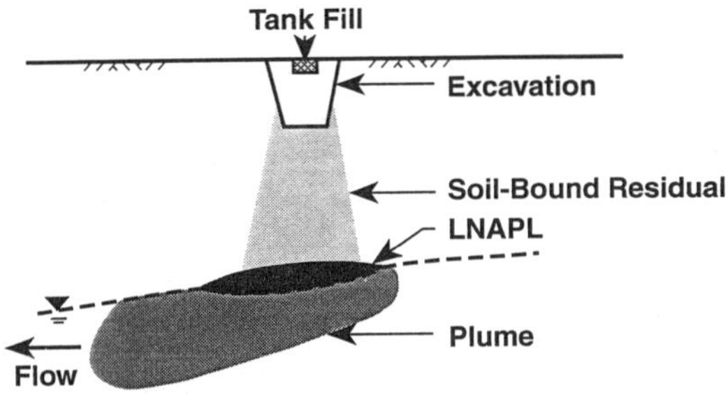

FIG. 3*c*—*This model considered the situation should a larger volume release have occurred. It shows the possible presence of a lens of liquid gasoline (LNAPL) floating on the water table.*

filter sands and gravels, grout) installed down a borehole by a driller. This is a fundamental requisite of STERDI. Another requisite is that engineering experience is sufficient to allow for generic, rather than site-specific, remedy designs.

Presumptive remedy designing involves planning remedial systems suitable for the contaminants and their hypothetical distributions, as depicted in the conceptual site models. Limited site-specific data are necessary to successfully apply certain remediation technologies. For example, soil vapor vacuum extraction systems have wide applicability to the cleanup of volatile organic compounds in the vadose zone, and their design is not highly sensitive to every peculiarity of the site setting; uncertainties can be compensated for as part of the design. Off-the-shelf equipment and "rule-of-thumb" design parameters can be sufficient to provide for effective remedial systems.

Consider the earlier illustration. Should the contaminant distribution shown in Fig. 3*a* actually have occurred, then a vapor extraction well (VEW) installed as shown in the presumptive remedy design of Fig. 4*a* would have been effective. If the contaminant distributions shown in Fig. 3*b* or 3*c* had occurred, then the monitor well (MW), recovery well (RW), and VEW array shown in Fig. 4*b* would have been effective.

The in-ground components of the remedial systems shown are typical driller installations, and it is obvious that such equipment could be easily sized for, and placed in, an exploratory borehole based upon chemical testing data which revealed the extent of contamination. To allow such installations during the one and only field effort in a STERDI project, however, the chemical testing data must be generated simultaneously with the contaminant-delineation explorations as described in the following.

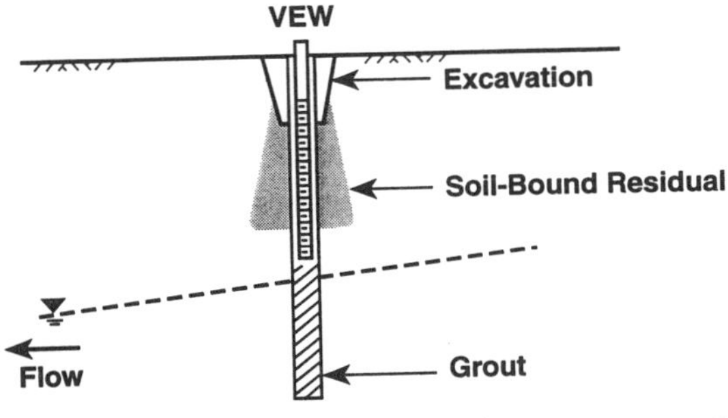

FIG. 4*a*—*An effective presumptive remedy for this gasoline contaminant distribution would be to operate a vapor extraction well (VEW) with sufficient radius of influence.*

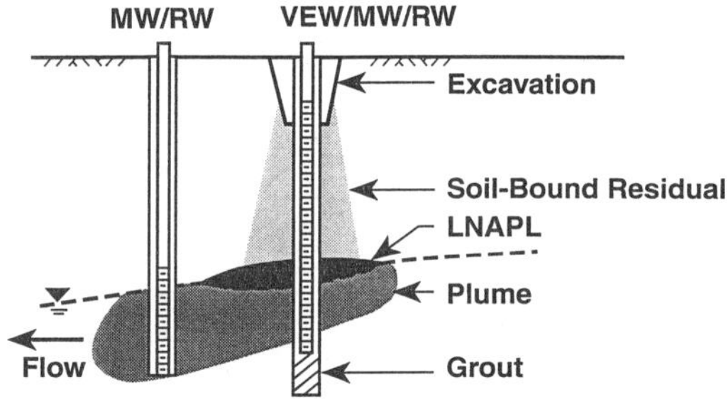

FIG. 4*b*—*If this gasoline contaminant distribution had occurred, or if the LNAPL lens were not present, then the presumptive remedy depicted would be appropriate. A well through the spill zone could serve multiple purposes, for vadose-zone vapor extraction (VEW), ground-water monitoring (MW), and ground-water or LNAPL recovery (RW) or both. A downgradient well could serve two purposes, ground-water monitoring or recovery or both.*

Exploration, Sampling, and Analysis Plan (ESAP)

General—The goal of the ESAP is to lay out and sequence the exploratory boreholes to allow recognition of which of the hypothetical contaminant distributions (that is, which of the conceptual site models) actually prevails, as early in the exploration program as possible. Samples collected from the borings are analyzed in the field in a procedure called "chemical logging" (described below), so that the actual contaminant distribution is revealed as explorations advance and can be compared immediately with the hypothesized distributions. Such "real-time" chemical testing is encouraged by EPA for certain programs [such as RCRA Facility Investigations (U.S. EPA 1989)], although it does not appear to be standard procedure yet in projects done outside of these programs.

In the case used for illustration (Fig. 3), the first soil boring was drilled through the known heart of the spill zone and soil samples were collected and analyzed continuously, recognizing that the findings of this single exploration would distinguish between the three hypothesized contaminant distributions. As discussed in the case history later, the findings indicated that the model depicted in Fig. 3*b* prevailed.

It is necessary also to plan the borehole layout and sequence to be able to recognize each borehole's suitability to be completed as a remedial system component while the drill rig still occupies the hole. In other words, the efficiency of STERDI is diminished if poor sequencing is followed, such that the need for an installation in a now-collapsed borehole is recognized only after a later boring is completed and chemically logged.

Soil Sampling and Boring Logs—In Connecticut, the usual practice is to drill soil borings through the overburden using hollow-stem augers. Soil samples are collected with standard split spoons at appropriate depths (Storer 1992), taking heed of the conceptual models to provide for a test of the hypothesized conditions. Soils are visually examined and their characteristics are described in accordance with the Unified Soil Classification System (Howard 1986, U.S. Army Corps of Engineers 1953). Standard test boring logs are completed. Observations concerning the depth to ground water are made during drilling.

The following specific soil samples are collected from each sample depth-interval:

- Two 1⅓-oz (40 mL) volatile organic analysis (VOA) vials, filled three-quarters full, and fitted with polytetrafluoroethylene-lined septum caps (used for in-field sample analysis, as discussed below);

- One 2-oz (60 mL) glass jar, filled completely, and fitted with a polytetrafluoroethylene-lined septum cap. (The glass jar sample is stored at 4°C for possible future certified-laboratory analysis to provide confirmation of field-test findings.)

Chemical Logging: Sample Analysis, Instrumentation, and Calibration—The field testing equipment is set up in a van or in a nearby building. The equipment consists of a gas chromatograph (GC), a lap-top computer for in-field data management, a constant-temperature water bath, syringes, and calibration standards. A 6-ft-long (1.8 m) folding work table is standard equipment, as it provides sufficient room for equipment setup. The equipment is run off a portable generator if another electrical supply is not convenient.

Soil samples are analyzed using a static, heated, headspace method [modified EPA Method 3810 (U.S. EPA 1986)]. The technique is simple, but provides rapidly obtained analytical results and allows the accurate determination of relative VOC concentrations in the soil samples. For similar soils from a site, under elevated constant-temperature conditions, the contaminant vapor concentrations induced in the jar headspaces above the samples are proportional to the original concentrations of the contaminant in the sample; in a jar, the ratio of contaminant concentration in the headspace to its concentration still associated with the sample is a function of the effective partition coefficient (Ettre et al. 1983). The bath temperature chosen for the method is low enough to prevent condensation of the sample in the injection port of the GC, but high enough to drive most of the VOCs from the sample into the headspace. Hence, sample-to-sample variations in the extent of partitioning into the vapor phase, which could result from soil-type variation if the samples were kept at ambient temperature, are minimized.

The 1⅓-oz (40 mL) VOA vials are placed in a constant-temperature water bath held at 35°C. One of the VOA vials is removed from the bath approximately 30 min later, shaken vigorously to encourage partitioning of the VOCs into the vapor phase and to uniformly distribute the vapors in the jar headspace prior to analysis. The sample is analyzed using a Photovac 10S50 gas chromatograph under the following instrument conditions:

Detector	—10.6 eV photoionization detector
Oven temperature	—104°F (40°C), isothermal
Carrier gas and flow rate	—Hydrocarbon-free air, approximately 0.00018 cfm (5 mLmin)
Column	—CPSil 5CP, capillary

An aliquot of the headspace from atop the sample is directly injected into the GC using a 50-μL or smaller volume gastight syringe.

If sample re-analysis is required for QA/QC purposes, or if chromatographic peaks are outside of the range of the instrument calibration curve, the second VOA vial of the sample set is analyzed. Injection volumes are adjusted as appropriate. Duplicate injections from a single VOA vial are not made since loss of VOCs occurs rapidly following puncture of the vial septum.

Samples can be analyzed for some or all of the following compounds, with the listed approximate limits of detection:

Compound	Detection Limit (ppb)
1,2-dichloroethylene (cis and trans isomers)	1
1,1-dichloroethylene	1

Compound	Detection Limit (ppb)
1,1,1-trichloroethane	10
trichloroethylene	1
tetrachloroethylene	1
methyl tert-butyl ether	100
benzene	1
ethylbenzene	2
toluene	1
xylene (ortho, meta, and para isomers)	2
acetone	100
methyl ethyl ketone	100
methyl iso-butyl ketone	100

The GC is calibrated by injecting headspace samples from atop aqueous standards. The analytical results for soil samples are not equivalent to those of EPA's purge-and-trap or TCLP methods, but they do correlate well, as other authors have shown statistically (Balcerzak and Wilbur 1994). The purpose of chemical logging is not so much to measure the absolute concentration of a contaminant in a sample, but rather to determine the relative distribution of contamination; the latter is achieved using this method.

Recognizing the Prevailing Conceptual Site Model and Appropriate Presumptive Remedy Design

The contaminant concentrations are plotted on an *x-y* graph against sample depth for each soil sample alongside the soil boring log having the same vertical scale. Also plotted is the approximate depth to ground water. Examples of chemical logs are shown in Figs. 5*a* and 6*a*; details of the logs are discussed in the following section.

The chemical log and soil boring log are useful in evaluating whether the borehole penetrated a spill zone (that is, high levels of contaminants in the vadose zone) or a ground water contaminant plume, and in recognizing preferential ground-water migration pathways and preferential soil-adsorbed contaminant layers. Which of the conceptual site models actually prevails is determined by comparison of the chemical logs with the various hypothetical distributions. In accordance with the presumptive remedy design for the realized site condition, the appropriate in-ground portions of the remedial system are installed. Of course, if the conceptual models are comprehensive and well thought out, the procedure will preclude the iterations typical of the traditional approach.

Case Studies

Gasoline Spill

Gasoline-contaminated near-surface soil was discovered during the removal and replacement of an underground gasoline storage tank system at a facility in Connecticut. The contaminated soil had resulted from leakage from underground piping. Obviously contaminated soil in the heart of the release was excavated out to the apparent edge of the spill zone and down to a depth of about 8 ft (2.4 m) below ground surface, at which depth further excavation became impractical.

Existing data on the spill and partial cleanup were provided; the results of laboratory soil analyses indicated the excavation sidewalls were not significantly impacted by gasoline re-

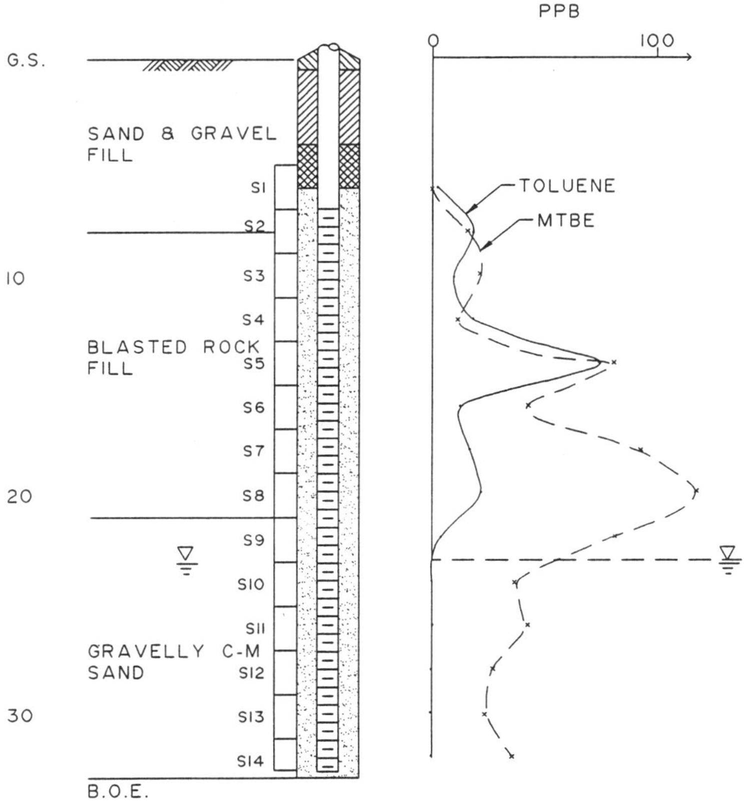

FIG. 5a—*Chemical log of a soil boring drilled at a gasoline spill, showing split-spoon sampling intervals (such as S1), soil texture descriptions, and contaminant concentrations (in parts-per-billion, ppb) for toluene and methyl tert-butyl ether (MTBE). The figure also shows the construction details of the combined vapor extraction/ground-water monitoring/ground-water recovery well (VEW/MW/RW) that was installed (10 ft equals 3.0 m).*

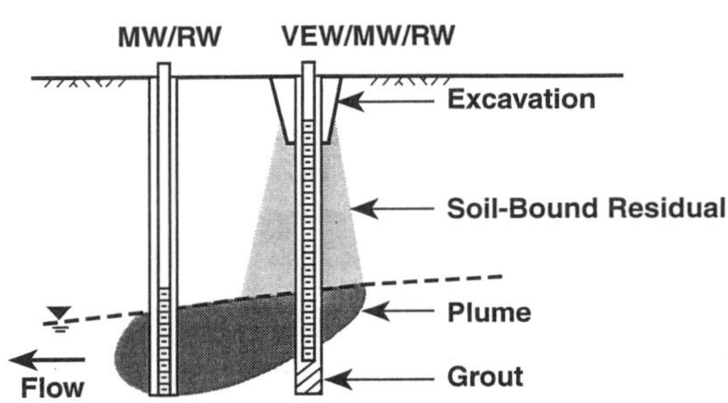

FIG. 5b—*Conceptual site model and presumptive remedy design for a gasoline spill. This hypothetical contaminant distribution was recognized as actually existing, and this presumptive remedy as appropriate, based on the chemical logs for the two borings (one of which is shown in Fig. 5a).*

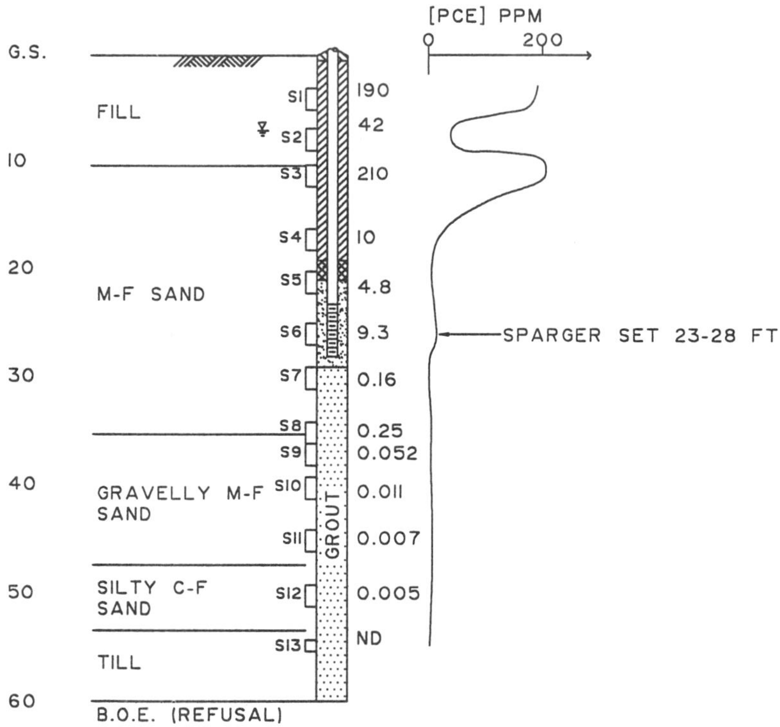

FIG. 6a—*Chemical log of a soil boring drilled in a tetrachloroethylene (PCE) spill zone, with concentrations in parts-per-million (ppm). An air sparging well ("bubbler") was installed so that dissolved- and soil-adsorbed-phase contaminants below the water table could be purged (10 ft equals 3.0 m).*

siduals, but the bottom soils had VOC concentrations between 1 and 13 parts-per-million (ppm).

The conceptual site models, presumptive remedy designs, and an exploration, sampling and analysis plan were based on the information relating to conditions around the excavation and on historic geotechnical engineering soil logs for the facility, as discussed earlier. The fieldwork to implement the plan and install the in-ground portions of the remedial system was completed in three days.

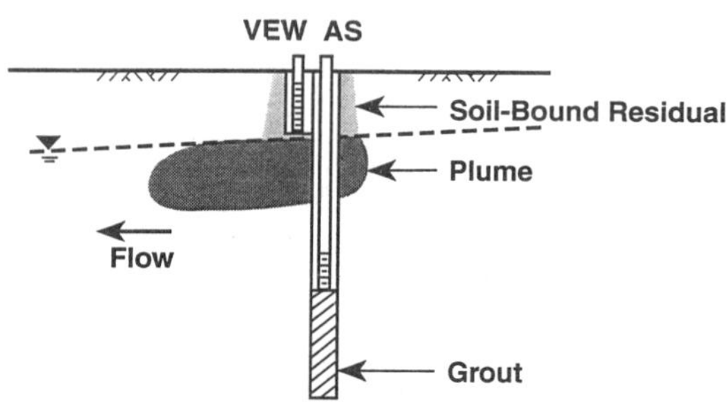

FIG. 6b—*This conceptual model and presumptive remedy prevailed for a PCE spill, as supported by the chemical log shown in Fig. 6a.*

The chemical log shown in Fig. 5*a* resulted from the first borehole drilled into the known spill zone. Sand and gravel fill material, placed following the excavation, was present to about 8 ft (2.4 m) deep. Below, blasted rock fill was encountered to a depth of about 21 ft (6.4 m); native gravelly sand was present beneath the fill. The depth to ground water, based on observations made during drilling, was about 23 ft (7.0 m).

Gasoline constituents [including benzene, toluene, ethylbenzene, xylenes, and methyl tert-butyl ether (MTBE)] were detected in soil samples collected from the borehole at concentrations up to 100 parts-per-billion (ppb). Figure 5*a* shows toluene and MTBE were present above the water table. MTBE was the only contaminant detected in soil samples collected below the water table, likely resulting from the high mobility (solubility) of that compound as compared with the other gasoline constituents. The log confirmed that the contaminant distribution hypothesized in Fig. 3*b* prevailed.

In accordance with the presumptive remedy designs, a vapor extraction well (VEW) was installed in the spill zone borehole as shown in Fig. 5*b*. The screened length of the well was extended into the water table so that the installation could also serve as a ground-water monitoring well (MW), and as an emergency ground-water recovery well (RW) if it ever became necessary to contain the contaminated ground water. The monitoring well function was desired so that the recovery well operation could be triggered early if contaminant levels increased and threatened a downgradient drinking water resource.

A second borehole was drilled approximately 40 ft (12.2 m) from the first, in the inferred downgradient direction from the spill zone; the spacing reflected a calculated minimum radius of influence of the VEW. The ground-water flow direction was readily estimated, given the marked local topographic relief. (If this had not been the case, then additional boring locations could have been specified, along an arc at this distance away from the first borehole, to accommodate the uncertainty.) The boring was chemically logged to evaluate the lateral extent of contamination above the water table, so that the need for additional VEWs to clean up beyond the expected radius of influence of the initial VEW could be determined. Gasoline constituents were not detected in soil samples collected between 5 and 30 ft (1.5 and 9.1 m) deep; hence the conceptual model of Fig. 3*b* was further confirmed.

A monitoring well was installed in the second boring, again to evaluate possible migration of the contaminant plume toward the drinking water well, located some 500 ft (152 m) downgradient of the spill zone. In addition, the well was constructed to be suitable for use as an emergency recovery well if the need arose.

The remedial investigation and installation required only a few days and cost about $19 000. In the authors' experiences, similar projects conducted in the traditional way would have taken several months to complete, at about twice the cost.

The remediation system was shown to have an adequate area of influence, relative to the extent of soil contamination, in performance monitoring during the life of the system. Like many applications of vapor extraction remediation systems, the gasoline VOC contamination was reduced to nondetectable in about 18 months.

Tetrachloroethylene Spill

A number of small releases of tetrachloroethylene (PCE), over a period of years, occurred during solvent transfer to and from aboveground bulk storage tanks at a chemical distribution facility in Connecticut. During a two-day drilling program following a STERDI plan, we identified residual soil contamination above and below the water table, delineated the vertical extent of the contamination in the aquifer beneath the spill zone, and installed a well screened below the contaminant zone that was suitable for use as an air sparging well for cleanup of dissolved PCE in ground water and soil-adsorbed PCE in the saturated zone.

The chemical log shown in Fig. 6*a* was developed from a borehole completed in the PCE spill zone. Our conceptual models had been based on preliminary previous sampling and testing conducted by others.

PCE occurred at high levels at depths between 3 and 36 ft (0.9 and 11 m), then levels decreased systematically to nondetectable below that depth [36 ft (11 m)]. The highest concentrations were detected in the shallowest sample, collected from the vadose zone [3 to 5 ft (0.9 to 1.5 m)], and just below the water table in the 10-to-12-ft-deep (3.0 to 3.7 m) sample (S3). The depth of S3 corresponded to the top of the natural floodplain soil. The distribution suggested that the PCE spilled at ground surface migrated downwards with percolating precipitation, and concentrated where a change in soil type and permeability was encountered (that is, the top of natural material). The decrease in contamination with depth indicated that liquid PCE had not penetrated through a significant portion of the saturated zone.

Once the borehole had clearly advanced beneath the bottom of the contaminated aquifer interval, the lower portion of the spill zone borehole was sealed with cement/bentonite grout up to a depth of 30 ft (9.1 m) below grade. An air-injection (sparging) well screen was installed between 23 and 28 ft (7.0 and 8.5 m) deep, below the zone of significant PCE contamination. The STERDI plan then called for a vapor extraction well to be installed in a boring adjacent to the air sparging well to collect contaminants from above the water table in the spill zone, and those purged from beneath the water table, as shown in Fig. 6*b*. Although the operational history of this remedial system is beyond the scope of this paper, the system is expected to achieve the cleanup goals straightforwardly.

Conclusions

STERDI procedures can allow completion of spill characterization and in-ground remediation component installation in a few days, whereas traditional approaches commonly take months. Because the work goes so quickly, and because some normally repetitive steps (such as reviews and approvals, and driller mobilizations) can be avoided, a STERDI project can also cost significantly less than a comparable traditional project.

STERDI can be used where prior knowledge of site geologic conditions and the nature of a release exists. It relies on the ability to predict the various possible contaminant distributions which may have resulted from a release. Where geologic or contaminant conditions are so complex that they cannot be reliably hypothesized by conceptual models, then the effectiveness of STERDI procedures are diminished. However, experience has shown that there are many instances of spill-zone investigations, leading to remediation of overburden soils and ground water, that are amenable to STERDI.

STERDI also relies on the ability to conduct in-field, "real-time" quantitative chemical analyses. Chemical logging for volatile organic compounds, such as oft-encountered petroleum and solvent constituents, is now a well-developed art. Moreover, recent advances in reliable real-time analyses for other classes of contaminants (such as heavy metals) may make STERDI useful for other types of releases. In this regard, the applicability of STERDI is limited only by whether there are remedial technologies for those other contaminants which utilize driller-installed monitor well construction materials or in-ground remediation components.

References

Balcerzak, P. and Wilbur, D., 1994, "On-Site VOC Analysis Gains Favor," *Pollution Engineering,* Jan., pp. 62–64.

Ettre, L. S., Kolb, B., and Hurt, S. G., 1983, "Techniques of Headspace Gas Chromatography," *American Laboratory,* Oct., pp. 76–83.

Howard, A., 1986, *Soil Classification Handbook, Unified Soil Classification Systems,* Training Manual Nos. 4, 5, and 6, U.S. Department of the Interior, Bureau of Reclamation, Division of Research and Laboratory Science, Geotechnical Branch, Denver, CO.

Storer, R. A., Ed., 1994, "Practice for Soil Investigation and Sampling by Auger Borings, D1452-80," *Annual Book of ASTM Standards,* Vol. 4.08, American Society for Testing and Materials, Philadelphia, pp. 110–111.

U.S. Army Corps of Engineers, 1953, "The Unified Soil Classification System," *Technical Memorandum No. 3-357,* Vol. 1, Waterways Experiment Station, Vicksburg, MS.

U.S. Environmental Protection Agency, 1986, *Test Methods for Evaluating Solid Waste, SW-846,* Sept. pp. 3810-1–3810-5.

U.S. Environmental Protection Agency, 1988, *Guidance for Conducting Remedial Investigations and Feasibility Studies (RI/FS) Under CERCLA,* EPA/540/G-89/004, Oct.

U.S. Environmental Protection Agency, 1989, *RCRA Facility Investigation (RFI) Guidance, Interim Final,* EPA/530-SW-89-03, May.

Barry C. O'Melia[1] *and Dean R. Parson*[2]

Dual-Phase Vacuum Extraction Technology for Soil and Ground-Water Remediation: A Case Study

REFERENCE: O'Melia, B. C. and Parson, D. R., **"Dual-Phase Vacuum Extraction Technology for Soil and Ground-Water Remediation: A Case Study,"** *Volatile Organic Compounds in the Environment, ASTM STP 1261,* W. Wang, J. Schnoor, and J. Doi, Eds., American Society for Testing and Materials, 1996, pp. 272–286.

ABSTRACT: Soil vapor extraction (SVE) has become a common, effective method to remediate soils contaminated by volatile organic compounds (VOCs). The technology induces a soil gas flow through the vadose zone, which extracts VOCs into the soil gas being extracted. The Dual-Phase Vacuum Extraction (DVE) system uses a high vacuum source to extract both soil gas and ground water from the same extraction well. The creation of a two-phase flow in the extraction piping eliminates the need for a separate pumping system for ground-water recovery. The high vacuum can also induce a higher recovery of ground water than typical downhole pumping methods can provide.

An Air Force Base (AFB) site, located in the United States, previously used an underground storage tank to contain waste solvents, primarily methylene chloride and perchloroethylene, from paint stripping and application operations prior to disposal. Leakage from the tank system resulted in subsurface soils and ground water with elevated levels of these chlorinated solvents. The soils at the site consist of silty clays and clayey silts with low permeabilities. The DVE system was pilot-tested to evaluate performances for site remediation.

Two extraction wells were tested separately. The average soil gas extraction rate per well was 1.05 m^3 per minute with radius of influence of 10.7 m. For ground water, the average recovery rate per well was 3 L/min with a radius of influence of 23.2 m. These radii influences were significantly higher than those estimated for traditional soil vapor extraction or ground-water removal technologies. Removal rates ranged from 11.4 kg/day for an area with high levels of VOCs to 2.3 kg/day for an area with lower levels of VOCs.

KEYWORDS: soil vapor extraction (SVE), dual vacuum extraction (DVE)

Soil vapor extraction (SVE) has become a common, effective method to remediate volatile organic compounds (VOCs) contaminated soils. The technology induces a soil gas flow through the vadose zone, which extracts VOCs into the soil gas being extracted. The Dual-Phase Vacuum Extraction (DVE) system uses a high vacuum source to extract both soil gas and ground water from the same extraction well. The creation of a two-phase flow in the extraction piping eliminates the need for a separate pumping system for ground-water recovery. The high vacuum can also induce a higher recovery of ground water than typical downhole pumping methods can provide. This paper summarizes a pilot test to evaluate the application and feasibility of DVE technology at Air Force bases in the United States.

[1] Project engineer, Woodward-Clyde Federal Services, 1500 Century Square, 1501 4th Ave., Seattle, WA 98101.

[2] Senior project manager, Terracon Environmental, 1609 Oak Ridge Drive, Suite 101, Fort Collins, CO 80525.

Based on the results of a remedial investigation (RI), the potential existed for migration of VOCs away from the site. The feasibility study (FS) identified in-situ remediation, using the DVE technology, as the preferred remedy. The FS recommended the development and implementation of an on-site pilot study prior to any full-scale remediation. The focal point of the site is the paint stripping and equalization tank (PSET), which was used to collect spent solvent, wash water, and paint solutions from aircraft maintenance activities in the adjacent building.

The DVE pilot test system was installed and operated from January 4, 1993 through January 24, 1993. The scope of work included:

1. Installing two vapor extraction wells and three soil gas probes.
2. Installing and operating the pilot DVE unit.
3. Collecting soil, extracted soil gas, and ground-water samples.
4. Evaluating the remediation performance of the DVE pilot unit.

Applicability to Site

Site History—As part of past maintenance operations at the site, aircraft were periodically repainted at the adjacent building. Mechanics would chemically strip paint from the aircraft with a methylene chloride solution and then repaint the aircraft. During stripping operations, the paint and spent solvent were allowed to settle on the concrete floor of the building. The floor was then washed with water. The solution of wash water, spent solvent, and paint was collected at the south side of the building and conveyed by a gravity drain line to the PSET, where the sludge was separated from the liquid waste. The sludges were then pumped from the tank and disposed of as required. The liquid waste was discharged to an adjacent sanitary sewer. Subsurface soil sampling and ground-water sampling have been performed at this location to assess the impact of any past spills or leaks of hazardous chemicals.

Local Geology/Hydrogeology—Soils at the site are generally silty clays and clayey silts with discrete layers of silt or sand underlain by approximately 36.6 m of shale interbedded with siltstone. Previous field investigations at the site have encountered bedrock from approximately 3.66 to 6.71 m below the surface. The bedrock consists of gray to reddish-brown siltstone underlaid by a reddish-brown sandstone. This sandstone unit is referred to as the lower water bearing unit. The sandstone unit in the vicinity of MW-12-13 was 10 m thick.

A preliminary investigation encountered water at depths ranging from 2.4 to 2.9 m below the surface. During the most recent investigation, ground water was encountered between depths of 3 and 5 m. Based upon measurements taken on March 2, 1990, the gradient of the potentiometric surface was calculated to be 0.019 m/m to the northwest. Three additional rounds of measurements taken between November 1990 and January 1991 indicate that the gradient varies from northwest to southwest. Slug tests were performed on shallow wells during December 1990 and February 1991. The test results and subsequent analysis of the data indicated hydraulic conductivities ranging from 5.7×10^{-6} cm/s to 1.3×10^{-4} cm/s.

Nature and Extent of Contamination—The soil and shallow ground water at the site in the vicinity of the PSET contain elevated levels of contaminants. Contaminants consist primarily of solvents from the previous paint stripping operation waste stream. Compounds associated with paint stripping wastes include methylene chloride, chloroform, trichloroethene (TCE), tetrachloroethene (PCE), acetone, 2-butanone, 1,1,1-trichloroethane, toluene, and ethylbenzene. However, methylene chloride is the predominant contaminant.

Soil—Methylene chloride, at concentrations as high as 10 000 mg/kg, has been detected in the soil near the northwest lower corner of the PSET. Concentrations in the soil are estimated to diminish radially outward as shown in Fig. 1. Other primary solvent constituents

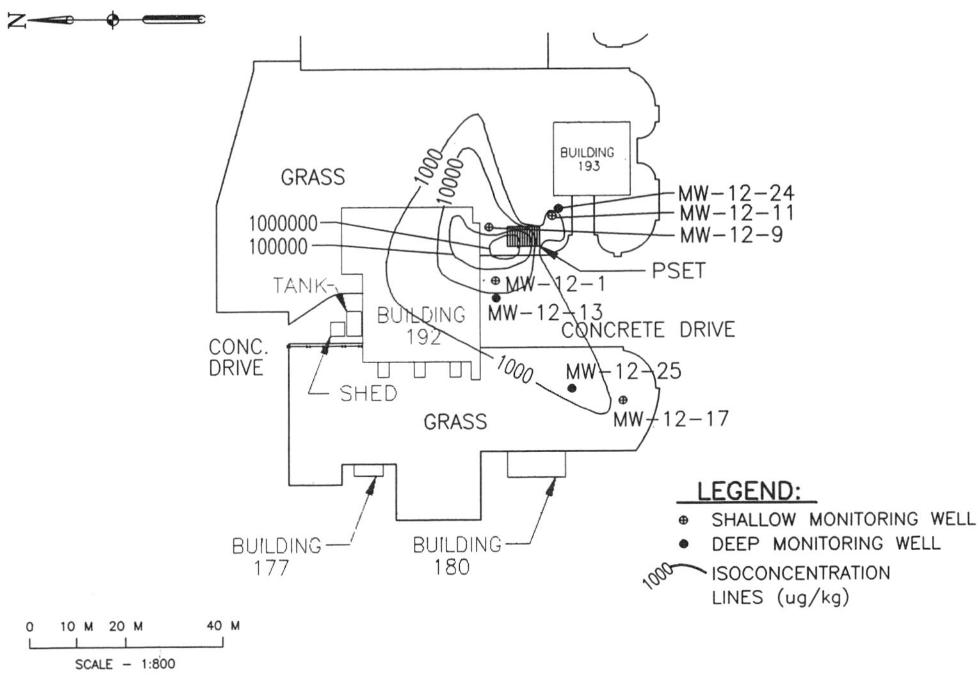

FIG. 1—*Dual-phase vapor extraction well and soil gas probe locations.*

identified from previous soil sampling programs include acetone, 2-butanone, TCE, and PCE. These other contaminant concentrations are typically lower in relative concentrations than the methylene chloride concentrations.

The lateral extent of soil contamination from VOCs is typified by the methylene chloride isoconcentration lines in Fig. 1. The 1000 μg/kg line should not be interpreted as an outer limit of the VOC plume but rather a boundary where nondetection levels were approached.

Vertically, the presence of elevated VOC concentrations in the site soils appears to be limited to between the base of the PSET (2.7 to 3.0 m below grade) and the bedrock surface (about 6.1 m below grade).

Shallow Ground Water—Methylene chloride, TCE, and PCE were the most frequently detected VOCs in shallow ground-water samples at the site. Methylene chloride concentrations up to 3 200 000 μg/L (3200 mg/L) were detected adjacent to the PSET. The levels of 1,2-dichloroethylene (1,2-DCE), TCE, PCE, ethylbenzene, xylene, and toluene in ground-water samples exceed established maximum contaminant levels (MCLs) for ground water.

Deep Ground Water—In 1991, deep ground-water monitoring wells penetrating the lower aquifer zone were sampled and analyzed for VOCs. MW-12-13 alone showed significant concentrations of contaminants, with concentrations of methylene chloride and PCE of 136 μg/L and 268 μg/L, respectively.

Technical Description

Dual-Phase Vapor Extraction

Dual-Phase Vapor Extraction (DVE) systems function by applying vacuum at a wellhead sufficient to extract both soil gas and ground water through the same pipe. In low-permeability soils, a DVE system can achieve ground-water extraction and recovery rates that can significantly exceed those accomplished by conventional well pumping systems. Volatile contaminants are stripped from surrounding soils into the soil gas phase and from water entrained in the extraction well flow. The DVE technique can also volatilize and extract free product, if present. DVE has been demonstrated successfully under these conditions in the United States and Europe (Roy 1991). This paper presents an evaluation of DVE at a site in the United States.

DVE Pilot Testing Equipment—The mobile DVE pilot test unit used in this study has a liquid ring vacuum pump capable of extracting up to 3.7 actual cubic meters per minute (acmm) of air at 254 mm of mercury (mm Hg) and developing a maximum vacuum of up to 660 mm Hg at a pumping rate of 1.6 acmm. The unit is electric powered with a 7.5 kW motor and is capable of running on an external, three-phase, alternating current power supply of 230 or 460 V.

The DVE unit is designed to extract soil gas and ground water through the same extraction well. A small-diameter pipe (3.8 cm) called a drawndown pipe, is installed in the extraction well within 30.5 cm of the bottom of the well casing (Fig. 2). The pipe is connected to the DVE pilot unit using piping or hoses that lie on the ground surface. The initial step is to lower the ground-water level in the extraction well. This is accomplished by using the vacuum pump on the DVE unit, an additional vacuum pump, or actually placing a ground-water pump in the bottom of the well. Once the well is dewatered, the vacuum pump on the DVE pilot unit begins to extract soil gas through the small-diameter pipe. As the ground-water table starts to recover, the water level rises toward the mouth of the small-diameter pipe. The high velocity of soil gas entering the pipe (typically greater than 30.5 m/s) entrains the water as droplets and as a water layer on the inside of the pipe. This produces a two-phase flow of water and soil gas flowing upward in the small-diameter pipe. The flow of soil gas through the pipe carries the extracted ground water into the DVE unit. This technology is effective for aquifers that do not produce more than 4.0 L/min per well. If the ground-water extraction rates are too large, the entire small-diameter pipe fills with water, thereby extracting only ground water. Data indicate that the vacuum produces a widening radius of influence that results in microfracturing of the soil matrix. This mechanism increases the primary conductivity of the soil and allows for a greater radius of influence for both soil gas and ground-water extraction. As the microfractures propagate, both ground water and soil gas are more readily extracted; subsequently, lower vacuum is required to maintain the initial flow rate (Lewis et al. 1992).

The process and instrumentation diagram (P&ID) for the DVE system is shown in Fig. 3. The extracted soil gas and ground water flow into the primary water knockout tank to remove the entrained water. The soil gas is routed out the top of the knockout tank through the vacuum pump. The gas is then routed through a secondary water knockout tank to remove condensed water generated through the vacuum pump. The extracted soil gas exits the DVE unit and is discharged to the atmosphere. Depending upon regulatory requirements, off-gas treatment may be required for other sites and full-scale systems. Water accumulates in the primary knockout tank until an electronic float switch turns on the liquid discharge pump. Water is pumped out of the primary tank, through a flow totalizer, and into the water holding tank. Water that accumulates in the secondary knockout tank is pumped with the recirculating pump and cooled by an air-cooled heat exchanger for use as the liquid ring pump sealant. Typically, supplemental water is added to the secondary tank to provide makeup water to the pump seal. The system has a number of safety switches and automatic controls for identification of operation problems and for shutdown.

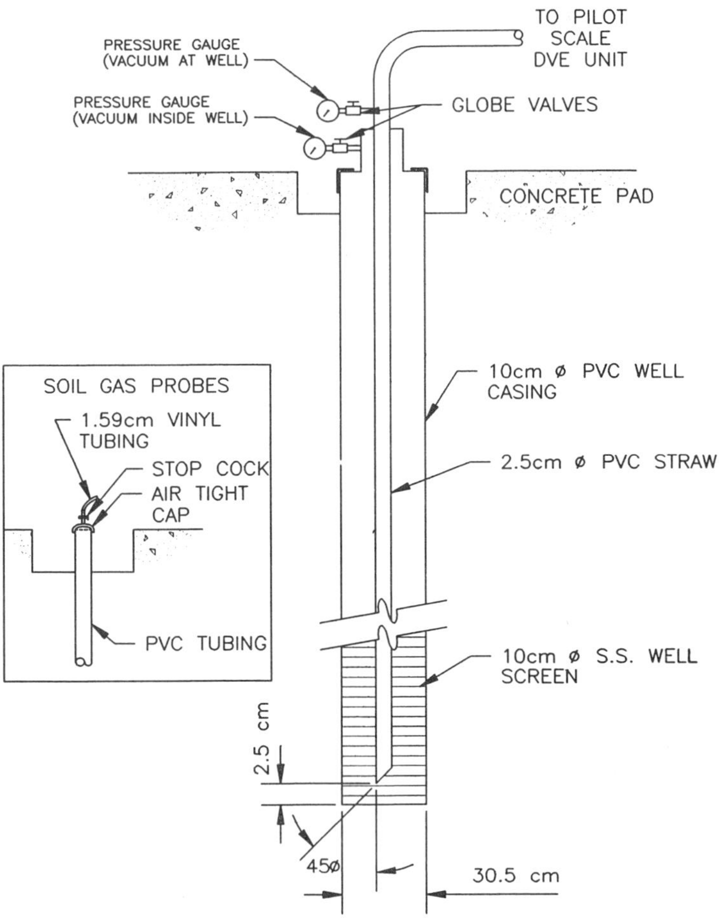

FIG. 2—*DVE extraction well configuration and piezometer configuration.*

Pilot Test Objectives—The following data objectives have been identified to evaluate the effectiveness of a DVE system:

1. Measure the extraction rates of ground water and soil gas from two selected wells.
2. Measure resulting ground-water drawdown.
3. Measure concentrations of compound VOCs in the extracted soil gas and ground water.
4. Estimate radius of influence in the vadose zone for given soil gas extraction rate and vacuum pressure at the vapor extraction well.

Field Program

Field Measurements

Two vapor extraction wells and three soil gas probes were installed as noted in Fig. 4. In January 1993, two separate pilot tests were run on Vapor Extraction Well 1 (VE-1) and

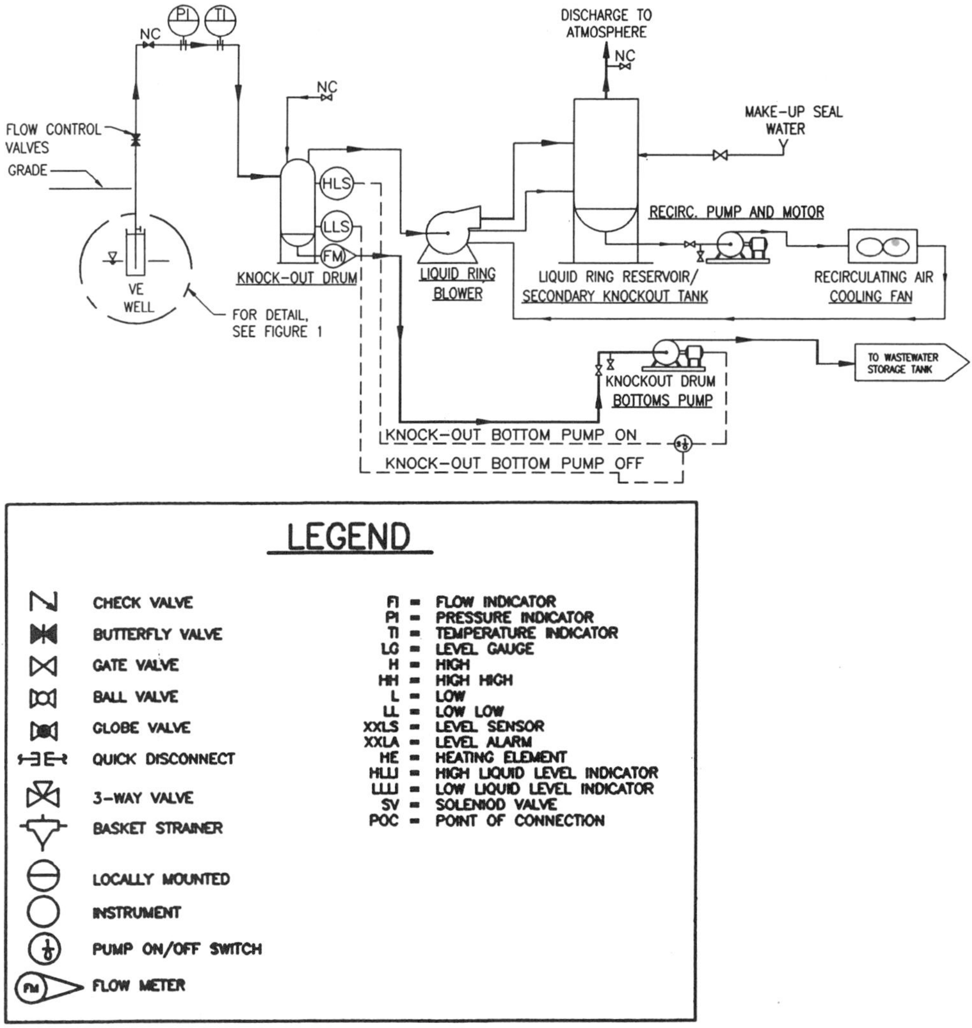

FIG. 3—*P&ID DVE pilot system (not to scale).*

Vapor Extraction Well 2 (VE-2), respectively. Three categories of field measurements were obtained during the pilot test, including soil gas extraction, ground-water extraction, and other field parameters.

Soil Gas—During the pilot test, several soil gas extraction parameters, including photoionization detector (PID) measurements of the exhaust gas and pressure readings at the soil gas probes, were measured.

During operation on VE-2, the soil gas probe Soil Gas 1 (SG-1) readings started at 8 mm of water column vacuum and declined by 85.3 mm of water column vacuum after one day of operation. The readings leveled out at approximately 86.4 mm of water column vacuum. Soil Gas 2 (SG-2) readings started at 0 mm of water column vacuum, decreased to 163 mm of water column vacuum after one day of operation, and stabilized at approximately 178

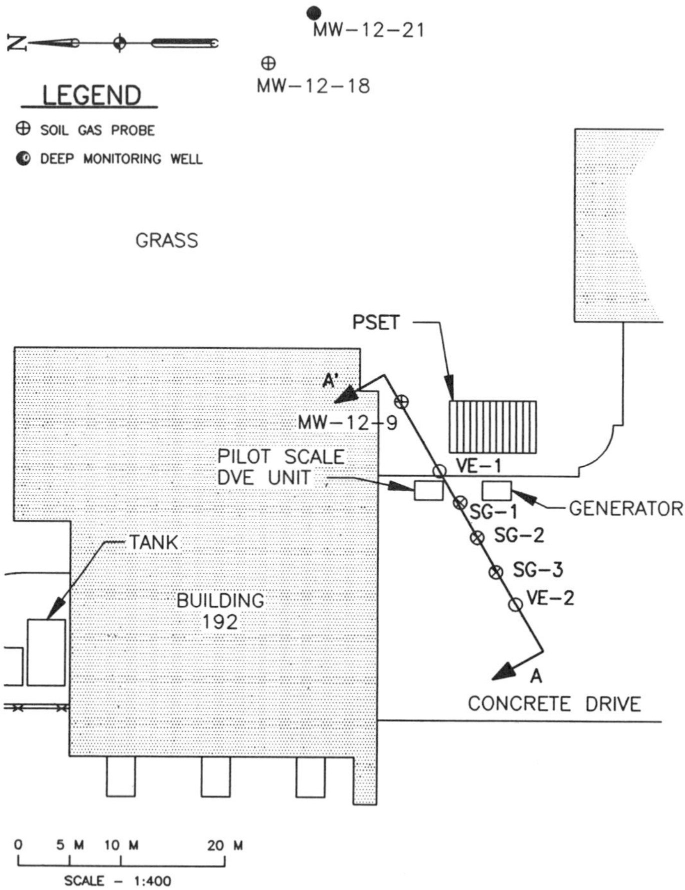

FIG. 4—*Dual-phase vapor extraction well and soil gas probe locations.*

mm of water column vacuum. Soil Gas 3 (SG-3) decreased to 288.3 mm of water column vacuum after one day and stabilized at approximately 305 mm of water column vacuum. As an additional measuring point, soil gas pressure at VE-1 was measured on 17 January 1993. At that time, the measurement was approximately 8.4 mm of water column vacuum. Airflow extraction rates were approximately 1.05 standard cubic meters per minute (scmm).

During the pilot test on VE-1, SG-1 readings started at 0 mm of water column vacuum and decreased to 212 mm of water column vacuum after 17 h of operation. The readings leveled out at approximately 216 mm of water column vacuum. SG-2 readings decreased to 114 mm of water column vacuum after 17 h of operation and averaged approximately 102 mm of water column vacuum. SG-3 readings decreased to 85.6 mm of water column vacuum after 17 h and averaged approximately 92.3 mm of water column vacuum thereafter. VE-2 readings decreased to 23.4 mm of water column vacuum after 17 h and then leveled out at approximately 25.4 mm of water column vacuum. Airflow extraction rates were approximately 100 scmm.

Ground Water—Ground-water levels were measured at MW-12-12, MW-12-17, and MW-12-18 (Fig. 4) to evaluate the radius of influence on ground water from the extraction wells. These results are discussed later.

The extracted ground-water volume was measured using a flow totalizer. For the VE-1 test, 10 310 L of ground water were extracted at a rate of approximately 2.65 L/min; for the VE-2 test, 19 420 L of ground water were extracted at a rate of approximately 3.4 L/min. The total volume of ground water extracted was 29 730 L.

Other Field Parameters—Other parameters measured in the field included VOC screening of soil samples, pH of extracted ground water, and distance between extraction wells and soil gas probes. Soil was collected from split-spoon samples during installation of the vapor extraction wells and screened with a PID to detect VOCs. The interval with the highest readings was submitted for VOC analysis using EPA Method 8240. For VE-1, the screening measurements ranged from 15 to greater than 1000 ppm. The 3.05 to 3.35 m sampler intervals, 4.57 to 4.9 m sample intervals, and 5.79 to 6.10 m sample intervals all exceeded the upper limit (1000 ppm) of the PID. The 4.57 to 4.9 m sample interval was submitted for VOC analysis. For VE-2, the screening measurements ranged from 38 to greater than 1000 ppm. The 4.57 to 4.19 m sample interval registered the highest reading (greater than 1000 ppm) and was submitted for VOC analysis using EPA Method 8240.

Analytical Testing Results

Soil—Soil samples were collected during the installation of the two VE wells. The sample intervals were 0 to 0.46 m and every 1.5 m thereafter up to 6.1 m. The soil samples from VE-1 and VE-2 for the 4.57 to 4.9 m interval were submitted to the laboratory for VOC analysis using EPA Method 8240. Soil samples collected from VE-1 contained acetone and tetrachloroethene at concentrations of 2000 and 45 000 μg/kg, respectively. Soil samples collected from VE-2 contained methylene chloride and tetrachloroethene at concentrations of 1.9 and 1.3 μg/kg, respectively.

Extracted Ground Water—Samples of extracted ground water were collected from the discharge pump of the first knockout tank during the operation of the DVE pilot test at approximately 24-h intervals. Three extracted ground-water samples were obtained during extraction from VE-1. Of the target VOCs of interest, only methylene chloride and tetrachloroethene were detected with concentrations ranging from 32 000 to 75 000 μg/L and 2800 to 8200 μg/L, respectively. Four extracted ground-water samples were collected during the pilot test for VE-2 at approximately 24-h intervals. Tetrachloroethene was detected in three samples ranging from 140 to 200 μg/L. Methylene chloride was found in all four samples at concentrations varying from 2200 to 4600 μg/L.

Soil Gas—Extracted soil gas samples were submitted to the analytical laboratory for chemical analysis of VOCs. At approximately 12-h intervals, soil gas samples were collected from VE-1. All seven samples contained methylene chloride ranging in concentrations from 5 100 000 parts per billion by volume (ppb v/v) declining to 1 300 000 ppb v/v. Tetrachloroethene was also detected in all seven samples at an initial concentration of 1 300 000 ppb v/v, decreasing to a final concentration of 310 000 ppb v/v. Nine soil gas samples were collected during DVE operation on VE-2. Analytical testing results did not show a consistent decline in the VOC concentrations as in the VE-1 test but varied throughout the test. Cis-1,2-dichloroethene was found in the first sample at a concentration of 2300 ppb v/v. Methylene chloride was detected in all nine samples at concentrations ranging from 200 000 to 320 000 ppb v/v. Tetrachloroethene was found at concentrations ranging from 11 000 to 37 000 ppb v/v.

Pilot Test Evaluation

Ground-water drawdown or airflow permeability testing had not been performed at the site previous to the DVE pilot test. However, in 1990 and 1991, ground-water drawdown and hydraulic conductivity slug testing were performed in the upper saturated zone approximately 0.8 km northeast of the site. The drawdown test performed outside the site was not completed because the water level in the pumping well was drawn below the pump intake at a rate of 1.4 L/min after 123 min of testing. Hydraulic conductivities in areas of that test were between 2.17×10^{-5} cm/s and 7.22×10^{-4} cm/s. A storativity constant of 1.07×10^{-2} was calculated. As noted earlier, hydraulic conductivities identified at the site range from 5.7×10^{-6} cm/s to 1.3×10^{-4} cm/s.

The DVE pilot test was performed in the upper saturated zone at the site. This zone consists of 4.88 to 5.33 m of clayey silt and extends to bedrock in the area of the vapor extraction wells. The saturated thickness was approximately 3.66 m at the vapor extraction wells.

Ground-water Extraction Radius of Influence

At the operating extraction rate of 3.41 L/min from VE-2, water level measurements dropped approximately 0.46 to 1.16 m during three days of operation in the two nearest observation wells (MW-12-9 and MW-12-1). These two wells are located approximately 7.62 and 14.6 m from VE-2, respectively. Water level measurements observed in MW-12-12, 48.8 m downgradient from the VE-2 test, indicated a decline of between 0.076 to 0.1 m after three days of operation. However, a water level decrease of this magnitude may not be directly attributed to the DVE test. No regional trend ground-water data were evaluated. A vacuum of between 229 and 254 mm Hg was observed at the extraction well during the VE-2 test.

At the operating extraction rate of 2.65 L/min from VE-1, water level measurements dropped approximately 1.04 to 1.37 m during pilot test operation in the two observation wells closest to VE-1, MW-12-1, and MW-12-9. Water level measurements from MW-12-12, 48.2 m hydraulically downgradient from the VE-1 test, did not positively indicate that the ground-water radius of influence due to pumping extended to that point. A vacuum between 229 and 330 mm Hg was observed in the extraction pipe during the VE-1 test.

Vapor Extraction Radius of Influence

At the operating vapor extraction rate of 1.047 scmm for the test at VE-2, vacuum readings between approximately 76.2 and 3.05 mm of water were recorded in the three soil gas probes 2.44 to 7.32 m away from VE-2. VE-1, located 9.75 m away from the VE-2 test, indicated a maximum vacuum reading of approximately 8.4 mm of water vacuum. At the operating vapor extraction rate of approximately 1 scmm from the test on VE-1, vacuum readings between approximately 76.2 and 22.9 mm of water vacuum were recorded in the three soil gas probes 2.4 and 27.3 m away from VE-1. VE-2, located 9.75 m away from the VE-1 test, indicated a maximum vacuum reading of approximately 25.4 mm of water vacuum.

The steady-state method of calculating the soil permeability to airflow (K_{air}) was selected based upon shallow depths to ground water and short duration to reach steady state (USAF 1992).

In order to calculate a K_{air} based on observed field readings, the radius of influence was assumed to be 10.7 m based on readings of 8.38 mm of water 9.75 m from the VE-2 test.

A calculation of K_{air}, based on a radius of airflow influence of 10.7 m and yielding a K_{air} of 1.07×10^{-8} cm² (1.1 Darcy), was performed using the following equation

$$K = \frac{Q\mu \ln (R_w/R_I)}{H\pi P_w[1 - (P_{Atm}/P_w)^2]}$$

where

k = soil permeability to airflow (cm²) or (Darcy),
Q = flow (cm³/s),
μ = viscosity of air = 1.8×10^{-4} g/cm-s or 0.018 cp,
P_w = absolute pressure at extraction well (g/cm-s²) or (atm),
P_{Atm} = absolute ambient pressure at the site, 1 atm = 1.01×10 g/cm-s²,
R_w = radius of vapor extraction well (cm),
R_I = radius of influence of vapor extraction well (cm), and
H = height of well screen (cm).

This field derived K_{air} compares to K_{air} in the literature of 5×10^{-8} cm² for a silty soil (Johnson et al. 1990). Evaluation of the influence of the silt layers on K_{air} was not performed. However, the presence of these layers may have impacted variables H and R_I.

Contamination Removal Rates

The DVE unit was operated on VE-1 and VE-2 for approximately 2.5 and 4 days, respectively. The average daily ground-water extraction rate for VE-1 ranged between 4.5 and 2.4 L/min with an overall average of 2.65 L/min. The average daily ground-water extraction rate for VE-2 varied from 7.08 to 3.10 L/min with an overall average of 3.37 L/min.

Extracted water and air samples were collected during the pilot test. The analytical results were presented earlier. The extraction rates of water and soil gas discussed previously were combined with analytical results to determine contaminant removal rates during the pilot test. These results are illustrated in Figs. 5, 6, 7, and 8.

The removal rate for methylene chloride in water at VE-1 was initially 0.486 kg/day. It decreased to 0.114 kg/day over the course of the pilot test. Tetrachloroethene was initially removed at 0.055 kg/day and decreased to 0.009 kg/day. No other target VOCs were detected. It should be noted that the reported concentrations of VOCs in water are not representative of concentrations observed in the ground water. During extraction, the water and soil gas are intimately mixed. VOCs can be stripped from the water and exhausted with the airstream.

The air contaminant removal rate for methylene chloride ranged from 29.2 kg/day initially and decreased to 7.73 kg/day when operations ceased. Tetrachloroethene in air was initially at 14.5 kg/day and decreased to 3.45 kg/day.

At VE-2, the removal rate for methylene chloride in water was initially 0.023 kg/day and decreased to 0.0205 kg/day. Tetrachloroethene was initially not detected but increased to 0.0091 kg/day when operations ceased. Methylene chloride removal in the air was initially 1.15 kg/day and increased to 1.4 kg/day with a maximum rate of 1.83 kg/day. Tetrachloroethene was initially produced in air at 0.122 kg/day and increased to 0.323 kg/day with a maximum of 0.402 kg/day. This upward trend of VOC removal between 0.122 and 0.323 kg/day may have been the result of higher-level VOC contamination in the vicinity of VE-1 being drawn into VE-2.

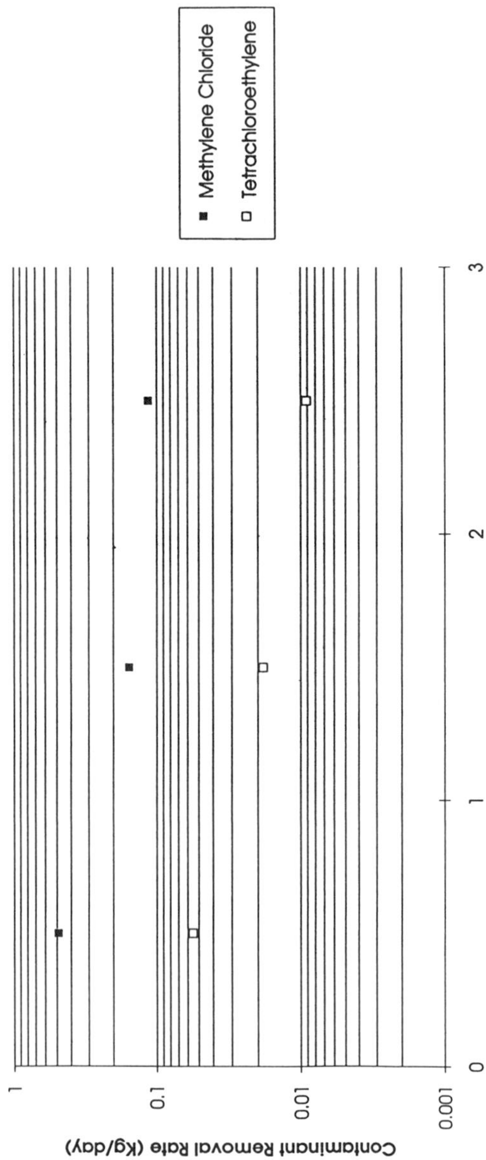

FIG. 5—*Contaminant removal rate in extracted ground water for VE-1.*

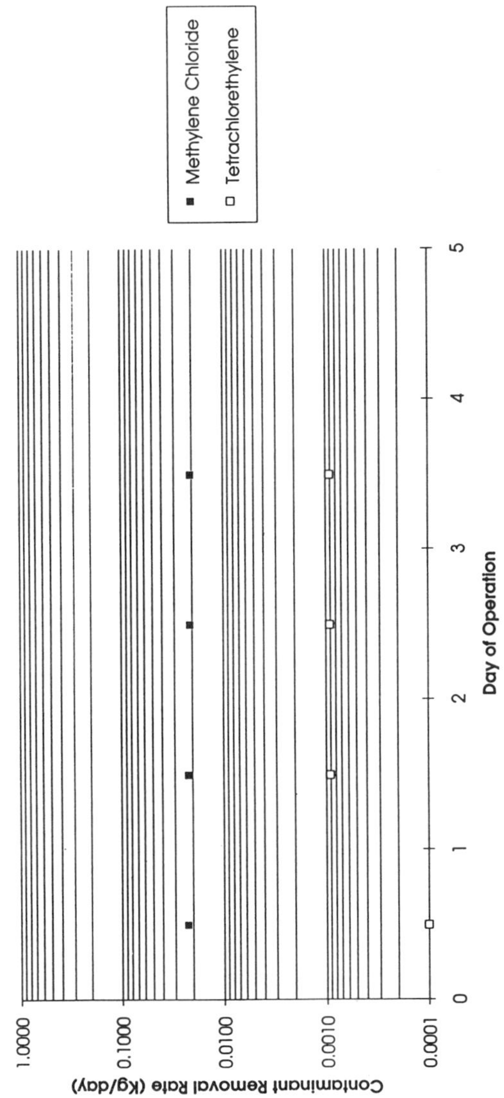

FIG. 6—*Contaminant removal rate in extracted ground water for VE-2.*

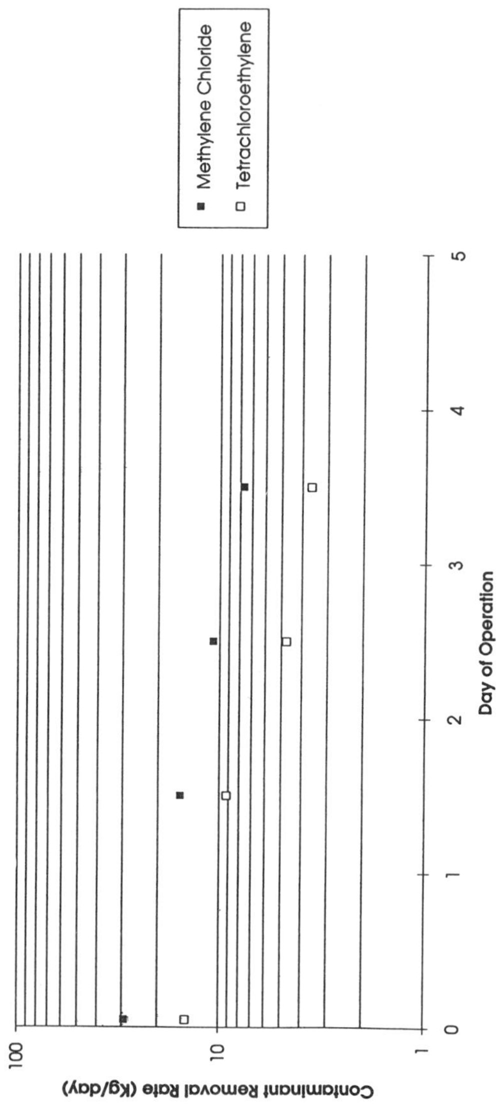

FIG. 7—*Contaminant removal rate in extracted soil gas for VE-1.*

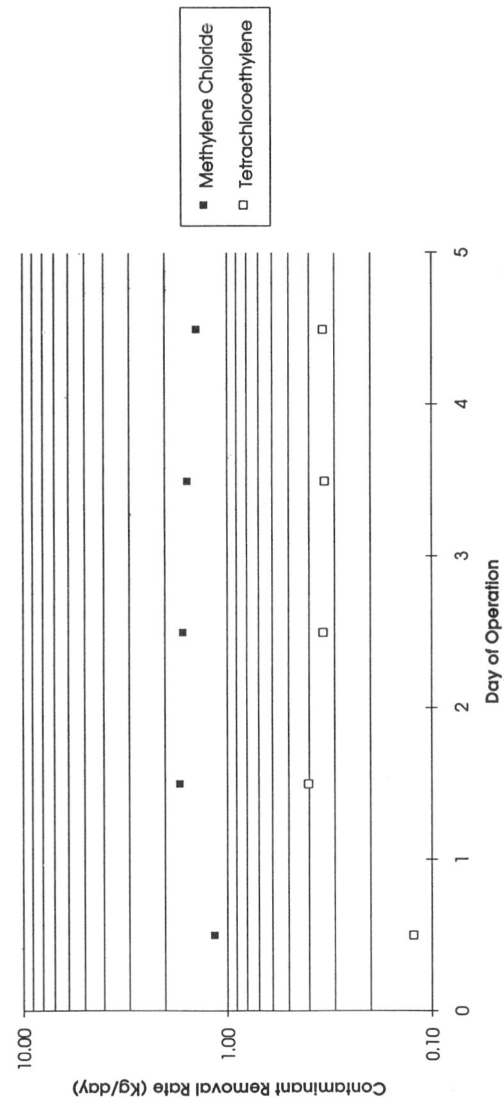

FIG. 8—Contaminant removal rate in extracted soil gas for VE-2.

Evaluation Summary

The DVE pilot test system performed well while recovering VOC in ground water (1.9 to 3.785 L/min per well) and extracted soil gas (greater than 1.0 scmm) from the source area. The following observations can be made based upon the data collected in the pilot study:

1. Ground-water extraction rates were significantly higher under vacuum than previously observed, possibly due to vacuum applied during testing.
2. Cumulative VOC removal rates on the order of 0.1 to 10 kg/day were identified per well in soil gas and ground water.

References

Johnson, P., Stanley, C. C., Kemblowski, M. W., Byers, D. L., and Colthart, J. D., 1990, "A Practical Approach to the Design, Operation, and Monitoring of In Situ Soil-Venting Systems," *Ground Water Monitoring and Remediation,* Spring.

Lewis, R., White, J., and Ryan, P., 1992, "Dual Phase Extraction Applied to Groundwater and Clayey Soils in the Removal of Chlorinated Hydrocarbons," U.S. EPA/A&WMA International Symposium.

Roy, K., 1991, "Dual Vacuum Extraction Sinks below the Surface in Groundwater Cleanup," *Hazmat World,* Nov.

USAF, 1992, "Test Plan and Technical Protocol for Field Treatability Test for Bioventing. Air Force Center for Environmental Excellence," Brooks AFB, Texas, May.

Author Index

Subject Index